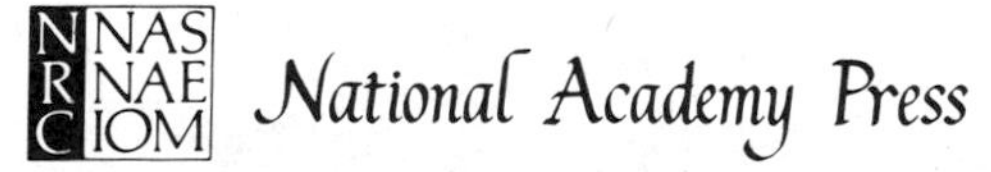


The National Academy Press was created by the National Academy of Sciences to publish the reports issued by the Academy and by the National Academy of Engineering, the Institute of Medicine, and the National Research Council, all operating under the charter granted to the National Academy of Sciences by the Congress of the United States.

# Safety of Existing Dams

## Evaluation and Improvement

# Safety of Existing Dams

## Evaluation and Improvement



Committee on the Safety of Existing Dams
Water Science and Technology Board
Commission on Engineering and Technical Systems
National Research Council



NATIONAL ACADEMY PRESS
Washington, D.C. 1983

**National Academy Press, 2101 Constitution Avenue, NW, Washington, DC 20418**

---

NOTICE: The project that is the subject of this report was approved by the Governing Board of the National Research Council, whose members are drawn from the councils of the National Academy of Sciences, the National Academy of Engineering, and the Institute of Medicine. The members of the committee responsible for the report were chosen for their special competences and with regard for appropriate balance.

This report has been reviewed by a group other than the authors, according to procedures approved by a Report Review Committee consisting of members of the National Academy of Sciences, the National Academy of Engineering, and the Institute of Medicine.

The National Research Council was established by the National Academy of Sciences in 1916 to associate the broad community of science and technology with the Academy's purposes of furthering knowledge and of advising the federal government. The Council operates in accordance with general policies determined by the Academy under the authority of its congressional charter of 1863, which establishes the Academy as a private, nonprofit, self-governing membership corporation. The Council has become the principal operating agency of both the National Academy of Sciences and the National Academy of Engineering in the conduct of their services to the government, the public, and the scientific and engineering communities. It is administered jointly by both Academies and the Institute of Medicine. The National Academy of Engineering and the Institute of Medicine were established in 1964 and 1970, respectively, under the charter of the National Academy of Sciences.

This report represents work supported by contract number EMW-C-0756, work unit number 6311F, between the Federal Emergency Management Agency and the National Research Council.

**Library of Congress Cataloging in Publication Data**

National Research Council (U.S.). Committee on the
Safety of Existing Dams.
Safety of existing dams.

Includes index.
1. Dam safety—United States. I. Title.
TC556.N37 1983 627'.8 83-12094
ISBN 0-309-03387-X



Printed in the United States of America
First Printing, September 1983
Second Printing, August 1984

## COMMITTEE ON SAFETY OF EXISTING DAMS

ROBERT B. JANSEN, Consulting Civil Engineer, Spokane, Washington, *Chairman*
HARL P. ALDRICH, Haley and Aldrich, Inc., Cambridge, Massachusetts
ROBERT A. BURKS, Southern California Edison Company, Rosemead, California
CLIFFORD J. CORTRIGHT, Consulting Civil Engineer, Sacramento, California
JAMES J. DOODY, Department of Water Resources, Sacramento, California
JACOB H. DOUMA, Consulting Civil Engineer, Great Falls, Virginia
JOSEPH J. ELLAM, Pennsylvania Department of Environmental Resources, Harrisburg
CHARLES H. GARDNER, North Carolina Department of Natural Resources, Raleigh
WILLIAM R. JUDD, Purdue University, West Lafayette, Indiana
DAN R. LAWRENCE, Department of Water Resources, Phoenix, Arizona
ROBERT J. LEVETT, Niagara Mohawk Power Corporation, Syracuse, New York
ARTHUR G. STRASSBURGER, Pacific Gas and Electric Company, San Francisco, California
BRUCE A. TSCHANTZ, University of Tennessee, Knoxville
ERIK H. VANMARCKE, Massachusetts Institute of Technology
HOMER B. WILLIS, Consulting Civil Engineer, Bethesda, Maryland

### Technical Consultant

CHARLES F. CORNS, Consulting Engineer, Springfield, Virginia

### NRC Project Manager

SHEILA D. DAVID, Staff Officer

## WORKSHOP PARTICIPANTS

GEORGE L. BUCHANAN, Tennessee Valley Authority, Knoxville
CATALINO B. CECILIO, Pacific Gas and Electric Company, San Francisco, California
LLEWELLYN L. CROSS, Chas. T. Main, Inc., Boston, Massachusetts
RAY F. DEBRUHL, North Carolina Department of Administration, Raleigh
JAMES M. DUNCAN, University of California, Berkeley
LLOYD E. FOWLER, Goleta Water District, California
VERNON K. HAGEN, U.S. Army Corps of Engineers, Washington, D.C.
JOSEPH S. HAUGH, USDA Soil Conservation Service, Washington, D.C.
DAVID LOUIE, Harza Engineering Company, Chicago, Illinois
J. DAVID LYTLE, U.S. Army Corps of Engineers, St. Louis, Missouri
MARTIN W. MCCANN, Stanford University, California
JEROME RAPHAEL, University of California, Berkeley
HARESH SHAH, Stanford University, California
THOMAS V. SWAFFORD, Fairfield Glade Resort Developers, Crossville, Tennessee
HARRY E. THOMAS, Federal Energy Regulatory Commission, Washington, D.C.
LAWRENCE J. VON THUN, U.S. Bureau of Reclamation, Denver, Colorado
JACK G. WULFF, W. A. Wahler & Associates, Palo Alto, Calfifornia

## FEMA Representative

WILLIAM BIVINS, Project Officer, Federal Emergency Management Agency, Washington, D.C.

## COMMITTEE PANELS

### Panel on Risk Assessment

Erik Vanmarcke, *Chairman*
James J. Doody
Joseph J. Ellam
Haresh Shah
Vernon K. Hagen
Lawrence J. VonThun
Joseph S. Haugh
Martin W. McCann, Jr.

### Panel on Hydraulic/Hydrologic Considerations

Homer B. Willis, *Chairman*
Robert J. Levett
Jacob H. Douma
Bruce A. Tschantz
Catalino B. Cecilio
Llewellyn L. Cross
David Louie

### Panel on Concrete and Masonry Dams

Arthur G. Strassburger, *Chairman*
William R. Judd
Robert A. Burks
Jerome Raphael
George L. Buchanan

### Panel on Embankment Dams

Harl P. Aldrich, *Chairman*
Charles H. Gardner
Clifford J. Cortright
James M. Duncan
Thomas V. Swafford
Jack G. Wulff
Ray F. DeBruhl

### Panel on Instrumentation

Dan R. Lawrence, *Chairman*
Robert A. Burks
Lloyd E. Fowler
J. David Lytle
Harry E. Thomas

### Panel on Geological/Seismological Consideration

William R. Judd, *Chairman*
Harry E. Thomas
Charles H. Gardner
Jerome Raphael
James M. Duncan

# Preface

Because of several disasters in recent years, the safety of dams has received increasing attention throughout the world. Governments at all levels have come to recognize and, in many cases, to accept their responsibilities in this area. In the United States, federal and state agencies have been active in inventorying and inspecting dams in the interest of improved safeguards. The results point to deficiencies that are widespread and to a problem of national importance. From the disasters and from the evaluations of thousands of dams, the message is clear that the threat to public safety is large and must be reduced. Although the danger is evident, its elimination will be difficult for at least two principal reasons: those responsible must be ready to take action and the funds for remedial programs must be found.

In recognition of the need for a nationwide initiative that would foster a cooperative approach to dam safety, the Committee on the Safety of Existing Dams was created under the auspices of the National Research Council at the request of the Federal Emergency Management Agency. Members of the committee and its work groups were enlisted on the basis of their expertise in the engineering of dams and in the related professional disciplines. To qualify, they had to be outstanding in their fields and willing to contribute their knowledge and their ideas to the common purpose. The committee was organized to include civil engineers, representatives of state agencies responsible for dam safety, private corporate dam owners, geologists, hydraulic engineers, risk analysts, and others knowledgeable about federal and state dam safety programs. The charge that was laid out for each of them was demanding, and they responded commendably without exception. In composing the work groups, care was taken to ensure a range of experience and viewpoint so that the product would be balanced. Rather than a committee and work groups in the usual sense, an assembly of authors contributed individually, while the work group chairmen, the consultant, and the committee chairman served as planners, coordinators, and editors providing their own technical input. In structuring the effort we were guided by the belief that a collection of individual works, properly integrated, would be worth more than a blended group offering, the development of which might be burdened by excessive oral exchange.

With such an array of special and dedicated talent, the opportunity for accomplishment was large. To maximize this potential, work assignments were made in advance, and the contributors were encouraged to volunteer freely from their experience. The challenge was unanimously accepted. Each member was asked to consider himself in the role of adviser to a responsible dam owner or to an engineer and to suggest practical ways to approach the analysis and remedy of a suspected or actual deficiency. Designated tasks were designed to cover the gamut of problems, while avoiding inefficient duplication of effort. This report thus presents the advice of experts on how to solve the puzzle of an inadequate structure and how to apply economical and professionally acceptable remedies.

We have been guided by the need to optimize benefits from a given level of expenditure. The basic premise is that improvement of deficient dams must begin without delay, even though initial funding may be insufficient for comprehensive solutions. In some cases this may entail a staged approach to corrective work, but this is regarded as better than no action. Some solutions based on the risk assessment methods discussed in this report may not fully comply with the highest current standards of some federal agencies. We emphasize that we do not advocate a lowering of such criteria. A high level of excellence must continue to be the ultimate goal of those who strive for improvement of dams.

The limitations inherent in an evaluation of existing dams must be acknowledged. Although those unschooled in the intricacies might expect it to be an exact science, it is in fact full of uncertainty and dependent on judgment. The total range and character of risk may not be predictable, due in large part to the unknowns of a site and a structure. The goal of preventive and remedial engineering is to reduce uncertainties, recognizing that absolute safety may not be ensured in every case.

Professionals experienced in the evaluation and improvement of dam safety know that their job is to lower risk to the minimum that is practically attainable. This requires incremental investment in removal of deficiencies in the order of the hazard that they present. No matter how much money is spent, some unknown risk may remain. Many problems are not amenable to inexpensive solution, but the committee believes that an inadequate dam should be examined without rigid adherence to convention, always searching for ways to lessen risk within the unavoidable limits that are present.

The idealist may think that this report does not confront all the problems faced by those who are responsible for dams but whose resources are limited. Of course, the remedies may not be as complete as they would be with the availability of abundant funds. Some states and owners cannot afford the preventive and corrective work that common standards would dictate. A perfectionist might suggest that states and owners should save their money until the job can be done completely. The message of this report is that there are ways to remedy deficiencies progressively, attacking the most serious problems first and economizing where possible but not to the extent that applicable guidelines are disregarded. The experienced professional knows that through years of perseverance the reduction of risks can in many cases be achieved only in this way—setting sensible priorities and recognizing that some remedies may have to await later action. Those responsible for repairs sometimes adopt the contrary view, that the corrective effort would cost too much and must therefore be postponed indefinitely. Total elimination of risk may not be attainable because of financial restraints. The practical objective is to reduce risk to a more tolerable level. Even if this falls short of highest standards, it is certainly preferable to waiting for money that may never arrive.

Inevitably, the question arises regarding the extent of deficiency that can be allowed while necessary funding is sought. If that limit is exceeded, an alternative that could be weighed would be to abandon the dam. However, the hazard might not be eliminated simply by abandonment. Breaching or removal of a dam also requires engineering, and such work can be expensive. In some cases such actions might cost more than correcting the inadequacies. The emphasis of this report is on keeping a dam in service by using preventive and remedial engineering techniques, which the committee regards as the most positive approach to dam safety.

Several reasons can be cited for lack of compliance with standards. A shortage of financial resources is common. Sometimes a dam owner needs to be convinced that the deficiencies are intolerable. This is best accomplished by practicing engineers and state officials rather than by the courts, although the judicial process remains as a last recourse.

In the United States each state must ensure that its dams are inspected and that their safety is evaluated (excluding federal structures). Some states do not provide enough money for this. Such problems are inseparable from the technical considerations that serve the primary purpose of this report. While their solution is largely beyond the scope of this report, there is adequate precedent for resolving such governmental dilemmas. For example, in some European countries the laws provide for periodic evaluations by consulting engineers retained by the owners. In the United States this has a successful parallel in the independent inspection programs required at water power projects licensed by the Federal Energy Regulatory Commission. These precedents suggest that the legislative barriers are not insurmountable. Of course, a prerequisite is that the governing bodies be concerned enough about public safety that they will take the necessary initiative.

Current dam safety standards have been developed on sound bases. The committee does not advocate their revision solely because of financial pressures; however, they must be justified by the benefits. In many jurisdictions standards are adjusted to suit individual conditions. For instance, a small dam in an unpopulated area would be required to withstand less severe tests than the probable maximum flood or the maximum credible earthquake. Most state safety programs make allowances for the wide ranges of dam characteristics, locations, and consequences of failure. They recognize the need for stricter rules in metropolitan areas than in isolated rural environments, including requirements for closer surveillance.

Engineers experienced in the field of dam safety may question the emphasis that this report gives to risk-based decision analysis. They know that the relative degree of risk at a dam is difficult and often impossible to quantify. Necessary approximations rely heavily on judgment, which comes from working with many kinds of dams and a myriad of conditions. In the evaluation of existing dams, numerical methods sometimes are less useful than empirical approaches. This report therefore may appear to give undue prominence to sophisticated decision analyses based on assessments of probabilities. Since such analyses have been used successfully in only a few dam safety reviews, they could have been relegated to secondary status. Despite their lack of acceptance by practicing engineers, however, the committee believes that they hold promise and that they should be more fully explored by the profession. Such methods have generally been judged as being too theoretical, lacking input from practical experience, and producing after much study results that seem to be intuitively apparent to experienced engineers. Furthermore, it has been argued that, although the calculations may tend to be complex, the concepts on which they are based are overly simplistic, evidencing minimal recognition of the uncertainties intrinsic in the study of existing dams. Despite these criticisms, which are shared to some extent by some of the contributors to this report, it was concluded that the underlying principles need to be expanded and developed with participation by those actually responsible for dams and that there is a need for merging apparently academic concepts with those of engineers who have firsthand knowledge of dams and their problems. It is hoped that the case histories presented in this report may serve to demonstrate the potential of such cooperative effort. In this difficult field the search for new methods to keep dams safe must be continuous.

The members of the committee, the workshop participants, and the committee's technical consultant devoted much time and enthusiasm to their tasks. Their reward will be measured by acceptance and use of this report to ensure better dams. We owe much to the strong support provided by the National Research Council and the Federal Emergency Management Agency. The staffs of these institutions facilitated all aspects of the work program, including the very important phase of review and publication of the report. Our hope is that we have made a useful contribution to public safety.

Robert B. Jansen, *Chairman*
Committee on the Safety of Existing Dams

# Contents

# List of Figures and Tables

FIGURES

## TABLES

# Safety of Existing Dams

## Evaluation and Improvement

# 1
# Introduction

## PURPOSE

The goal of this report is the enhancement of dam safety. It was prepared by the Committee on the Safety of Existing Dams, National Research Council (NRC), to present in a single volume all essential aspects of dam safety. A major objective of the report is to provide guidance for achieving improvements in the safety of existing dams within financial restraints. Many dam owners are faced with problems of such a nature and extent that they are unable to finance remedial measures. In May 1982 the U.S. Army Corps of Engineers reported that no remedial measures had been instituted at 64% of the unsafe dams found during its 4-year inspection program, principally because of the owners' lack of resources. To these owners, as well as to regulatory agencies and others concerned with the engineering and surveillance of dams, the committee presents its suggestions and guidance for assessing and improving the safety of existing dams. The contents of this report are intended to be informational and not to advocate rigid criteria or standards. In no instance does the committee intend to recommend the lowering of existing dam safety standards.

## SCOPE

The scope of the committee's study and the conclusions of this report concern technical issues pertinent to dam safety. The study includes examinations of risk assessment techniques; engineering methodologies for stability and hydrologic evaluations; and methods and devices to identify, reduce, and/or eliminate deficiencies in existing dams. Included are case histories demonstrating economical solutions to specific problems and also possible nonstructural approaches.

The Committee on the Safety of Existing Dams, operating under a June 1983 completion deadline, arranged a relatively brief but intensive study of its assigned task. It conducted a 2-day meeting in Washington, D.C., on June 2–3, 1982, to initiate the effort; a meeting of the panel chairmen in Spokane, Washington, on August 10, 1982, to plan a workshop; and a 3-day workshop meeting in Denver, Colorado, on October 5–7, 1982. A committee meeting was also held on March 7 and 8, 1983, to complete the draft of the report.

Participants in the workshop included committee members, members of NRC staff, and other experts with a broad range of experience in dam engineering and dam safety. Task assignments for the workshop were divided among five working groups: (1) Risk Assessment, (2) Stability of Embankment Dams and Their Foundations, (3) Stability of Masonry Dams and Their Foundations, (4) Hydraulic/Hydrologic Considerations, and (5) Instrumentation. Advance assignments were made to individual participants for specific contributions to the workshop to ensure complete coverage of all issues. During the workshop a separate task group was designated to address the general subjects of geology and seismology.

The participants discussed the various technical aspects of enhancing dam safety with a view to reaching a consensus on desirable approaches whenever possible. From these presentations and discussions, the committee reached the conclusions presented in this report.

## BACKGROUND

This study by the Committee on the Safety of Existing Dams is the second phase of a comprehensive study concerning policy and technical issues related to the safety of dams. In October 1981 the Federal Emergency Management Agency (FEMA) asked the NRC to undertake such a study. For the first phase, FEMA asked the NRC to identify impediments to state-run programs for dam safety, to suggest federal actions to remove or mitigate those impediments, and to define how the U.S. government could help make nonfederal dams safer. In response, the NRC created the Committee on Safety of Nonfederal Dams to review and discuss the issues involved. The efforts of that committee were completed in February 1982 and reported in the 1982 publication entitled *Safety of Nonfederal Dams, A Review of the Federal Role*.

Unlike the first phase, the second phase concerns the technical considerations relating to dam safety and is applicable to all existing dams, federal as well as nonfederal. The NRC created a new Committee on the Safety of Existing Dams in May 1982 to examine the technical issues of dam safety and to develop guidance on how to achieve improvements in the safety of dams, with due recognition of financial constraints. The committee's task as defined by FEMA was as follows:

- To inventory and assess risk techniques and formulate guidelines on their use to rectify problems faced by dam owners and states with limited financial resources.
- To review and evaluate methods and devices that can be applied, along with risk assessments, to identify, reduce, and/or eliminate deficiencies in existing dams (includes development of a glossary of terms; evaluation of hydrologies and stability parameters; and formulation of guidance for mitigation of such problems as overtopping, weak foundations, piping, and seismicity).
- To examine methodologies for assessing the potential impact of adverse conditions (e.g., maximum credible earthquake, probable maximum flood) on existing dams and potential modifications in order to set limits of acceptable damage to a dam. The methodologies must support the assumption of nonfailure of the structure. Additionally, the methodologies are not to be applied to major structures where failure is catastrophic. Guidance here is intended to be offered on how to achieve improvements in the safety of existing dams within financial constraints.

## REFERENCES

National Research Council, Committee on Safety of Nonfederal Dams (1982) *Safety of Nonfederal Dams—A Review of the Federal Role*, National Academy Press, Washington, D.C.

U.S. Army Corps of Engineers (1982) *National Program for Inspection of Nonfederal Dams—Final Report to Congress*.

# 2
# The Safety of Dams

For centuries, dams have provided mankind with such essential benefits as water supply, flood control, recreation, hydropower, and irrigation. They are an integral part of society's infrastructure. In the last decade, several major dam failures have increased public awareness of the potential hazards caused by dams.

In today's technical world, dam failures are rated as one of the major "low-probability, high-loss" events. The large number of dams that are 30 or more years old is a matter of great concern. Many of the older dams are characterized by increased hazard potential due to downstream development and increased risk due to structural deterioration or inadequate spillway capacity.

The National Dam Inspection Program (PL 92-367) developed an inventory of about 68,000 dams that were classified according to their potential for loss of life and property damage (U.S. Army Corps of Engineers 1982b). About 8,800 "high hazard" dams (those whose failure would cause loss of life or substantial economic damage) were inspected and evaluated. Specific remedial actions have been recommended, ranging from more detailed investigations to immediate repair for correction of emergency conditions. The responsibility for the subsequent inspections, investigations, and any remedial work rests with the owners of the dams. In most states the actions or inactions of the dam owners will be monitored by a state agency responsible for supervision of the safety of dams.

The National Dam Inspection Program provided a beginning to what is hoped to be a continuing effort to identify and alleviate the potential hazards presented by dams. Essential to the success of such an effort are understandings of the causes of dam failures and the effects of age; competent inspection and maintenance programs; thorough knowledge of individual site conditions as revealed by design, construction, and operating records, in addition to inspections and investigations; and an emergency action plan to minimize the consequences of dam failure. The remainder of this chapter presents a discussion of these elements.

## CAUSES OF DAM FAILURES

### Dam Failure Surveys

A number of studies have been made of dam failures and accidents. The results of one survey, by the International Commission on Large Dams (ICOLD), were reported in its publication *Lessons from Dam Incidents, USA*. N. J. Schnitter (1979 Transactions of ICOLD Congress, New Delhi) summarized the survey data in the form illustrated by Figures 2-1 through 2-5. These data pertain only to dams more than 15 meters in height and include only failures resulting in water releases downstream.

Figure 2-1 shows the relative importance of the three main causes of failures: overtopping, foundation defects, and piping. Overall, these three causes have about the same rate of incidence.

Figure 2-2 gives the incidence of the causes of failure as a function of the dam's age at the time of failure. It can be seen that foundation failures occurred relatively early, while the other causes may take much longer to materialize.

Figure 2-3 compares the heights of the failed dams to those of all dams built and shows that 50% of the failed dams are between 15 and 20 meters high.

Figure 2-4 shows the relation between dams built and failed for the various dam types from 1900 to 1969. According to the bottom graph, gravity dams appear the safest, followed by arch and fill dams. Buttress dams have the poorest record but are also the ones used least.

Figure 2-5 shows the improvement of the rate of failure over the 1900–1975 period. The upper graph is in semilogarithmic scale and gives the percentage of failed dams in relation to all dams in operation or at risk at a given time. The lower graph gives the proportion of the built dams that later failed and shows that modern fill and concrete dams are about equally safe.

The United States Committee on Large Dams (USCOLD) made a survey of incidents to dams in the United States. Results of the initial study, which covered failures and accidents to dams through 1972, were published

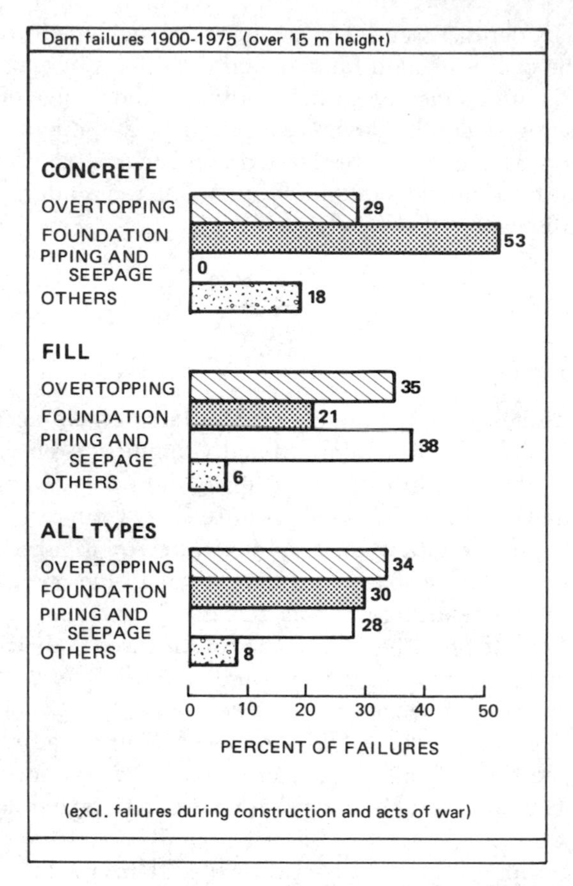


FIGURE 2-1 Cause of failure. SOURCE: ICOLD (1973).

jointly by the American Society of Civil Engineers (ASCE) and USCOLD in 1975 in *Lessons from Dam Incidents, USA*. These data were updated through subsequent USCOLD surveys of incidents occurring between 1972 and 1979. Table 2-1 was compiled from the information developed by the USCOLD surveys and includes accidents as well as failures. The USCOLD surveys pertained only to dams 15 or more meters in height.

Table 2-2 pertains only to concrete dams and lists the number of incidents in the USCOLD surveys for each principal type of such dams. Tables 2-1 and 2-2 list incidents by the earliest, or "triggering," principal cause as accurately as could be determined from the survey data. For instance, where failure was due to piping of embankment materials through a corroded outlet, the corrosion or deterioration was accepted as being the primordial cause of failure. Also, where a sliding failure was due to overtopping flows that eroded the foundation at the toe of a concrete dam, overtopping was listed as the cause of failure. Only one cause is listed for each incident. While only a few of the incidents were attributed to faulty construction, it is reasonable to expect that many of the other failures were due, at least in part, to inadequate construction or design investigations. However, the information on the specific cases is not sufficient to establish such inadequacies as the primordial causes.

## Failure Modes and Causes

Table 2-3 pertains to embankment dams. It is of particular interest because it correlates failure modes and causes. As indicated, the modes and causes of failure are varied, multiple, and often complex and interrelated, i.e.,

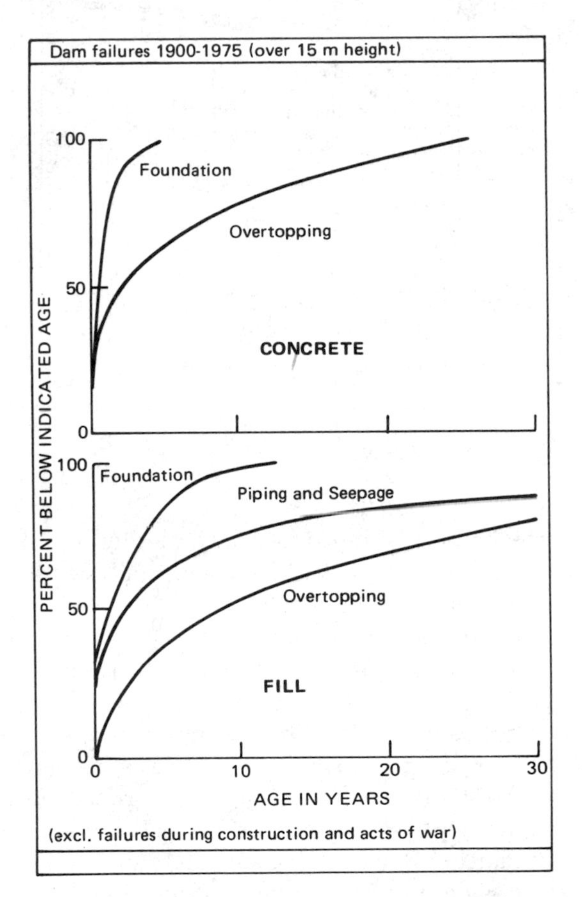


FIGURE 2-2 Age at failure. SOURCE: ICOLD (1973).

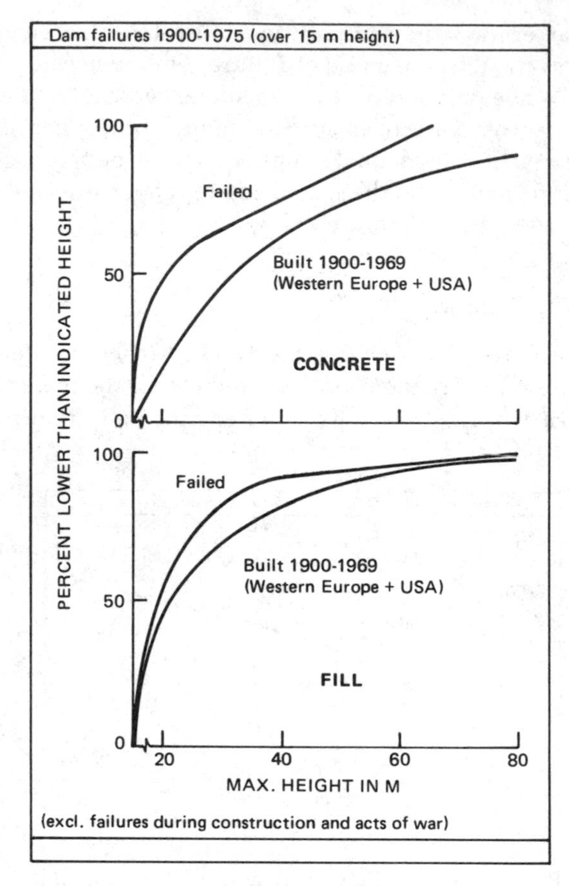


FIGURE 2-3 Height of dams. SOURCE: ICOLD (1973).

often the triggering cause may not truly have resulted in failure had the dam not had a secondary weakness. These causes illustrate the need for careful, critical review of all facets of a dam. Such a review should be based on a competent understanding of causes (and weaknesses), individually and collectively, and should be made periodically by experts in the field of dam engineering.

Many dam failures could be cited to illustrate complex causes and the difficulty of identifying a simple, single root cause. For example, the 1976 Teton failure may be attributed to seepage failure (piping) (Jansen 1980). But several contributing physical (and institutional) causes may be identified (Independent Panel to Review the Cause of Teton Dam Failure 1976). In another example, a dam in Florida was lost due to a slope failure, triggered by seepage erosion of fine sandy soils at the embankment's toe. The soils lacked sufficient cohesion to support holes or cavities normally associated with piping and were removed from the surface of the dam toe by excessive seepage velocities and quantities. This undermining of the toe by seepage resulted in a structural failure, but the prime cause was the nature of the foundation soils.

The complex interrelationship of failure modes and causes makes it extremely difficult to prepare summary tables such as Table 2-1. It also explains why different evaluators could arrive at different conclusions regarding prime causes. Certainly any such table should be accompanied by

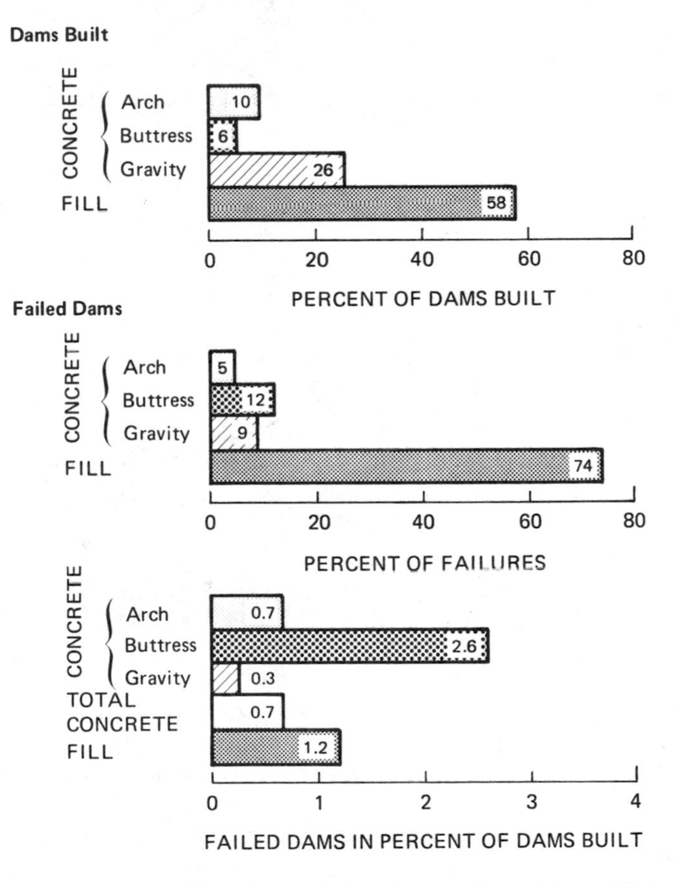


FIGURE 2-4 Dam types (Western Europe and USA, 1900-1969).
SOURCE: ICOLD (1979).

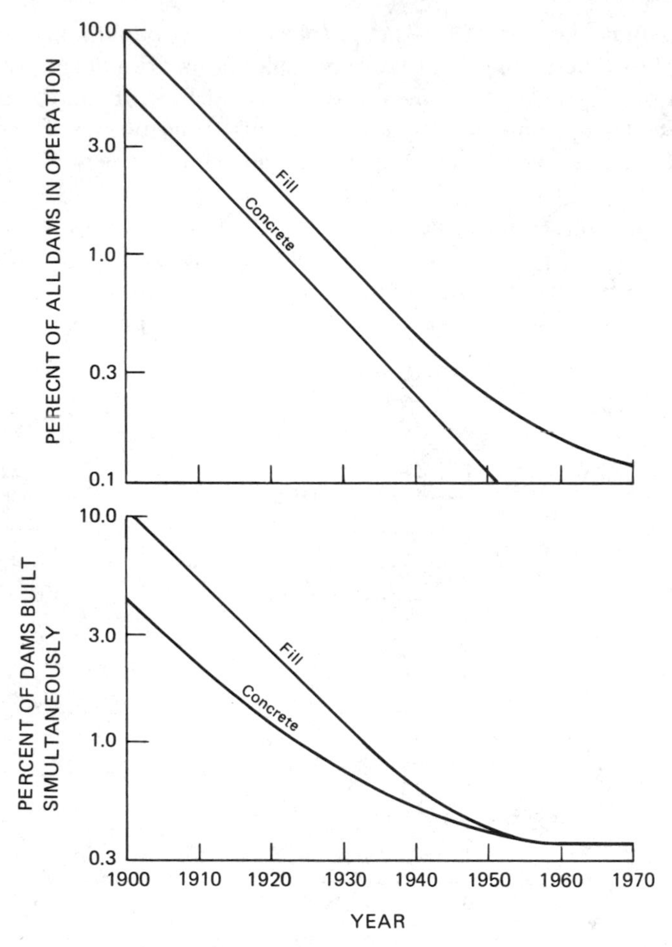


FIGURE 2-5 Probability of failure (Western Europe and USA). SOURCE: ICOLD (1979).

a commentary to provide the reader with a better understanding of the data. Thus in the following descriptions of each category of cause identified in Table 2-1, additional information is given about the involved incidents.

*Overtopping*

Overtopping caused about 26% of the reported failures and represents about 13% of all incidents. The principal reason for overtopping was inadequate spillway capacity. However, in 2 failure cases overtopping was attributed to blockage of the spillways and in 2 others to settlement and erosion of the embankment crest, thus reducing the freeboard. In 1 of the latter cases the settlement was great enough to lower the elevation of the top of embankment below that of the spillway crest.

Six concrete dams have failed due to overtopping and 3 others were involved in accidents. Two of the overtopping failures resulted from instability due to erosion of the rock foundation at the toe of dam, and 4 were due to the washout of an abutment or adjacent embankment structure. In 1 of these events a saddle spillway was first undermined and destroyed, and then the abutment ridge between the spillway and the dam was lost by erosion. In one instance of erosion of the rock at the toe of the dam, piping was suspected as a contributing cause.

The 3 overtopping accidents reported for concrete dams involved erosion of the downstream foundation in only 1 case. In another instance the powerhouse and equipment were damaged, but the dam sustained no damage.

**TABLE 2-1** Causes of Dam Incidents

| | Type of Dam | | | | | | | | |
|---|---|---|---|---|---|---|---|---|---|
| | Concrete | | Embankment | | Other* | | Totals | | |
| Cause | F | A | F | A | F | A | F | A | F&A |
| Overtopping | 6 | 3 | 18 | 7 | 3 | | 27 | 10 | 37 |
| Flow erosion | 3 | | 14 | 17 | | | 17 | 17 | 34 |
| Slope protection damage | | | | 13 | | | | 13 | 13 |
| Embankment leakage, piping | | | 23 | 14 | | | 23 | 14 | 37 |
| Foundation leakage, piping | 5 | 6 | 11 | 43 | 1 | | 17 | 49 | 66 |
| Sliding | 2 | | 5 | 28 | | | 7 | 28 | 35 |
| Deformation | | 2 | 3 | 29 | 3 | | 6 | 31 | 37 |
| Deterioration | | 6 | 2 | 3 | | | 2 | 9 | 11 |
| Earthquake instability | | | | 3 | | | | 3 | 3 |
| Faulty construction | 2 | | | 3 | | | 2 | 3 | 5 |
| Gate failures | 1 | 2 | 1 | 3 | | | 2 | 5 | 7 |
| TOTAL | 19 | 19 | 77 | 163 | 7 | | 103 | 182 | 285 |

*Steel, masonry-wood, or timber crib.

F = failure.
A = accident = an incident where failure was prevented by remedial work or operating procedures, such as drawing down the pool.

SOURCE: Compiled from *Lessons from Dam Incidents, USA*, ASCE/USCOLD 1975, and supplementary survey data supplied by USCOLD.

TABLE 2-2 Causes of Concrete Dam Incidents

| | Concrete Dam Type | | | | | | | | |
|---|---|---|---|---|---|---|---|---|---|
| | Arch | | Buttress | | Gravity | | Totals | | |
| | F | A | F | A | F | A | F | A | F&A |
| Overtopping | 2 | 1 | 1 | | 3 | 2 | 6 | 3 | 9 |
| Flow erosion | 1 | | 1 | | 1 | | 3 | | 3 |
| Foundation leakage, piping | 1 | 1 | 2 | | 2 | 5 | 5 | 6 | 11 |
| Sliding | | | | | 2 | | 2 | | 2 |
| Deformation | | 2 | | | | | | 2 | 2 |
| Deterioration | | 3 | | 2 | | 1 | | 6 | 6 |
| Faulty construction | | | | | 2 | | 2 | | 2 |
| Gate failures | | | | | 1 | 2 | 1 | 2 | 3 |
| TOTAL | 4 | 7 | 4 | 2 | 11 | 10 | 19 | 19 | 38 |

F = failure.
A = accident.

SOURCE: Compiled from *Lessons from Dam Incidents, USA*, ASCE/USCOLD 1975, and supplementary survey data supplied by USCOLD.

In the third, structural cracking was believed to have been caused by the overtopping load on the structure, resulting in subsequent reservoir leakage through the dam.

### *Flow Erosion*

This category includes all incidents caused by erosion except for overtopping, piping, and failure of slope protection. Flow erosion caused 17% of the failures and 12% of all reported incidents. Of the 17 failures, 14 were at embankment dams where, except in 2 cases, the spillways failed or were washed out. In 1 instance the gate structure failed due to erosion of its foundation, and in another the embankment adjacent to the spillway weir was washed out. In the latter case, overtopping and/or poor compaction of the spillway-embankment interface was suspected but not confirmed. With respect to the 3 concrete dam failures, the spillways were destroyed in 2 instances and in the other, a small buttress dam, the entire dam was destroyed.

The 17 reported accidents relating to flow erosion all involved embankment dams. In 1 case the downstream embankment slope was eroded, and in 2 other instances erosion of the outlets was involved. Two of the accidents actually were due to cavitation erosion in the tunnels. The remaining 12 accidents involved the loss or damage to spillway structures.

### *Slope Protection Damage*

Damage to slope protection was not reported to be involved in any failures; however, in 1 accident the undermining of riprap by wave action led to embankment erosion very nearly breaching the dam. The 13 reported accidents represent about 4% of all incidents. Of the 13 accidents, 6 involved concrete protection and the others riprap. In some of the latter cases the wave action pulled fill material through the riprap, and in the others riprap was either too small or not durable.

### *Embankment Leakage and Piping*

Embankment leakage and piping accounted for 22% of the failures and 13% of all reported incidents. In 5 of the 37 incidents piping is known to have occurred along an outlet conduit or at the interface with abutment or concrete gravity structure.

### *Foundation Leakage and Piping*

Foundation leakage and piping accounted for 17% of all failures and 24% of all reported incidents. It is the number one cause of all incidents. Six concrete dams, 1 steel dam, and 11 embankment dams were involved in the 18 failures. In at least 11 of the 49 accidents, which involved 6 concrete and 43 embankment dams, the leakage occurred in the abutments. Some reports cite inadequate grouting or relief wells and drains as causing the leakage and piping. In 1 event piping was caused by artesian pressures and not reservoir water.

### *Sliding*

This category covers instability as represented by sliding in foundations or the embankment or abutment slopes. Sliding accounted for 6% of all failures and 12% of all incidents reported. Of the 6 failures, 1 was a concrete gravity structure where, during first filling, the structure's slide downstream of about 18 inches was preceded by a downstream abutment slide, followed by large quantities of water leaking from the ground just downstream of the dam. The reservoir was emptied successfully, but before repairs were accomplished, the reservoir filled again causing large sections of the dam to "overturn or open like a door." The 5 embankment failures occurred in the downstream slopes, 1 due to excessively steep slopes and the others probably due to excessive seepage forces.

All of the 28 reported sliding accidents involved embankment dams. In 2 cases the slides occurred in abutment slopes, in 10 cases in the downstream

**TABLE 2-3** Earth Dam Failures

| Form | General Characteristics | Causes | Preventive or Corrective Measures |
|---|---|---|---|
| | | *Hydraulic Failures (30% of all failures)* | |
| Overtopping | Flow over embankment, washing out dam. | Inadequate spillway capacity. | Spillway designed for maximum flood. |
| | | Clogging of spillway with debris. | Maintenance, trash booms, clean design. |
| | | Insufficient freeboard due to settlement, skimpy design. | Allowance for freeboard and settlement in design; increase crest height or add flood parapet. |
| Wave erosion | Notching of upstream face by waves, currents. | Lack of riprap, too small riprap. | Properly designed riprap. |
| Toe erosion | Erosion of toe by outlet discharge. | Spillway too close to dam. | Training walls. |
| | | Inadequate riprap. | Properly designed riprap. |
| Gullying | Rainfall erosion of dam face. | Lack of sod or poor surface drainage. | Sod, fine riprap; surface drains. |
| | | *Seepage Failures (40% of all failures)* | |
| Loss of water | Excessive loss of water from reservoir and/or occasionally increased seepage or increased groundwater levels near reservoir. | Pervious reservoir rim or bottom. | Banket reservoir with compacted clay or chemical admix; grout seams, cavities. |
| | | Pervious dam foundation. | Use foundation cutoff; grout; upstream blanket. |
| | | Pervious dam. | Impervious core. |
| | | Leaking conduits. | Watertight joints; waterstops; grouting. |

| | | | |
|---|---|---|---|
| | | Settlement cracks in dam. | Remove compressible foundation, avoid sharp changes in abutment slope, compact soils at high moisture. |
| | | Shrinkage cracks in dam. | Use low-plasticity clays for core, adequate compaction. |
| Seepage erosion or piping | Progressive internal erosion of soil from downstream side of dam or foundation backward toward the upstream side to form an open conduit or "pipe." Often leads to a washout of a section of the dam. | Settlement cracks in dam. | Remove compressible foundation, avoid sharp changes, internal drainage with protective filters. |
| | | Shrinkage cracks in dam. | Low-plasticity soil; adequate compaction; internal drainage with protective filters. |
| | | Pervious seams in foundation. | Foundation relief drain with filter; cutoff. |
| | | Pervious seams, roots, etc., in dam. | Construction control; core; internal drainage with protective filter. |
| | | Concentration of seepage at face. | Toe drain; internal drainage with filter. |
| | | Boundary seepage along conduits, walls. | Stub cutoff walls, collars; good soil compaction. |
| | | Leaking conduits. | Watertight joints; waterstops; materials. |
| | | Animal burrows. | Riprap, wire mesh. |

**TABLE 2-3** Earth Dam Failures (*continued*)

| Form | General Characteristics | Causes | Preventive or Corrective Measures |
|---|---|---|---|
| | | *Structural Failures (30% of all failures)* | |
| Foundation slide | Sliding of entire dam, one face, or both faces in opposite directions, with bulging of foundation in the direction of movement. | Soft or weak foundation. | Flatten slope; employ broad berms; remove weak material; stabilize soil. |
| | | Excess water pressure in confined sand or silt seams. | Drainage by deep drain trenches with protective filters; relief wells. |
| Upstream slope | Slide in upstream face with little or no bulging in foundation below toe. | Steep slope. | Flatten slope or employ berm at toe. |
| | | Weak embankment soil. | Increased compaction; better soil. |
| | | Sudden drawdown of pond. | Flatten slope, rock berms; operating rules. |
| Downstream slope | Slide in downstream face. | Steep slope. | Flatten slope or employ berm at toe. |
| | | Weak soil. | Increased compaction; better soil. |
| | | Loss of soil strength by seepage pressure or saturation by seepage or rainfall. | Core; internal drainage with protective filters; surface drainage. |
| Flow slide | Collapse and flow of soil in either upstream or downstream direction | Loose embankment soil at low cohesion, triggered by shock, vibration, seepage, or foundation movements. | Adequate compaction. |

SOURCE: Sowers (1961).

slope, in 11 cases in the upstream slope, and in 2 cases both in the upstream and downstream slopes. In 1 instance the slide was reported to have occurred in the foundation and to be due to very steep embankment slopes. Three reports did not indicate the location of the slope slides. Of the 11 slides in the upstream slope, 6 occurred during or immediately following reservoir drawdown. In several cases heavy rains preceded the slides, and 3 of the slides were known to have occurred in clay foundation layers.

### *Deformation*

This category covers instability cases other than those involving sliding. Of the 6 failures, 3 involved timber crib dams where either the logs slipped out of their sockets or ice or flood flows breached the dam. The other 3 failures were embankment dams where, in one case, deformation of the outlet pipe permitted the outward leakage of the full-flowing pipe, causing piping of the embankment. In another case ice pressures displaced the intake riser of the outlet works. In the third embankment dam the concrete intake riser collapsed, with resulting leakage and piping along the conduit barrel.

Of the 31 reported accidents, 29 occurred at embankment dams. However, in 19 of these cases the outlet or spillway was involved. In 5 instances the accident occurred in tunnels where serious leakage developed in 4 instances; in the other a complete blowout occurred. Excessive cracking, shearing, or collapse of outlet pipes occurred in 7 cases, in some instances due to differential settlement of the embankment. Failures of a valve structure, drop structure, and intake structure, the latter due to ice forces, were reported at 3 other embankment dams. One dam, a rockfill structure with a masonry shell, developed serious cracking in the shell due to differential settlements. In 3 instances differential settlements damaged spillway structures.

In the 12 accidents where ancillary structures were not involved, differential settlement of the embankment led to transverse and/or longitudinal embankment cracks. Some leakage and piping occurred at the location of transverse cracks.

### *Deterioration*

Two of the failures and 9 of the accidents were caused by deterioration. The 2 failures involved corrosion of outlet pipes, which allowed leakage and piping of embankment material into the outlet. The 9 accidents involved 3 embankment dams and 6 concrete dams. At the 3 embankment dams, leakage with piping of embankment material into the conduit was caused by pipe corrosion in 2 cases and by concrete deterioration in the other.

At 3 concrete dams the accidents were due to concrete deterioration caused by freeze-thaw damage. At another, alkali reactivity was the cause. Corrosion of the penstock and deterioration of timber bulkhead were listed as causes of the accidents at 2 concrete dams.

### *Earthquake Instability*

Three incidents of earthquake instability were reported—all considered to be accidents. Two of these were the Lower and Upper San Fernando (Van Norman) dams that were damaged during the 1971 San Fernando earthquake (Seed et al. 1973). These incidents are listed as accidents because reservoir water was not released downstream; however, essentially complete reconstruction of the dams was required. The other was the Hebgen Dam in Montana, which was damaged by the 1959 Madison Valley earthquake.

### *Faulty Construction*

Faulty construction was listed as the cause of 2 failures and 3 accidents. The failures occurred in concrete gravity dams and in 1 case was attributed to the omission of reinforcing steel. The 3 accidents occurred at embankment dams and in 2 cases were caused by poor bonding between old and new embankment material, leading to seepage and slope failures (in 1 case during drawdown). At the third, poor concrete tunnel construction led to severe leakage through construction joints and spalled areas.

### *Gate Failures*

Spillway gate failure was listed as the cause of failure of the dam in 2 cases. Gate or valve failure was the cause of 5 of the reported accidents, resulting in damage to downstream structures and/or loss of reservoir pool.

## Effects of Age and Aging

Data published by ASCE/USCOLD (1975) and ICOLD (1973) show that older dams have failed or suffered serious accidents approaching failure more frequently than dams of recent vintage. This is largely attributed to better engineering and construction of modern dams, especially since about 1940. The records also show that failures and accidents have been more frequent during first filling and in the early years, mostly due to design or construction flaws or latent site defects. Then follows an extended period of gradual aging with reduced frequency of failure and accidents during midlife. The frequency of accidents, but not failures, has then increased during later life, although some failures have occurred even after more than 100 years of satisfactory service.

Weathering and mechanical and chemical agents can gradually lead to accident or failure unless subtle changes are detected and counteracted. The engineering properties of both the foundation and the materials composing a dam can be altered by chemical changes that occur with time. Dams constructed for the purposes of water quality control, sewage disposal, and for storing manufacturing and milling wastes, such as tailings dams, are particularly susceptible to changes from chemical action. Foundation shearing strengths and bearing capacities can be reduced and permeabilities can be increased by dissolution. Progression of solution channeling in limestone foundations is a widespread problem. The permeabilities of critically precise filter zones and drain elements can be reduced or obstructed by precipitates. The effectiveness of cement grout curtains can be reduced by softening, solutioning, and chemical attack.

These time-related changes occur not only where chemical and industrial wastes are present in the stored water but also where the foundations are gypsiferous or calcareous or where the embankment zones have been constructed using deposits similarly constituted. If the mineral content of the stored water is very low, these changes can occur more rapidly.

Concrete can gradually deteriorate and weaken from leaching and frost action. Alkali-aggregate reaction in concrete is irreversible and can gradually destroy the integrity of the structure (Jansen et al. 1973).

Cracking of concrete in masonry dams should never be disregarded. Most cracks caused by shrinkage and temperature during the early period after construction do not penetrate deeply enough to be a threat to the dam's stability. However, sometimes cracking to significant depth can endanger the stability. This is because the monolithic behavior of the dam is affected, causing higher stress concentrations, and water pressure has freer access to the interior of the dam, causing higher pore pressures (principally uplift). Also, freeze-thaw damage to concrete is accelerated by the presence of cracks.

The metal components of appurtenant structures, such as trash racks, pipe, gates, valves, and hoists, gradually corrode unless continuously maintained. Deterioration can be rapid in an acidic environment. Unless continuously wet in a freshwater environment, timber structures such as cribbing will eventually decay from water content cycling and insect infestation and attack by organisms. Low-quality riprap will soften and disintegrate, destroying its effectiveness for erosion and slope protection.

About one-half of the dams inventoried by the U.S. Army Corps of Engineers (1982b) under the National Dam Inspection Program (PL 92-367) were constructed prior to 1960. Many of these dams can be expected to possess some of the above symptoms of aging. Even the more recently built dams may show signs of deterioration. Therefore, it is essential that periodic inspections be made to detect such symptoms and that timely measures be taken to arrest and correct the deficiencies.

## FIELD INSPECTIONS

An effective inspection program is essential to properly maintain a project in a safe condition. The program should involve three grades, or types, of inspections: (1) periodic technical inspections, (2) periodic maintenance inspections, and (3) informal observations by project personnel. Technical inspections are those involving specialists familiar with the design and construction of dams and include an assessment of the safety of project structures. Maintenance inspections are those performed at a greater frequency than technical inspections in order to detect at an early stage any significant developments in project conditions and involve consideration of operational capability as well as structural stability. The third type of inspection is actually a continuing effort performed by onsite project personnel (dam tenders, powerhouse operators, maintenance personnel) in the course of performing their normal duties.

### Technical Inspections

#### *Frequency of Inspections*

The frequency of technical inspections should depend on a number of factors. A dam that has not been properly inspected by experts for some years or a new or reconstructed dam should be inspected rather frequently in order to establish baseline data, information, and general familiarity. Initially, semiannual inspections would be prudent. These inspections are in addition to more frequent (say daily or weekly) visits by the regular caretakers or operators. It is advisable to have inspections made under variable operating conditions such as:

- Reservoir level down, so that the upstream face and abutments as well as the reservoir rim can be inspected.
- Reservoir full or preferably spilling. This permits checking for leakage or piezometer pressure under maximum head conditions. It also allows the inspector to assess hydraulic conditions of the spillway and its energy dissi-

pator. Operations of gates and valves can be checked, and downstream flow conditions can be assessed. Spillway approach conditions and potential debris problems can be reviewed.

As the inspectors become more familiar with the dam, and adequate data have been compiled, frequency of inspections may be reduced to perhaps once per year, then in some cases to once in 2 or more years, depending on the hazard rating. It is recommended that these "expert" inspections never be extended beyond 5 years even under the best of dam conditions.

Special inspections should always be conducted following any major problems or unusual event, such as earthquake, flood, vandalism, or sabotage.

### *Inspection Staff*

The inspection staff should be multidisciplined and need not include each member of the team each time. Most critical to the inspection is a civil engineer with significant (10 or more years) experience in the design and evaluation of dams. An engineer whose sole experience has been in earth dams would obviously not be the best-qualified engineer to assess a concrete or masonry dam and vice versa. There is considerable value in having an independent (not a member of the owner's staff) civil engineer on the inspection team. This provides a peer review considered to be extremely valuable in obviating bias. This should not preclude the owner from having a staff civil engineer accompany the independent engineer. It is extremely important to have the operating staff member, preferably the normal caretaker, assist in the inspection. This gives him an opportunity to learn what to look for in his frequent visits to the dam and permits the "experts" to gain firsthand information from him. A geologist should be a member of the team on its initial inspection and at about 5-year intervals on others, particularly where problems relating to geology are suspected or known to exist. Some continuity of inspection personnel from year to year is important. The first inspection should, if at all possible, include interviews with the original designer, the owner, the constructor, and current as well as previous caretakers/operators.

### *Inspection Scope*

The field examination must be both systematic and comprehensive because very subtle changes or visual indicators can often be important in the evaluation of an existing or potential safety problem. Probably the greatest value of such an inspection is the direct and early disclosure of obvious, developing, or incipient conditions that threaten the integrity of the dam. Often, on the basis of this visual examination alone, an experienced engineer can judge the severity of any problems and determine the rapidity with which remedial measures should be taken to prevent a failure or a serious accident that might lead to a failure. Liberal notetaking and photographs of items of interest are important factors in documenting the conditions noted and establishing the basis for determining and evaluating subsequent changes that may occur.

The geology and topography of the surrounding area, including the reservoir, should be inspected in order to assess general features and the quality of the foundation and the reservoir. Inspectors should look for discontinuities, slides, artificial cuts and fills, and signs of erosion, particularly in the vicinity of the spillway and the dam/foundation contact.

Existing records, such as preceding inspection reports and notes, water levels, spill and leakage records, movement survey results, photographs, and piezometric and other instrumentation records, should all be reviewed. The adequacy of the existing records and their maintenance also should be reviewed.

A review of all past records should be made. These should, whenever available, include preconstruction investigation records, design criteria and design analysis records, and available construction records. Photos taken during initial construction, or subsequent photos, are often valuable.

The entire downstream face and, whenever feasible, the upstream face should be inspected for overall quality of materials, leaks, offsets, cracks, erosion, moisture, crazing, vegetation, and surficial deposits. Parapets, walls, spillway channels, galleries, and bridges also should be inspected for these items.

The spillway channel should be examined for erosion, condition of log booms, and susceptibility to blockage. The condition of gates and operating equipment, including motors, cables, chains and controls, should be noted and the gates operated if feasible.

Outlets, including conduits, gates, and machinery, should be inspected. Galleries should be checked for signs of seepage, leakage, internal pressures, and condition of drains or signs of blockage.

The evaluation of safety, which is a principal component of a technical inspection, is discussed in detail later in this chapter. Techniques and procedures for accomplishing technical dam inspections and making the associated safety evaluations are described in detail in *Recommended Guidelines for Safety Inspection of Dams* (Chief of Engineers 1975), *Safety Evaluation of Existing Dams* (U.S. Bureau of Reclamation 1980), and *Guide for Safety Evaluation and Periodic Inspection of Existing Dams* (Forest Service and Soil Conservation Service 1980).

### *Checklists and Inspection Forms*

It is extremely important that checklists and inspection report forms be used and completed for all inspections. These forms can be formal or informal. They should be completed during and immediately after the inspection, not the next day or later. The checklist should be developed prior to the inspection and should reflect the features peculiar to that particular project.

### *Photographs*

There is considerable value in aerial and close-up photographs, especially of areas of deterioration or those suspected of deterioration. Stereo-paired aerial photos are particularly valuable for reviewing possible progression of slides or of erosion of spillway dissipators or flow channels. They can also help in assessing downstream growth or habitation conditions. Infrared photos could be useful in locating wet areas that might otherwise not be obvious or detectable and that might indicate seepage.

## Maintenance Inspections

Formal maintenance inspections should be conducted on a semiannual-to-annual basis to monitor the behavior and condition of the structure and of all operating equipment. The inspection should be performed by an engineer or experienced supervisor of dam operations, who should note any adverse changes in physical conditions, such as erosion, corrosion, blockages of drains, blockages of spillway channels and other water passages, and subsidence. The condition and adequacy of all monitoring equipment and instruments also should be reviewed.

All gates and emergency power sources to operating equipment, including motors, cables, chains, and controls, should be inspected and operated if feasible. A principal objective of this inspection, besides offering an opportunity to check on aspects pertinent to the safety of project features, is to promote an efficient and effective maintenance program. It is desirable that such inspections be made by persons not directly involved in or responsible for the day-to-day operation and maintenance of the project.

## Informal Observations

Dam tenders and maintenance personnel should be charged with the responsibility of helping to monitor the behavior and safety of dams. If alert, such personnel could discover existing defects during routine operational and maintenance activities. For example, a mowing-machine operator cutting grass along the toe of an embankment may come across an upward bulging mass of freshly disturbed ground, which might indicate incipient slope instability. Or a gate tender approaching the gate controls might observe a small sinkhole on the embankment crest that could be developing from the piping of fines into a deteriorating outlet conduit. Rodent holes might also be discovered in this manner.

Personnel such as operators, maintenance crew members, and all others, including the owners of smaller dams, who are at a dam in the course of their normal duties should be watchful for any unusual events or strange conditions and should report them at once to those in authority. Alertness and inquisitiveness on the part of such individuals could afford an important surveillance program for the project.

## MAINTENANCE

Normal maintenance activities include the surveillance of the project's physical conditions and the timely correction of any deficiencies that might develop, as well as the preservation of the operating capability of the project. The inspections described in the section Maintenance Inspections should be an integral part of the maintenance program.

Maintenance activities should include the surveillance of all aspects of the structure pertinent to safety. For example, seepage or leakage through the foundation or abutment areas should be closely monitored. (This can take on greater importance, depending on the integrity of the material.) Also, uplift pressures are critical to stability. If instruments are not available to monitor this pressure, they should, if possible, be installed. Collecting, processing, and evaluating surveillance instrumentation data are ways to detect the development of defects in a dam and are helpful in the investigation of a specific or suspected defect. Often, records are collected, but processing and evaluation are delayed or long neglected. All data relating to dam safety should be promptly evaluated. Installations for the collection of precautionary types of surveillance data are usually made as a matter of course during construction but may also be installed after construction. Installations for investigating specific or suspected defects are usually made upon the appearance of new or changing conditions and events. Instrumentation for dams is discussed in Chapter 10.

Another important maintenance objective is to preserve the water-passing capability of the project. Heavy growth or landslides upstream or downstream of a spillway could reduce its ability to pass its design flow. Also, the electrical and mechanical operating machinery and the spillway gates should all be maintained to ensure their operability under all conditions. The failure of any one of these could lead to failure of the dam.

Maintenance is an ongoing process that should never be neglected and that should continue throughout the operating life of the dam. To provide proper maintenance services, all material regarding the design, construction, and operation of the dam should be available in a location where it is readily accessible for the inspection and maintenance programs.

## RECORDS

Complete records on each dam, including initial site investigations; preconstruction and final geologic reports; design assumptions and criteria; contract plans and specifications; construction history; descriptions of repairs or modifications; and documentation of conditions and performance after the dam is in operation, including history of major floods and instrumentation records, should be maintained. Construction photographs are extremely valuable. Experience has proven that many questions and concerns arise in the operating life of a dam for which thorough records are vital to assess such situations properly. This documentation should, of course, be continuously supplemented throughout the life of a dam by periodic inspection reports.

At present, some dams have adequate records, while many have little or none at all. An important objective of a periodic inspection program is to collect and develop such data. The inspection reports should eventually provide most of the information relating to safety of the dam, as they should usually contain a summary of major preoperational information and a documentation of all observations, assessments, damage, and repairs during operation. The importance of keeping such information well organized and readily available cannot be overemphasized.

In extracting and assimilating record data, the quality and accuracy of the records must be carefully assessed. If certain types of existing information, such as exploration and materials testing reports, are overlooked or are questionable, exploration and testing may have to be repeated, at considerable cost. It is imperative that all of these records be made available, not simply filed, to those involved in the evaluation of a dam's safety.

Comprehensive descriptions of all types of records and their utilization are contained in *Safety Evaluation of Existing Dams* (U.S. Bureau of Reclamation 1980), *Guide for Safety Evaluation and Periodic Inspection of Existing Dams* (Forest Service and Soil Conservation Service 1980), and *Feasibility Studies for Small Scale Hydropower Additions, Vol. IV, Existing Facility Integrity* (U.S. Army Corps of Engineers 1979).

## EVALUATION OF SAFETY

An evaluation of the stability and safety of an existing dam is a principal component of all technical inspections. Such an evaluation must be carefully and thoroughly performed by experienced personnel. It should consider all data of record, including design, construction, and operating history, and the results of a field inspection and any analyses necessary to determine the safety of project structures and the adequacy of maintenance and operating procedures.

A safety evaluation is generally amenable to a staged approach. The basic idea behind such an approach is to attempt to establish the integrity of a structure or to resolve a problem associated with it at the least possible cost. For example, if an adequate determination can be made from review and analysis of existing data and field observations, then that is all that should be done. If the review or observations indicate that additional special investigations are required to determine the condition of a facility or to evaluate and correct a problem, then these investigations can most effectively be planned on the basis of what the data review or observations show to be required. For the foregoing reasons the first stage of the evaluation should include, as a minimum, three basic steps: (1) review of existing data, (2) site inspection, and (3) evaluation of data and formulation of conclusions.

### Review of Existing Data

A thorough knowledge must first be gained on the basis of a dam's original design and its performance history and records, to provide a basis for judgments that will be made later. Whatever data are available for review can normally be obtained from the owner's files or from the files of the state agency regulating the safety of dams, if such an agency exists and takes an active role in the particular state in which the facility is located.

#### *Design Data*

The review should reveal whether the original design criteria and assumptions are satisfactory based on the current state of the art and, if not, whether they are acceptable. Original design assumptions may have been inconsistent with construction conditions or with subsequent events and conditions. The review should include hydrology and spillway capacity, materials investigations and specifications, criteria for outlets and other appurtenances, all geological and seismological reports, and all design analyses.

Original design methodologies and techniques can be very meaningful. Equally important are data on any analyses or reviews subsequent to original design. The same applies to any modifications or alterations in design.

### *Construction Data*

Construction data (relating to original construction or alterations) are as important, if not more so, than design data. Inspections and engineers' reports relating to foundation cleanup; grouting; concreting operations, such as strengths, placing methods, cleanup, cement, water-cement ratio; and aggregate source are important. For example, in one case knowledge of the method of placing concrete aided in determining the cause of deterioration of the dam.

Confirmation of compliance with specifications or information on changes to suit field conditions are important. Often, in the past, field changes were made to save money without the designers' knowledge or to mitigate damages to the construction resulting from unexpected events, such as flooding or accidents. Very often, in the case of very old dams, calculations or records cannot be found. In these cases construction records are all the more important. They are often documented in old publications or photos. All of the above can be invaluable in reviewing the safety of a dam as well as in establishing investigational programs.

### *Operating Records and Maintenance*

Records of operation and maintenance activities are often more readily available to the reviewer than are design or construction records. Any dam owner should make a concerted effort to compile such records regardless of whether past records exist. In other words, a late start is better than no start. These records should be compiled whether the dam is presently considered to be in excellent condition, in good or fair condition, or a hazard or a high risk.

The reviewer (evaluator) should consider the frequency and quality of inspections. Infrequent inspections and casual maintenance should alert the evaluator to potential problems. At the very least the reviewer should undertake a more intensive evaluation unless inspections and maintenance have been frequent and of good quality. The housekeeping level at a dam is often a good indication of the level or quality of care given to the dam.

In the absence of suitable records the evaluator should establish a program of frequent inspections and should initiate survey and data collection procedures in order to establish baseline data. The level of this program depends on the hazards and risks presented by the dam.

Often an owner will have a lot of scattered, irregular, incomplete, or unreliable data. Data often need to be plotted. There is then a need for data analysis, in addition to collection. All data should be reviewed and challenged for reliability and completeness. Upon first evaluation the evaluator should gather information from all feasible sources. Information provided by a regular dam caretaker, for example, is probably better and more reliable than that provided by a nearby powerhouse operator. A water superintendent may be a reliable source.

Of particular value are data on leakage, water levels, deflections, flood levels at the dam and downstream thereof, oral and written comments about repairs, and reports of operating problems with equipment. All of the above should be plotted on a time scale to permit evaluation of interrelationships of the data. A dam's age, quality of maintenance, presence of operating records, and apparent deterioration all have a bearing on its safety.

It is important to acknowledge that often the worst problem at a dam may not be the dam itself but a lack of knowledge thereof. It is impossible to evaluate a dam's safety without knowledge of the structure. The greater the knowledge, the better the evaluation. Instrument data (from the dam) should be programmed for computer storage and analysis. This method of data handling will permit development of a computer analysis program that will 'red-flag' critical data points. For example, piezometer data can be programmed to indicate with a special symbol when the readings approach a critical level of uplift or pore pressure. Thus, when the evaluator scans the computer listing his attention will be drawn to potentially dangerous conditions.

## Site Inspection

Once the evaluator has thoroughly reviewed the existing data, a site inspection should be performed to observe pertinent visual evidence. This inspection should conform to all the requirements of a formal technical inspection and should provide the evaluator with intimate knowledge of project conditions and problems. It should also provide the evaluator the opportunity to resolve any discrepancy or question that may exist concerning record data such as drawings, instrumentation data, or operating procedures.

## Hazard Potential

A safety evaluation should include a review of the dam's hazard potential and should determine whether new developments in the downstream area substantiate a change in the hazard level. (See section on Classification of Inundation Areas.) In this connection the emergency action plan (discussed later) should be reviewed to ensure that it reflects the current hazard potential status.

## Evaluation and Conclusions

After all available data have been reviewed and the site has been examined in detail, the evaluator should analyze all pertinent information revealed by the record, all conditions observed at the site, and the results of any engineering calculations.

One of the most useful techniques to apply in the process of evaluating the safety and stability of an existing dam and its appurtenant works is to compare performance, as indicated by field observations, instrumentation measurements, and the results of any required special investigations performed to evaluate a specific problem, with the assumptions and calculations made in the original design of the facility. In doing this the engineer will often be made aware of criteria that were in vogue at the time the design was originally accomplished. In many cases such criteria will still be appropriate. However, it is important to bear in mind that the state of the art is not static. It changes as engineering knowledge and technology advance and as natural events occur that deviate from prior experience. As a result the reliability of designs based on the state of the art that existed when the dam was designed and constructed must always be compared with existing practices.

Following the analyses of all the data the evaluator should prepare a report detailing his findings. If sufficient information is available to make a judgment regarding the project's stability and safety, the report should include such a conclusion with any associated recommendations. If not, the report should detail the additional information needed and recommend the investigations required to develop the needed data.

It is a well-known fact that in the initial assessment or reassessment of an existing dam, particularly an old one, the unknown is the principal determinant for an investigation or exploratory program. Lack of knowledge makes it virtually impossible to determine the potential hazard of a dam and makes a viable risk assessment impossible. It is also axiomatic that the more dams that are investigated, or the more a single dam is investigated, the more likely that a potential or real problem will be discovered. Only then can the problem be assessed and corrected if necessary.

## Additional Investigations

The need and nature of additional investigations, exploratory programs, or monitoring programs will depend on the potential for problems as determined by the inspection and on the availability of good information and records. They will also depend on the evaluator's assessments of risks and consequences of failure. Such investigations may involve theoretical studies as well as field investigations. Additional studies are sometimes needed to better define the stress and stability conditions and to evaluate alternative remedial measures. The investigative program can consist of a wide variety of tasks, depending on the nature of the known or suspected problem. The type of supplemental information and numerical data needed will concern structural, geologic, and performance features unobtainable by direct visual examination. Some kind of exploration may be required for sample extraction; for providing access for direct observation; and for instrumental measurements of deformation, hydrostatic pressures, seepage, etc. Data may also be obtained by nondestructive testing. Laboratory tests may be required to determine engineering properties of the materials of the dam and appurtenances and of the foundation for use in analyses and to assess their general condition. Performance instrumentation may be required. Applicable techniques of subsurface exploration, geologic mapping, laboratory testing, and instrumentation are described in numerous excellent references, such as the *Handbook of Dam Engineering* and various other references listed in Chapters 5 through 10 of this volume.

After additional data have been obtained, the engineering analyses and methods employed are generally similar to those that would have been conducted in the initial evaluation had the data been available. Particular care should be taken to study suspicious or questionable features and conditions. The engineering data and information to be used in the analyses are those specifically obtained for that purpose during the investigations. For example, unless available data on spillway design indicate conclusively that the spillway meets present-day design standards, a new flood estimate should be made, and the existing spillway should be analytically tested for its ability to safely handle the updated flood. Or, as another example, if the stability of an embankment dam appears marginal for any reason (such as apparently over-steep slopes, unusual saturation patterns, low-strength soils, or indications of high foundation pore pressures), a stability analysis and companion seepage analysis should be made using soil strengths and permeability rates obtained by sampling and testing for use in those specific analyses.

As valuable as they are, numerical analyses cannot provide total and absolute answers upon which to base the evaluation. Many physical conditions and reactive mechanisms cannot be mathematically analyzed. Therefore, after all the objective factors that may influence the evaluation have been gathered, interpreted, analyzed, and discussed, the investigator must still exercise judgment as to whether the dam is adequate in its present condition or requires remedial or other measures. There are no clear-cut rules by which the decisions can be made. Instead, the investigator may need to employ empirical reasoning and objective assessments, compare the case with successfully performing similar dams, and apply criteria in common use by the profession. When the perceived problems involve areas of specialized engineering practice and there would be significant losses from failure of the structure, experts in the pertinent specialties should be brought in as consultants.

## EMERGENCY ACTION PLANNING

### Current Policies and Practices

While the intent of dam design, construction, operation, maintenance, and inspection of dams is to minimize the risk of dam failures, it is recognized that the possibility of dam failures still exists. Even though the probability of such failures is usually small, preplanning is required to (1) identify conditions that could lead to failure, in order to initiate emergency measures to prevent such failures as a first priority, and (2) if this is not possible, to minimize the extent and effects of such failures. The operating and mobilizing procedures to be followed upon indication of an impending or postulate dam failure or a major flood should be carefully predetermined.

Following the failure of Teton Dam in 1976, President Carter directed the appropriate federal agencies to develop guidelines for dam safety. Subsequently, in June 1979, *Federal Guidelines for Dam Safety* was published by the Federal Coordinating Council for Science, Engineering and Technology. While these guidelines were developed to encourage high safety standards in organizational and technical management activities and procedures of federal agencies, they are also considered applicable to state dam safety agencies and private and nonfederal dam owners.

A basic tenet of these guidelines is that an emergency action plan, commensurate with the dam size and location (i.e., hazard classification), should be formulated for each dam. The guidelines require an evaluation of the emergency potential created from a postulated dam failure by use of flood inundation maps; development of an emergency action plan, coordinated with local civil preparedness officials; and a formal procedure to detect, evaluate, and mitigate any potential safety problem. Owners of private dams should evaluate the possible modes of failure of each dam, be aware of indicators or precursors of failure for each mode, and consider the possible emergency actions appropriate for each mode and the effects on downstream areas of failure by each mode. Evaluation should recognize the possibility of failure during flood events as well as during normal operating conditions and should provide a basis for emergency planning actions in terms of notification and evacuation procedures where failure would pose a significant danger to human life and property. Plans should then be prepared in a degree of detail commensurate with the hazard, and instructions should be provided to operators and attendants regarding the actions to be taken in an emergency. Planning should be coordinated with local officials, as necessary, to enable those officials to draw up a workable plan for notifying and evacuating local communities when conditions threatening dam failure arise.

Some states and several federal agencies have already developed their own emergency action planning guidelines and have implemented plans at dams consistent with the major elements contained in the *Federal Guidelines*. Among the federal agencies, the U.S. Army Corps of Engineers (1980) has published *Flood Emergency Plans, Guidelines for Corps Dams*. The Corps publication merits consideration by private dam owners for its detailed procedure and case study examples. Additionally, the Corps (1982a) has recently published a manual, *Emergency Planning for Dams, Bibliography and Abstracts of Selected Publications*, for assisting planners with relevant materials and references to emergency planning for dams and preparation of flood evacuation plans.

Finally, the user is directed to technical guidelines and recommendations on emergency action planning for federal agencies that have been published by a subcommittee of representatives from federal agencies having responsibilities for dam safety for the Interagency Committee on Dam Safety (ICODS) (FEMA 1982).

### Evaluation of Emergency Potential

Prior to development of an emergency action plan, consideration must be given to the extent of land areas and the types of development within the areas that would be inundated as a result of dam failure and to the probable time available for emergency response.

#### *Determination of Mode of Dam Failure*

There are many potential causes and modes of dam failure, depending on the type of structure and its foundation characteristics. Similarly, there are degrees of failure (partial vs. complete) and, often, progressive stages of failure (gradual vs. sudden). Many dam failures can be prevented from reaching a final catastrophic stage by recognition of early indicators or precursor conditions and by prompt, effective emergency actions. While emergency planning should emphasize preventive actions, recognition must be given to the catastrophic condition, and the hazard potential must be evaluated in that light. Analyses should be made to determine the most likely mode of dam failure under the most adverse condition and the resulting peak water outflow following the failure. Where there is a series of dams on a stream, analyses should include consideration of the potential for progressive "domino effect" failure of the dams. Appendix A of Chapter 4 provides an example of guidelines on estimating modes of dam failure for formulating emergency action plans by an investigator-owned utility.

### *Inundation Maps*

To evaluate the effects of dam failure, maps should be prepared that delineate the area that would be inundated in the event of failure. Inundation maps should account for multiple dam failures where such failures are possible. Land uses and significant development or improvements within the area of inundation should be indicated. The maps should be equivalent to or more detailed than the United States Geological Survey (USGS) 1:24,000-scale quadrangle maps, 7.5-minute series, or of sufficient scale and detail to identify clearly the area that should be evacuated if there is evident danger of failure of the dam. Copies of the maps should be distributed to local government officials for use in the development of an evacuation plan. Figure 2-6 is a sample inundation map.

A 1980 dam break flood study of 50 dams located in Gwinnett County, Georgia (prepared by cooperating state and federal agencies for the county) reported flood inundation study results on 1:12,000 (1 inch = 1,000 feet) maps, which were scaled from the USGS maps (Georgia Environmental Protection Division 1980).

### *Classification of Inundation Areas*

To assist in the evaluation of hazard potential, areas delineated on inundation maps should be classified in accordance with the degree of occupancy and hazard potential. The potential for loss of life is affected by many factors, including but not limited to the capacity and number of exit roads to higher ground and available transportation.

Hazard potential is greatest in urban areas. Since the extent of inundation is usually difficult to delineate precisely because of topographic map limitations, the evaluation of hazard potential should be conservative. The hazard potential for affected recreation areas varies greatly, depending on the type of recreation offered, intensity of use, communication facilities, and available transportation. The potential for loss of life may be increased

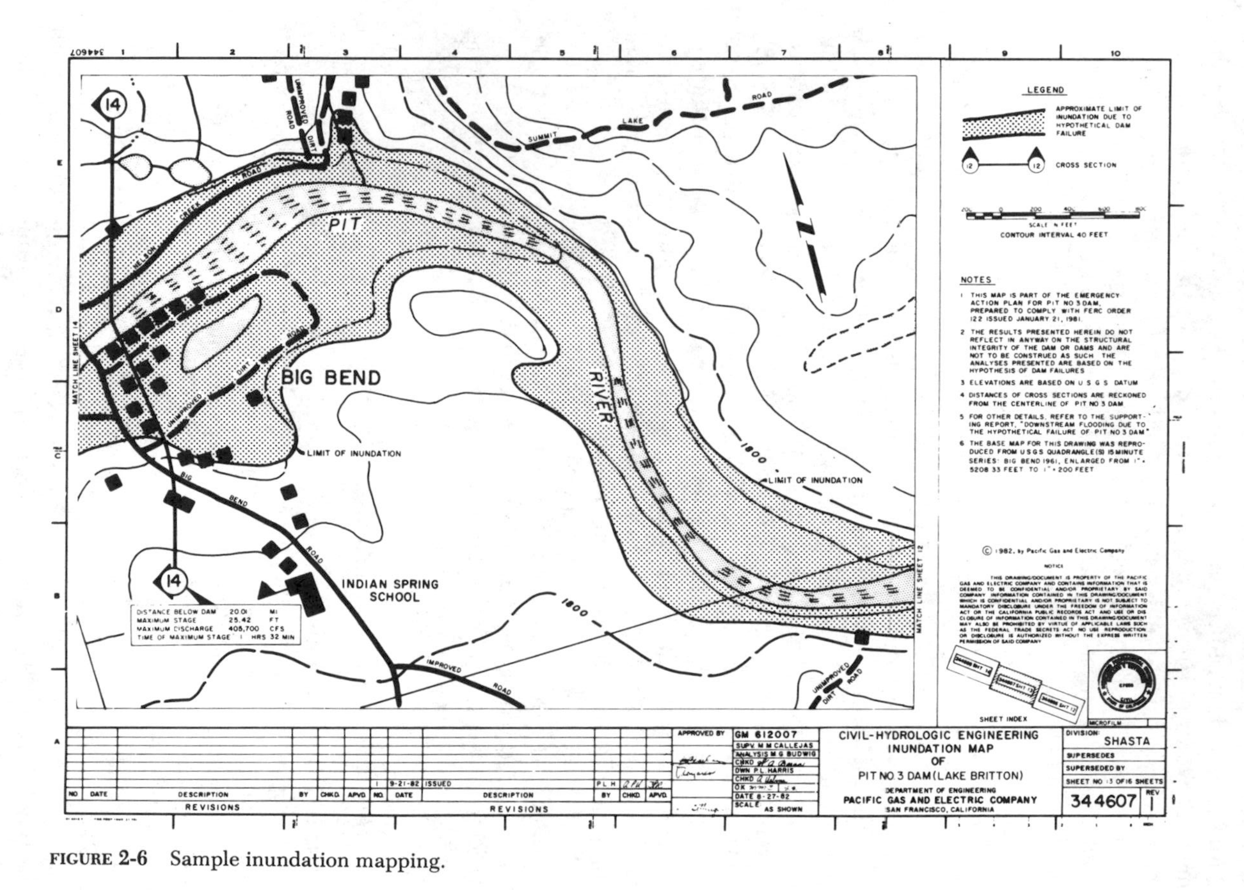


FIGURE 2-6 Sample inundation mapping.

where recreationists are widely scattered over the area of potential inundation, since they would be difficult to locate on short notice.

Many industries and utilities requiring substantial quantities of water for one or more stages in the manufacture of products or generation of power are located on or near rivers or streams. Flooding of these areas and industries can (in addition to causing the potential for loss of life and damage to machinery, manufactured products, raw materials, and materials in process of manufacture) interrupt essential community services.

Rural areas usually have the least hazard potential. However, the potential for loss of life exists, and damage to large areas of intensely cultivated agricultural land can cause high economic loss.

### *Time Available for Response*

Analyses should be made to evaluate the structural, foundation, and other characteristics of the dam and to determine those conditions that could be expected to result in slow, rapid, or practically instantaneous dam failure. Wave travel times, as discussed in Chapter 4, should also be established to help determine the time available for response.

## Actions to Be Taken to Prevent Failure or to Minimize Effects of Failure

### *Development of an Emergency Action Plan*

An emergency action plan should be developed for each dam that constitutes a hazard to life and property, incorporating preplanned emergency measures to be taken prior to and following assumed dam failure. The plan should be coordinated with local governmental and other authorities involved in public safety and should be approved by the appropriate top-level agency or owner management. To the extent possible, the emergency action plan should include notification plans, which are discussed in the section Notification Plans.

Emergency scenarios should be prepared for possible modes of failure of each dam. These scenarios should be used periodically to test the readiness capabilities of project staff and logistics.

A procedure should be established for review and revision, as necessary, of the emergency action plan, including notification plans and evacuation plans, at least once every 2 years. Such reviews should be coordinated among all organizations responsible for preparation and execution of the plans.

*Notification Plans*

Plans for notification of key personnel and the public are an integral part of the emergency action plan and should be prepared for slowly developing, rapidly developing, and instantaneous dam failure conditions. Notification plans should include a list of names and position titles, addresses, office and home telephone numbers, and radio communication frequencies and call signals, if available, for agency or owner personnel, public officials, and other personnel and alternates who should be notified as soon as emergency situations develop. A procedure should be developed to keep the list current.

Each type of notification plan should contain the order in which key owner supervisory personnel or alternates should be notified. At least one key supervisory level or job position should be designated to be manned or the responsible person should be immediately available by telephone or radio 24 hours a day. A copy of each notification plan should be posted in a prominent place near a telephone and/or radio transmitter. All selected personnel should be familiar with the plans and the procedures each is to follow in the event of an emergency. Copies of the notification plans should be readily available at the home and the office of each person involved.

Where dams located upstream from the dam for which the plan is being prepared could be operated to reduce inflow or where the operation of downstream dams would be affected by failure of the dam, owners and operators of those dams should be kept informed of the current and expected conditions of the dam as the information becomes available.

Civil defense officials having jurisdiction over the area subject to inundation should receive early notification. Local law enforcement officials and, when possible, local government officials and public safety officials should receive early notification. (In some areas such notification will be accomplished by civil defense authorities.)

The capabilities of the Defense Civil Preparedness Agency's National Warning System (NAWAS) should be determined for the project and utilized as appropriate. Information can be obtained from state or local civil defense organizations.

Potentially affected industries downstream should be kept informed so that actions to reduce risk of life and economic loss can be taken. Coordination with local government and civil defense officials would determine responsibility for the notification. Normally, this would be a local government responsibility.

When it is determined that a dam may be in danger of failing, the public officials responsible for the decision to implement the evacuation plan should be kept informed of the developing emergency conditions.

The news media, including radio, television, and newspapers, should be utilized to the extent available and appropriate. Notification plans should define emergency situations for which each medium will be utilized and should include an example of a news release that would be the most effective for each possible emergency. Use of news media should be preplanned insofar as is possible by agency and owner personnel and the state and/or local government. Information should be written in clear, concise language. Releases to news media should not be relied on as the primary means of notification.

Notification of recreation users is frequently difficult because the individuals are often alone and away from any means of ready communication. Consideration should be given to the use of standard emergency warning devices, such as sirens, at the dam site. Consideration should also be given to the use of helicopters with bullhorns for areas farther downstream. Vehicles equipped with public address systems and helicopters with bullhorns are capable of covering large areas effectively.

Telephone communication should not be solely relied on in critical situations. A backup radio communication system should be provided and tested at least once every 3 months. Consideration should be given to the establishment of a radio communication system prior to the beginning of construction and to the maintenance of the system throughout the life of the project.

### *Evacuation Plans*

Evacuation plans should be prepared and implemented by the local jurisdiction controlling inundation areas. This would normally not be the dam agency or owner. Evacuation plans should conform to local needs and vary in complexity in accordance with the type and degree of occupancy of the potentially affected area. The plans may include delineation of the area to be evacuated; routes to be used; traffic control measures; shelter; methods of providing emergency transportation; special procedures for the evacuation and care of people from such institutions as hospitals, nursing homes, and prisons; procedures for securing the perimeter and for interior security of the area; procedures for the lifting of the evacuation order and reentry to the area; and details indicating which organizations are responsible for specific functions and for furnishing the materials, equipment, and personnel resources required.

The assistance of local civil defense personnel, if available, should be requested in preparation of the evacuation plan. State and local law enforcement agencies usually will be responsible for the execution of much of the plan and should be represented in the planning effort. State and local laws and ordinances may require that other state, county, and local government agencies have a role in the preparation, review, approval, or execution of the plan. Before finalization, a copy of the plan should be furnished to the dam agency or owner for information and comment.

### *Stockpiling Repair Materials*

Where feasible, suitable construction materials should be stockpiled for emergency use to prevent failure of a dam. The amounts and types of construction materials needed for emergency repairs should be determined based on the structural, foundation, and other characteristics of the dam; design and construction history; and history of prior problems.

### *Locating Local Repair Forces*

Arrangements should be made with, and a current list maintained of, local entities, including contractors, and federal, state, and local construction departments for possible emergency use of equipment and labor.

### *Training Operating Personnel*

Owners of large impoundments should have technically qualified project personnel who are trained in problem detection, evaluation, and appropriate remedial (emergency and nonemergency) measures. These personnel should be thoroughly familiar with the project's operating manual. This is essential for proper evaluation of developing situations at all levels of responsibility that, initially, must be based on at-site observations. A sufficient number of personnel should be trained to assure adequate coverage at all times. If a dam is operated by remote control, arrangements must be made for dispatching trained personnel to the project at any indication of distress.

### *Increasing Inspection Frequency*

Frequency of appropriate surveillance activities should be increased when the reservoir level exceeds a predetermined elevation. Piezometers, water-level gauges, and other instruments should be read frequently and on schedule. The project structures should be inspected as often as necessary to monitor conditions related to known problems and to detect indications of change or new problems that could arise. Hourly or continuous surveillance may be mandated in some instances. Any change in conditions should be reported promptly to the supervisor for further evaluation.

The owner or his supervisor should issue additional instructions, as necessary, and alert repair crews and contractors for necessary repair work if developing conditions indicate that emergency repairs or other remedial measures may be required.

### Actions to Be Taken Upon Discovery of a Potentially Unsafe Condition

#### *Notification of Supervisory Personnel*

It is essential, if time permits, to notify the proper supervisory personnel since development of failure could vary in some or many respects from previous forecasts or assumptions and advice may be needed.

#### *Initiation of Predetermined Remedial Action*

At least one technically qualified individual, previously trained in problem detection, evaluation, and remedial action, should be at the project or on call at all times. Depending on the nature and seriousness of the problem and the time available, emergency actions can be initiated, such as lowering the reservoir and holding water in upstream reservoirs. Other actions to be taken include notifying appropriate highway and traffic control officials promptly of any rim slides or other reservoir embankment failures that may endanger public highways.

#### *Determination of Need for Public Notification*

To the extent possible, emergency situations that will require immediate notification of public officials in time to allow evacuation of the potentially affected areas should be predefined for the use of management and project personnel. If sufficient time is available the decision to notify public officials that the dam can be expected to fail will be made at a predetermined supervisory level within the agency or owner organization. If failure is imminent or has already occurred, project personnel at the dam site would be responsible for direct notification of the public officials. The urgency of the situation should be made clear so that public officials will take positive action immediately.

## REFERENCES

ASCE/USCOLD (1975) *Lessons from Dam Incidents, USA*, American Society of Civil Engineers, New York.

Chief of Engineers (1975) *Recommended Guidelines for Safety Inspection of Dams*, National

Program of Inspection of Dams, Vol. I, Appendix D, Department of the Army, Washington, D.C.

Federal Coordinating Council for Science, Engineering and Technology (1979) *Federal Guidelines for Dam Safety*, Federal Emergency Management Agency, Washington, D.C.

Federal Emergency Management Agency (1982) Interagency Committee on Dam Safety (ICODS), Subcommittee on Emergency Action Planning, *Dam Safety, Emergency Action Planning.*

Forest Service and Soil Conservation Service, U.S. Department of Agriculture (1980) *Guide for Safety Evaluation and Periodic Inspection of Existing Dams.*

Georgia Environmental Protection Division (1980) Georgia Soil and Water Conservation Committee, et al., *Dam Breach Flood Maps for Gwinnett Co., Georgia.*

Golze, A. R., ed. (1977) *Handbook of Dam Engineering*, Van Nostrand Reinhold Co., New York.

ICOLD (1973) *Lessons from Dam Incidents*, Abridged Edition, USCOLD, Boston, Massachusetts.

ICOLD (1979) *Transactions of New Delhi Congress.*

Independent Panel to Review the Cause of Teton Dam Failure (1976) *Failure of Teton Dam*, Report to U.S. Department of Interior and State of Idaho.

Jansen, R. B. (1980) *Dams and Public Safety*, U.S. Bureau of Reclamation, Government Printing Office, Washington, D.C. (reprinted in 1983).

Jansen, R. B., Carlson, R. W., and Wilson, E. L. (1973) *Diagnosis and Treatment of Dams*, Transactions of 1973 Congress, ICOLD, Madrid, Spain.

Seed, H. B., Lee, K. L., Idriss, I. M., and Makdisi, F. (1973) *Analysis of the Slides in the San Fernando Dams During the Earthquake of February 9, 1971*, Earthquake Engineering Research Center, University of California, Berkeley.

Sowers, G. F. (1961) "The Use and Misuse of Earth Dams," *Consulting Engineering*, July.

U.S. Army Corps of Engineers (1979) *Feasibility Studies for Small Scale Hydropower Additions, Vol. IV, Existing Facility Integrity.*

U.S. Army Corps of Engineers (1980) *Flood Emergency Plans, Guidelines for Corps Dams*, Hydrologic Engineering Center, Davis, Calif., June, 47 pp.

U.S. Army Corps of Engineers (1982a) *Emergency Planning for Dams, Bibliography and Abstracts of Selected Publications*, Hydrologic Engineering Center, Davis, Calif.

U.S. Army Corps of Engineers (1982b) *National Program for Inspection of Non-Federal Dams—Final Report to Congress.*

U.S. Bureau of Reclamation (1980) *Safety Evaluation of Existing Dams*, Government Printing Office, Washington, D.C.

U.S. Department of Interior Teton Dam Failure Review Group (1980) *Failure of Teton Dam, Final Report.*

# 3
# Risk-Based Decision Analysis

## SUMMARY

Engineering is inherently based on a weighing of risks. Traditionally, this has been drawn to a large extent from judgment reinforced by experience. As techniques of risk analysis offered in the literature have become increasingly sophisticated, practical engineers and related professionals have preferred to apply time-tested judgmental approaches rather than new techniques. Yet there is a need to improve methods of risk analysis in the engineering of dams and other structures whose safety is important to the public interest. This especially applies where funding for remedial work is limited and expenditures must be directed to achieve an optimum reduction of risk.

Those who advocate the use of advanced risk-based techniques must communicate, in understandable terms, the merits of their systems. Too often these have been presented on a general and overly technical statistical basis. There is a need to apply these numerical approaches to site specifics by merging theory with realistic appraisal of local conditions. Probability analysis is logically applied to natural events that affect projects, such as in calculation of the frequency and intensity of rainfall that may recharge the water in an incipient landslide or of an earthquake that may trigger movement of an earth or rock mass. These applications are well accepted, as are procedures for estimating floods. However, there is a largely unexplored potential for extension of risk analysis into other aspects of dam safety.

The role of risk assessment is to provide a formal, consistent approach to evaluate the likelihood of occurrence of various adverse outcomes. In a decision analysis approach, actions are optimized in the face of uncertain adverse outcomes. Optimization could be to achieve minimum loss of life or property damage or to maximize risk reduction benefits at minimum cost. Contrary to the reasoning often given for the inadequacy or inappropriateness of risk analysis methods, uncertainty about events is the primary basis for using a formal probabilistic approach, not the reason to disavow its usefulness. This holds regardless of the sources of uncertainty, whether they are due to our limited modeling capabilities, scarcity of observational data, or the inherent randomness of the process. Another misconception is that the probabilistic risk analysis replaces engineering judgment and intuition. Far from being mutually exclusive, methods of risk analysis and engineering judgment complement and strengthen each other. The accountability and consistency of judgmental procedures can be improved by risk-based procedures.

Risk analysis helps decision makers summarize available information and quantify associated uncertainties of the available information. These procedures in themselves do not make decisions. Needless to say, when there is clear evidence of unsafe conditions at a dam, it is better to initiate remedial action (if possible) than to initiate extensive engineering studies, including formal risk analysis.

In prioritizing dams for safety evaluation, it is appropriate to use an approximate risk-based screening process. At this level of analysis, only relative risk evaluation is needed. If information-gathering and -analysis procedures are consistent for all the dams under investigation, the priorities obtained will be relatively insensitive to the decision criteria used for prioritizing. In risk analysis aimed at prioritizing dams, it is not necessary to do extensive probabilistic studies for hydrologic, geotechnical, or seismic aspects.

In conducting a more detailed probabilistic risk assessment for a given dam, it is necessary to gather and analyze as complete a package of information as is economically and technically possible. It is also necessary to evaluate risk of failure due to all external and internal load conditions. Before a decision is reached about the safety level of a dam, all economic, social, and political constraints should be incorporated.

The most important recommendation to a new user of risk analysis procedures is to overcome the reluctance to employ probabilistic procedures. It is true that more examples and case studies are needed, so that users can understand the simplicity and rationality of the available procedures. Groups that have developed the probabilistic risk-based studies should hold workshops and courses to apply the "academic" and "analytical" procedures to practical problem situations involving existing dams.

The work done by the U.S. Bureau of Reclamation and by certain university groups on probabilistic risk analysis for dams should be made available to the user community. This could provide further impetus to simplify and improve these methods. We believe that the limited acceptance of formal risk analysis comes more from lack of knowledge of the tools than from their complexity. Consistent and continuous use and improvements will increase acceptance in the long run. The methods will be adopted more widely as familiarity with them and appreciation of their value grow.

## INTRODUCTION

The resolution of dam safety problems requires an understanding not only of technical questions but also of complex financial and institutional problems. In the competition for limited public funds, a dam safety program is often seen as one of many worthwhile but expensive hazard mitigation programs.

In many cases the effective rehabilitation of an existing dam would impact severely on the financial resources of the dam owner. Increasingly, therefore, dam safety program managers, owners, and their technical staffs must be prepared to support and justify engineering decisions by the use of an analysis of the trade-off between cost and risk. For an owner of a large number of dams, such as a large utility, city, or water district, the problem may be to prioritize the dams for remedial measures and to budget appropriate funds in order to achieve the greatest safety for the least money. Methods of decision and risk analysis are helpful for making these decisions in as consistent a manner as possible.

The methodology for risk assessment has to be evaluated in the context of specific decision situations. The owner of a single dam with multiple deficiencies has to decide what priority to assign to his work plan in order to accomplish his objective. Should he remedy a stability problem or increase the spillway capacity, or reduce seepage, or consider a combination of remedial actions? An owner with several dams is confronted with similar decisions concerning the deficiencies of each dam.

Each of the decision situations mentioned involves a fundamental trade-off between (certain) expenditures and (uncertain) future gains and losses. It is useful to identify one of the alternatives in a decision situation as the "reference alternative." In decision making about existing dams, the obvious reference alternative is to "do nothing," i.e., to accept the risk and consequences associated with the status quo.

Perhaps the most critical step in the decision analysis process is to conceive alternatives for providing added protection or for providing it more cheaply. For each alternative the engineer must evaluate the added cost in relation to the "do nothing" alternative as well as its effectiveness in reducing the probability of failure or the consequences or both. Based on this evaluation and guided by appropriate decision criteria, the engineer must select the most favorable alternative or, when acting in an advisory capacity, present all the facts on cost and risk to the decision maker.

There is no single best methodology of risk assessment and risk management for dams. Evidently, the amount and quality of information available to the engineer will differ greatly depending on the nature of the decision, the resources available to analyze existing information or to seek new information, and the time available to reach the decision. It is therefore appropriate to use different methodologies requiring different levels of sophistication and different types of information about failure consequences and risk where the appropriate methodology is selected according to the decision situation.

The sequential nature of engineering decisions has important implications in risk assessment and management for dams. In general, assuming adequate records are kept, the amount and the quality of information about site conditions and structural properties increases with time. In this context it is useful to distinguish between decisions that have a "one-time" or "terminal" character (such as in design or rehabilitation) and those that do not (such as site exploration or dam inspection). Decisions in the latter category are expected to be followed either by other "nonterminal" decisions (more extensive exploration or inspection) or by a "terminal" decision. In the context of the U.S. Army Corps of Engineers' Dam Safety Program, decisions taken during "Phase I" dam inspection are nonterminal, while "Phase II" decisions involving repair and rehabilitation may be thought of as one-step or terminal decisions. The idea is that in analyzing a nonterminal decision it is necessary to consider follow-up actions, while one-step decisions can be analyzed in isolation. In the first step of a sequential decision process the engineer is usually concerned about data acquisition, while in the final step he seeks the actual realization of the benefits of proposed protective measures.

It is also useful to categorize safety-related decisions involving dams according to the number and type of structures involved. The decision situation may involve:

- a single dam,
- a group of dams in a given jurisdiction,
- a system of dams located in series on a waterway,
- a system of dams affecting a common area, or
- all dams of a certain type (e.g., concrete arch dams) affected by a particular set of design criteria or regulatory requirements.

Many considerations go into the formulation of the criteria for decisions affecting dam safety. It may be necessary to distinguish between the types of dams involved and their purpose (e.g., flood control vs. recreation), types of ownership (public vs. private), types of decisions (modification of operations vs. rehabilitation), and kinds of protective measures (structural vs. nonstructural).

Lest the merits of methods of risk assessment and risk management be overstated, it should be mentioned that questions involving hazard mitigation are often controversial. Different parties (owner, downstream resident, builder) are affected differently by the outcome of the decision. Hence, they tend to assess costs and risks differently and may select different criteria upon which to base the decision. In light of these conflicts the main value of decision analysis methodology may be that it provides a framework for organizing factual information about costs and risk, for structuring the decision-making process, and for promoting and uplifting communication among the opposing parties (Vanmarcke 1974).

## RISK ASSESSMENT: ALTERNATIVE METHODS

The detail to which risk assessment is carried out depends on the intended application of its results. If comparisons are to be drawn among dams or among alternative treatments for a given dam, it is important that the assessment procedures be consistent. Three broad levels of risk assessment are currently in practice and may be categorized as subjective, index-based, and formal (quantitative).

A subjective assessment is one in which all relevant factors are not systematically accounted for. The engineer or owner considers those factors that appear most important to the case and uses this assessment to identify a solution to the problem. Such an assessment may often result in a good decision and may be all that is needed, but it will only rarely lead to an optimum solution, and it will be difficult either to document, respect, or fully account for.

An index-based risk assessment is a systematic evaluation of the factors affecting dam safety that allows a ranking, rating, or scoring of a number of dams. It is more general and complete than a subjective assessment but does not permit numerical comparison of likelihood or expected cost. Specific site conditions are often difficult to factor in, except subjectively. Several examples of such a qualitative risk assessment are presented later in this section.

In a formal risk assessment one estimates occurrence frequencies, relative likelihoods of different levels of response and damage, and the various components of cost and consequences. Although an actual value of risk cost is determined, this value often need not be considered in absolute terms but as a number suitable for comparison among alternative risk reduction measures. An integral part of a risk assessment is that one should vary assumptions about any of the study parameters to determine their effect on the risk cost and (when used on risk management) on the optimal choice.

The most common format of a risk assessment involves the following steps:

- Identification of the events or sequences of events that can lead to dam failure and evaluation of their (relative) likelihood of occurrence.
- Identification of the potential modes of failure that might result from the adverse initiating events.
- Evaluation of the likelihood that a particular mode of dam failure will occur given a particular level of loading.
- Determination of the consequences of failure for each potential failure mode.
- Calculation of the risk costs, i.e., the summation of expected losses (economic and social) from potential dam failure.

A more detailed discussion of these steps is given below.

Step 1. Identify what loading conditions operate on the dam with the potential to cause a dam failure, and estimate the frequency of occurrence of these events.

The loading conditions that usually need to be considered are static reservoir load, seismic load, and hydrologic (flood) load. However, for a specific dam other loads (see next section) may need to be considered. In the course of the risk assessment, some loads may be ruled out as not having the potential for dam failure. It is recognized that the estimation of the frequency of hydrologic and seismic events with the potential to cause dam failure is difficult to make with confidence because of the lack of historic data. The evaluation of the likelihood of "internal" or "passive" initiating events (foundation instability due to strength deterioration or piping) is even more difficult. Nevertheless, such an estimate is a necessary part of any risk assessment.

Step 2. Given the configuration, characteristics, and condition of the dam-foundation-spillway system and the loading conditions to which it will be exposed, identify the potential *modes* of failure that may result from the loading events.

This is the first step in determining the response of the structure as well as the consequences of failure. The detail to which the modes need to be identified is very much site- and problem-specific. In some cases the assessment of risk and consequences may be satisfactorily made by assuming a complete and instantaneous dam breach, while in other cases this assumption would yield a very unrealistic estimate of damages (and hence the benefits of remedial measures).

Step 3. Estimate the likelihood that a particular mode of dam failure will occur for a range of levels of potential adverse loadings.

The third step is perhaps the most difficult in the risk assessment process. It is necessary to account for the entire range of loads through which it is exposed and for how a dam responds to these loads. An estimate of response is required for all types of loads over the entire damage potential range of loads such that the "unconditional" risk of failure may be realistically assessed. This step is one of the fundamental differences between a risk-based safety assessment and the "maximum event" analysis that is in common practice.

Step 4. Determine the consequences of failure for each potential failure mode.

Once a mode of failure has been defined, the flood caused by the dam break is routed through the flood plain. Inundation maps (and flow rates) are used to estimate the potential for loss of life, the level of property damage, and environmental impact. Uncertain factors such as season of year, time of day, reservoir elevation, and antecedent precipitation might have to be taken into account at this stage of the analysis. (These factors may also be important in identifying appropriate remedial measures.)

Step 5. Determine the risk costs or "expected losses" associated with the existing dam in its present condition.

The total economic risk cost is obtained by summing the product of the likelihood of the loading condition, the likelihood of dam failure in different modes given the loading condition, and the cost of the damages resulting from that failure mode over the entire range of load levels and failure modes.

Similar calculations can be made for the expected losses of life or the "social risk cost." These require expressing the life loss consequences for each potential failure mode in Step 4.

## Risk-Based Methods for Prioritizing Dams

The Stanford risk-based screening procedure (McCann et al. 1983b,c) is divided into three phases. The objective in the first phase is to identify the expected losses given dam failure. The work involved includes tasks for gathering data, an evaluation of the likelihood of dam failure due to a number of initiating events, routing of a flood wave as a result of instantaneous dam failure, and an estimate of the direct losses. The data collected will include information on the dam, the spillway, downstream topography and development, design criteria, inspection programs, etc. This corresponds essentially to the Phase I inspection of the National Dam Safety Inspection Program.

In the second phase the expected loss due to dam failure is determined. The probability of failure will be a function of the present structural capacity of the dam and the likelihood of occurrence of loading conditions that might induce failure.

The objective in the third phase is to consider the mitigation alternatives that might be available, what their associated costs are, and the additional level of safety they would achieve. An analysis of this type will provide the decision maker with information on the cost-effectiveness of upgrading a dam. The actual tasks include identification of a mitigation program to upgrade the dam, determination of the cost of implementation, and a reevaluation of the expected loss for the dam in an upgraded condition. The result of this phase is a ranking of the dams that can be based on a cost per unit of increased benefit.

In several MIT Research Reports (1982) an analogous framework is presented for decision analysis of hazard mitigation measures for existing dams. Several common decision situations are examined, including the problem of allocating limited funds for remedial work to a large number of dams. The proposed procedure is illustrated by means of a case study involving 16 actual dams located in rural Vermont. The basic sources of information are the Inspection Reports issued under the National Dam Inspection Program, where it is shown how costs, risks, and consequences can be estimated for the status quo as well as under different alternatives for upgrading.

A critical component in the analysis is of course the risk of dam failure, which is estimated by an updating procedure that permits combining different sources of information. Based on general background information on the group of dams under study (type, age, location), a subjective prior risk is assigned. Using specific information on each dam (from the Inspection Reports), the prior risk is then updated using Bayes's theorem. Emphasis is on demonstrating the flexibility of the model as it includes carrying out a sensitivity analysis with respect to the decision criteria and the input data.

As with all ranking and allocation procedures, an important step in the implementation of risk-based methods is the uniformity with which the method is applied. Consistency will be required in inspection reporting practice, interpretation of inspection reports (specific wording is useful), application of analysis procedures, etc. Regardless of the method adopted, whether probabilistic or deterministic, unless a degree of consistency in application can be ensured, achieving a reliable and consistent ranking will be jeopardized.

### Methods of Index-Based (Qualitative) Risk Assessment

The development of a methodology to conduct a fast, reliable evaluation of the safety of dams in a jurisdiction has not been limited to formal risk-based procedures. In a preliminary risk assessment required to screen or prioritize a large number of dams for inspection or rehabilitation, it may be appropriate to substitute an index-based procedure for formal quantitative risk analysis. Several organizations have developed such index-based procedures to provide a ranking or prioritizing of a system of dams, including the U.S. Bureau of Reclamation; the U.S. Army Corps of Engineers; and the states of Idaho, California, Pennsylvania, and North Carolina. Two of these methods are discussed below, in particular the procedure suggested by Hagen (1982) of the U.S. Army Corps of Engineers and the Safety Evaluation of Existing Dams (SEED) program used by the U.S. Bureau of Reclamation. These procedures consider the same factors as does a quantitative risk analysis, but they are in terms of a set of ranking parameters that take integer values between 1 and 5. A score of 1 is most favorable, a score of 5 least favorable. According to Hagen's method, the overall "relative risk" index for a dam equals the sum of an "overtopping failure score" and a "structural failure score," i.e.,

$$R_r = O_t + S_t,$$

where $O_t$ = overtopping failure score = $O_1 \times O_2 \times O_3$ and $S_t$ = structural failure score = $S_1 \times S_2 \times S_3$. The factors depend on the following considerations:

Factor $O_1$: Number of homes endangered by failure. (Based on difference in area inundated without failure and with failure, assuming water surface at the top of dam. Dam failure hydrograph superimposed on discharge prevailing at the time of failure.)

Factor $O_2$: Project flood capability in the percentage of current design flood standard. (Assuming the probable maximum flood is the current design flood standard.)

Factor $O_3$: Project capability to resist failure by overtopping. (Based on inspection of the structure and review of design and construction records.)

Factor $S_1$: Number of homes endangered by failure. (Area inundated is obtained from dam failure with water level at the top of flood control storage or normal maximum pool excluding surcharge used to pass design flood.)

Factor $S_2$: Evidence of structural distress. (Based on inspection of the structure and review of design and construction records.)

Factor $S_3$: Potential seismic activity. (Based on dam location on seismic zone map, knowledge of faults, recent earthquake epicenters, and dam design procedures.)

The maximum rating score for a dam by the selected rating scales would be 250 (i.e., 125 associated with "overtopping failure" and 125 with "structural failure"). A dam with a smaller score than some other dam should generally pose less of a risk.

### Safety Evaluation of Existing Dams (SEED)

The U.S. Bureau of Reclamation has developed a program to evaluate the safety of existing dams. The evaluation process includes a review of available data on a dam and a field inspection. The data review covers all aspects of the dam from geologic and seismic conditions to a review of the construction experience and operations. A detailed evaluation report is prepared on each dam, and a site rating (SR), a numerical measure of the dam's condition and damage potential, is assigned. A score is given on a scale of 0 to 9 to various elements shown in Table 3-1. The SR is a sum of the element scores.

During a field investigation a checklist of items is examined. For this purpose the Bureau has prepared a handbook to assist the examiner in identifying areas of potential distress in the dam (SEED 1980). On the basis of the site investigation, recommendations for upgrading are made, and their significance is measured by a weighting system that considers a categories' overall importance as well as its degree within each category. The sum of the weights for all recommendations is added to the SR to give a SEED value. The results for all dams are used to develop a SEED rank, where the highest SEED value has a rank of 1.

Other features of the SEED program include information on estimates of costs to carry out the recommended upgrades, scheduling information, status of different upgrades, and key personnel involved in the project. In addition, the information is stored on a computer and is continually updated.

## METHOD OF RISK ASSESSMENT FOR SPECIFIC CONDITIONS

It has been noted in previous discussions that an effective risk-based decision analysis must incorporate site-specific conditions related to the most likely failure modes, hazard conditions, and possible remedial measures. The technical elements that are common to different procedures of risk-based dam safety as-

TABLE 3-1 Hazard Rating Criteria in Hagen's Procedure

| | Condition | | | |
|---|---|---|---|---|
| Age (years) | Under 5<br>(0) | 5–24<br>(3) | 25–29<br>(4) | 50–<br>(9) |
| General condition | Excellent<br>(0) | Good<br>(3*) | Fair<br>(6) | Poor<br>(9) |
| Seepage problems | None<br>(0) | Slight[a]<br>(3) | Moderate<br>(6) | High<br>(9) |
| Structural behavior measurements current and within acceptable | Yes<br>(0) | — | Partial<br>(6) | No<br>(9) |

| | Damage Potential | | | |
|---|---|---|---|---|
| | Low | Moderate | High | Extreme |
| Capacity (acre-feet) | 0–999<br>(0) | 1,000–49,999<br>(3) | 50,000–499,999<br>(6) | 500,000–<br>(9) |
| Hydraulic height (feet) | 0–39<br>(0) | 40–99<br>(3) | 100–299<br>(6) | 300–<br>(9) |
| Hazard potential | (0) | (4) | (8) | |
| Hydrologic adequacy | Yes<br>(0) | — | — | No<br>(9) |
| Seismic zone | 0–1<br>(0) | 2<br>(3) | 3<br>(6) | 4<br>(9) |

NOTE: Number in parenthesis is the weighting factor.

[a]Assumed if not given.

sessment are estimation of the frequency of occurrence of loading events, evaluation of the response of the structure to the loading, and prediction of damage downstream.

The approach to the solution of these technical problems is best examined in the context of the major loading conditions to which the dam is exposed: static loading, hydrologic loading, and earthquake-induced dynamic loading.

## Risk of Dam Failure Due to Static Loading

To determine the risk of dam failure due to static loading, each of the failure modes relevant to a particular dam needs to be identified. The items in the list must preferably be all-inclusive and mutually exclusive. If the likelihood of certain failure modes is judged negligible compared with that of other modes, these may be omitted from formal consideration. Once the potential failure modes have been identified, there are three basic approaches for estimating the corresponding failure probabilities.

1. *Analytical (probability) approach*. In a common version of this method a factor of safety is computed for the condition. The uncertainty in the calculation is quantified by examination of the ranges and the variation of the individual input parameters to the analysis. Based on this uncertainty analysis the probability that the factor of safety will fall below 1 is determined.
2. *Empirical (historical frequency) approach*. In this method the number of failures for similar dams for the same failure condition is determined and divided by the total number of dam-years of operation for dams of this type as a crude estimate of the probability of failure. The obvious limitation of the approach is the scarcity of information in each narrowly defined category of dam type, age, loading type, failure mode, etc.
3. *Judgmental approach*. In this method the investigator attempts to quantify his judgment based on all available information. The judgmental statement may be made directly in terms of annual probability of failure of the dam due to a particular condition (e.g., probability of failure due to internal erosion $= 1 \times 10^{-3}$ annually), in terms of the chance of failure over a specified remaining operational life of the dam, or as a fraction of the probability associated with other modes (e.g., about twice the risk attributable to flooding and overtopping).

The analytical-probabilistic approach is the most elegant of these methods; however, adequate data to support or justify such studies are often not available. Some important potential failure modes do not lend themselves to a factor of safety formulation as is common in stability analysis. Also, the calculated probability of failure is very sensitive to the tails of probability functions describing the various parameters, and these are not well known at all. Therefore, it is not yet practical to incorporate these approaches in a comprehensive program of assignment of risks to a number of potential failure mechanisms from static loading.

A combination of the empirical and judgmental approaches appears to be most practical at the present time. Historical failure probabilities can be obtained for specific conditions and types of structures, but they need to be adjusted based on the conditions at a particular dam. This adjustment is based on the inspection, analysis, and judgment of the engineers performing the safety evaluation of the dam. The two estimates may be combined by means of a Bayesian updating procedure in which a weight is assigned to the relative confidence in each of the estimates (historical and engineer's judgment). Such a procedure is easily understandable, consistent with the level of data available, and practical to implement.

**Flood Risk Assessment**

Flood risk assessment has been practiced for many years in the study of water resources. Procedures used have been reviewed, evaluated, criticized, and extolled by experts in probability theory, hydrology, meteorology, mathematics, and other disciplines. Because of lack of agreement on the most appropriate methodology for estimating flood probabilities, the U.S. Water Resources Council recently directed the Interagency Hydrology Committee, under the council's auspices, to select a unique technique for computing flood probability for *gaged* locations. The committee responded by declaring that a unique technique does not exist that will provide the best answer for all locations and conditions. Nevertheless, the committee recommended that the Pearson Type III distribution with log transformation of the flood data be used as an uniform technique. The most recent publication describing the results of the Hydrology Committee's effort is Bulletin 17B. No similar publication has been developed for *ungaged* locations. This does not mean that adequate methods are not available but only that experts cannot agree on a single best method for every condition. The procedure recommended by Bulletin 17B covers flood events with return periods of 1:500 years only. Extrapolation to extreme rare events is not covered.

Traditionally, flood probability has been expressed in terms of annual exceedance probability. This means that specific flood magnitudes have an assigned probability of being exceeded during any given year. The use of exceedance probability for conveying the risk of the design flood being exceeded during the useful life of a project is extremely important in judging inflow design floods for sizing spillways and establishing crest elevations. For example, if a designer expects to establish an inflow design flood that would have a .01 probability (1% chance) of being exceeded during a 200-year useful life, a flood event with an annual exceedance probability of .00005 would need to be selected. This corresponds to an average exceedance interval (AEI) of 20,000 years and would be near the magnitude of a probable maximum flood. While analytical procedures (such as the binomial distribution) can be used to estimate such an event, the historical record is rather short, and such extreme extrapolations are difficult. Extrapolation becomes increasingly uncertain the further it is carried beyond the length of the period of record. Historical records of flood events are rather limited in the United States; therefore, in general, extrapolations beyond an annual exceedance probability of .01 (100 years AEI) will have a limited degree of reliability. In any case the analyst should quantify the uncertainty in the estimated exceedance probabilities.

There has been much debate regarding the wisdom of attempting to find an acceptable means for estimating the probability of extreme flood events. Mathematical formulas recommended for this purpose are always suspect because of the data sample available. Regional processing of data is used to improve the reliability of results. Historical records are searched to obtain evidence regarding past occurrences of large floods. Because of better data on historical storms, precipitation data are converted to runoff and probability of floods estimated from the larger data samples. Tree rings, sun spots, and other indicators of significant climate changes have also been proposed for estimating the probability of extreme floods. Stochastic hydrology uses random number generators to simulate long periods of flood record of any length desired. However, the generated floods are constrained by the statistical parameters of the relatively short actual data sample used to derive the parameters. For the probable maximum flood (PMF) the best approach for purposes of conducting a risk-based decision analysis may be to extend peak flow-probability relationships to conventionally derived values of PMFs by some reasonable mechanism. A PMF is defined as the largest flood considered reasonably possible for a specific location and its probability of exceedance should be close to zero. The function describing annual exceedance probability could be extended in a smooth curve from the limit of the relation obtained by historical records of flood events until the curve becomes asymptotic to the PMF value. The extended curve can then be varied in sensitivity studies accompanying the risk-based decision analysis.

### Earthquake Risk Assessment

Earthquakes pose a multitude of hazards to dams, either by direct loading of the structure or by initiating a sequence of events that may lead to dam failure. For example, strong ground shaking or fault offset at the dam foundation are direct loads on the structure, while an upstream dam failure, seiche, or landslide into the reservoir are earthquake-generated events that can lead to overtopping and failure.

A comprehensive probabilistic seismic risk assessment involves these two steps: (1) evaluation of the likelihood of occurrence of levels of seismic loading at the dam site (seismic hazard analysis) and (2) evaluation of the conditional probability of the different modes of dam failure given the occurrence of seismic loads (conditional reliability analysis). Overall seismic safety assessment requires combining the information generated in these two steps. The hazards typically associated with earthquakes are ground shaking, faulting, seiche, landslide into the reservoir, and upstream dam failure.

It is the job of the risk analyst to identify those failure scenarios that are the most significant contributors to the chance of an earthquake-induced dam failure. For example, slope instability of the upstream face of an earth embankment and liquefaction of the foundation may be postulated as potential failure modes when the dam is subjected to strong earthquake ground shaking. Once the hazards and corresponding modes of dam failure have been identified in a preliminary way, the risk analysis proceeds to the evaluation of the probabilities in each step of load-response-failure sequence. In the case of ground shaking and faulting the techniques available to evaluate occurrence probabilities are well established. However, the comment made in reference to the evaluation of the risk of extreme hydrologic events also applies here. The uncertainty in key input parameters leads to considerable variability in the estimates of probability of occurrence of earthquakes that have the potential of causing dam failure. It is therefore necessary to quantify the uncertainty in estimated seismic hazard.

A variety of opportunities are available to conduct probabilistic analyses, the number depending on the failure under consideration. As an example, various approaches exist to perform probabilistic slope stability analyses. The result of the probabilistic analysis expresses the chance of failure as a function of load level. These conditional probabilities usually increase monotonically from 0 to 1 as loading increases. Among the sources of uncertainty are the variability of material properties in the dam, the response prediction for known values of the input parameters, and in defining the failure state.

## CONSIDERATION OF REMEDIAL ACTIONS

The dam owner and engineering staff should become aware of and evaluate remedial measures that are available either to reduce the likelihood of dam failure or to reduce its consequences. These remedial measures can be structural or nonstructural. They are available individually or in combination with other remedial measures. The structural measures generally obtain their effectiveness from reducing the likelihood of the dam failure, whereas nonstructural measures may either reduce the likelihood of failure or reduce the consequences. The value of remedial measures is measured by the benefits obtained, the costs associated with the measure, and the adverse effects of implementation of the measure. Benefits may involve reduction in the number of lives lost, reduced damages, improved project outputs, compliance with regulatory or design requirements, etc. Costs should include future expenditures as well as the initial expenses required to make the measure effective. Indirect costs may be involved in some of the measures. The action may require a party other than the owner to spend money.

Discussed below are the most common alternatives available to an engineer or owner facing the decision of how to remedy problems with existing dams. The emphasis in the discussion of each alternative is on its impact on failure risk and consequences. Of course, in a formal risk assessment this qualitative evaluation is quantified in terms of fractional risk reduction and reduced losses in the event of a failure. A number of examples are presented in the section Examples of Risk Assessment and Decision Analysis.

### *The Status Quo Alternative*

*Do Nothing.* This is the reference alternative to which other measures are compared. If it is selected, there will obviously be no change in the consequences or the likelihood of failure.

### *Structural Measures*

Structural remedial measures can usually be classified as modifications of the dam, modifications to the spillway, corrective maintenance, or construction of new facilities, as follows:

*Modifications of the Dam.* A dam may be modified to reduce consequences of failure, to reduce the likelihood of overtopping, to enhance the structural integrity, or to resist erosion and external damage. In each case these improvements act to reduce the likelihood of dam failure. The measurement of change in the likelihood of dam failure is often difficult; however, it is usually accomplished by describing the effectiveness in reaching design standards. The cost of improvements to comply fully with design standards is often prohibitive. Therefore, evaluation of many alternative measures may be needed to determine which measures are the most cost-effective in reducing the risk of dam failure.

Reservoirs may be lowered or dams removed to reduce or eliminate the consequences of their failure. Such measures are not attractive to dam owners because the benefits from the dam are diminished or eliminated. Similarly, an owner's liability from the consequences of dam failure is also reduced or eliminated. These measures generally involve high costs and can have large adverse effects. Sediment accumulated in the reservoir area will be subject to transport downstream if the dam is removed. The environmental aspects of the reservoir area are also difficult to resolve when a dam is removed. Removal of a dam can affect those secondary users who have become accustomed to its presence and have learned to rely on it.

Increasing the height of a dam to reduce the likelihood of overtopping is most effective when it will fully accommodate the requirements for the inflow design flood. The increased dam height will result in increasing the consequences of its failure when full. It will also increase the potential level of the reservoir and flood lands previously unaffected by the project. Thus, this measure is not usually a viable alternative when there is concentrated development around the reservoir rim. When raising the top elevation of the dam, care must be taken to ensure that the foundation is capable of supporting the additional load. The cost associated with this measure may be high. However, when only a few feet of height are needed to comply with design standards, a wall or dike possibly may be constructed along the top of the dam within cost constraints.

*Modification of the Spillway.* Spillway modifications may be needed to enhance its ability to pass greater floods or improve its ability to resist erosive action during higher velocity flow. Greater capacity will reduce the likelihood of failure by larger floods, while better erosion resistance will reduce the likelihood of failure during the passage of the spillway's current capacity. It is a possibility that enlarging a spillway may increase the rate of discharge for more frequent storms and cause downstream flooding to increase in frequency.

Spillways may be widened or their crests lowered to increase flow capability. Additional spillways may also be provided to obtain greater flow capacity. These types of improvements can vary from a quick fix with a bulldozer to an elaborate new gated facility. If the crest is lowered without the addition of gates, the ability of the project to perform its intended functions may be seriously compromised. Also, new lands may be needed to accommodate the additional flow downstream from the spillway. Emergency spillways or fuse plug type spillways may be considered as an economical means of avoiding dam failure and reducing risk.

In some cases there is serious concern about an unlined spillway's ability to pass intended flows without rapid erosion followed by sudden discharge of the reservoir, which would be tantamount to a dam failure. In dam safety the objective is to avoid the sudden release of the reservoir. Damage to the spillway and to the dam are usually anticipated during the inflow design flood; however, maintenance of spillway crest is necessary to reduce the likelihood of sudden release of water. Types of protection may range from grass cover to an expensive concrete lining. Adverse effects are normally not severe.

*Construction of Upstream Facilities.* Reduction of the magnitude of flow reaching a dam of concern can be accomplished by construction of an upstream dam or diversion facility. Depending on the nature of the facility, the likelihood of dam failure by overtopping may be slightly reduced or essentially eliminated. The failure consequences would also be reduced. Generally, costs associated with these measures are large. Adverse effects may also tend to discourage their implementation.

*Corrective Maintenance.* Whenever apparent deficiences are observed at a dam, measures can be taken to reduce or eliminate the source of deficiency. The types of deficiency can be associated with internal or external features of the dam, or they can be associated with the stability of the spillway. Some of these methods include replacing concrete, bringing the top of the embankment up to grade, outlet repairs, replacing riprap, etc. These methods may become expensive but are normally not when taken care of in a timely manner. The effect of performing this maintenance is largely beneficial.

*Leveling the Top of an Earth Embankment.* This may be an applicable measure in cases where an allowance for settlement was made but has not taken place as anticipated (and is not expected to), resulting in a "crown" effect.

Because earth fills and their foundations are expected to consolidate and settle over time, it is common practice to camber or overfill earth embankments. Although reasonable allowances for this settlement are made, it is impossible to predict accurately the amount of settlement that will occur. The result is that in many cases, after essentially all settlement has occurred, the crest of the dam is not level, thus leaving an opportunity for concentration of flow if overtopping were to occur. This can be especially critical for dams not capable of withstanding floods as large as the PMF (e.g., low hazard dams). This concentrated flow has been specifically cited as a major contribution to some severe damage in actual overtoppings.

Implementation of this remedial measure would involve "shaving" off the camber and providing a level weir. For example, this could provide overtopping the entire length of dam by 1.5 feet instead of a concentrated flow at the abutments of perhaps twice that amount. Cost would be in the moderate range in this case, and the procedure would reduce the likelihood of failure due to overtopping. It is recognized that overtopping flows will concentrate in groins and, therefore, that special erosion protection would be needed there.

*Other Remedial Measures.* The above listing gives some common possible remedial measures. However, remedial actions are often quite site-specific and relate to the particular modes of failure (seepage, piping, liquefaction, deformation, etc.) that may be most relevant to a specific dam. Repair or strengthening of a parapet wall, plugging an access opening in a parapet wall, constructing a small dike downstream of the existing dam, relocating the dam to a new site, densification of the foundation, installation of relief wells downstream of the dam, and construction of stability berms are some examples of constructed or proposed remedial structural measures that can be implemented at a specific site or in a specific problem situation.

### *Nonstructural Remedial Measures*

Nonstructural remedial measures, such as those listed below, can be used to reduce the likelihood of dam failure and the consequences of dam failure.

*Intensive Surveillance.* This can be done to monitor dams with suspected or known problems. Depending on the specific site, various features may need inspection surveillance at different frequencies. Special inspections may be called for during or immediately following significant natural events, such as large floods or earthquakes. This alternative measure would usually need to be coupled with other structural or nonstructural measures, such as a lower operating pool, or may be used temporarily until positive measures are taken. Inspections themselves neither reduce the likelihood nor the consequences of dam failure; however, they can be effective in reducing risk if appropriate follow-up action is taken.

*Reservoir Regulation.* This provides a means to reduce the likelihood of failure. This may be a reduced likelihood of overtopping through lowering a reservoir to provide additional storage capacity for floods or it may be related to increased upstream slope stability through reducing the rate of drawdown. Reservoir restriction may be permanent, seasonal, or based on flood forecasting. The latter regulation can greatly reduce the probability of failure, but it may have no effect on the consequences of failure. Reservoir restriction by which the maximum reservoir elevation is limited year-round below the design level can reduce both the probability of failure and the consequences of failure. A case history of an example of the use of risk-based decision making to reduce risk by reservoir restriction is presented in the section Examples of Risk Assessment and Decision Analysis.

The cost of this alternative is generally low. However, other impacts can be significant by reducing benefits of the project and also by adverse environmental and economic effects in and around the reservoir area of the dam.

*Emergency Action Plans.* An emergency action plan (see Chapter 2) is a nonstructural measure that can be used as a temporary alternative until more positive remedial measures can be implemented. Such a plan does not reduce the likelihood of failure; however, it can be effective in reducing the consequences of one.

*Flood Plain Management.* A flood plain management program can be applied to all or only a portion of the area inundated by dam failure. Similarly, the program can be applied only to modification of uses of the flood plain or can be expanded to include existing uses. The preparation of inundation maps, as has been done in California, is a relatively conservative measure of modest cost. Such maps will not reduce the probability of failure but will provide a basis for determining the impact of a flood, which enhances evacuation and reduces liability.

Included are such measures as permanent relocation of downstream structures, change in existing land use, floodproofing of structures, installation of individual or group levees to protect a damageable area from flooding, and purchase or other land-use controls to keep future development from being subjected to flooding. Since these alternatives do not affect the dam itself, there is no change in the likelihood of failure.

The cost of this alternative can be quite high, depending on the nature of development or property values in the affected area. It should be noted that some measures, such as levees, may need to be higher (to contain flows from a breach) and therefore may be more expensive than they would if no dam existed. Social impacts could also be high, particularly in some areas involving relocation.

## EXAMPLES OF RISK ASSESSMENT AND DECISION ANALYSIS

In this section a number of case studies involving the application of risk analysis to dams will be described. The emphasis in each write-up is on the problem formulation and the results. The risk assessment methodology is briefly summarized, but its details are not presented and may be found in the case study reports that are referenced.

### Case I: Jackson Lake Dam

During the U.S. Bureau of Reclamation's review of its existing structures for adequacy of performance under current earthquake criteria, Jackson Lake Dam was identified as being located in an area with the potential for strong earthquake shaking. The fine-grained soils on which the embankment portion of the dam was founded and the hydraulic fill methods that were used to place a portion of the earth embankment presented the potential for liquefaction at the site under strong earthquake loading. A drilling, sampling, and laboratory testing program was initiated to define the location and physical properties of the various materials in the embankment and foundation. Once this information was obtained, a dynamic analysis of the dam with any proposed modifications or treatments could be carried out.

Because it was anticipated that it would be several years for any permanent remedial measures to be completed, temporary restrictions on the water level at Jackson Lake were considered. A risk analysis was performed to assess the probability of an overtopping condition as a function of the restriction level and to assess the level of downstream damage as a function of various modes of failure and restriction levels. The risk of overtopping was computed as a function of the probability of occurrence of various levels of earthquake, the probability of various structural responses due to these earthquakes, and the probability that the reservoir would be at an elevation that would lead to overtopping. The analysis showed that at elevation 6756.5 the risk of overtopping due to the primary mode of failure under consideration (liquefaction at base level of 6750) was reduced by about 50 percent (see Figure 3-1). Furthermore, it was seen by analysis of the outflow hydrograph that even in the event of overtopping (from the primary mode of failure hypothesized) the flood produced with the water level restricted to this level would be greatly reduced (see Figure 3-2). This level of reduction in risk appeared appropriate, but the magnitude of the benefits still needed to be incorporated into the decision analysis.

The overtopping risk analysis for restricted elevations below 6760 assumed the reservoir would be maintained at the restricted elevation on a year-round basis. Actual operations could produce a variable water level at about the mean elevation and yet minimize impact on recreational interests affected by the actual lake level as well as the timing and amount of releases. An operational plan that satisfied these objectives was developed by regional and project personnel. The plan called for water levels at elevation 6760 for 1 month out of the year but below elevation 6755 for about 9 months. These criteria provide a lower risk of overtopping than a constant restriction level of 6756.5 but present the potential for larger floods during the 1-month period when elevations reach elevation 6760 if overtopping were to occur during this period; however, the total benefits of this operating plan appeared to outweigh this short-term increased hazard.

The question of a level of acceptable risk is brought to bear at Jackson Lake as the twin goals of maintaining dam safety and providing maximum benefits from the project are brought together. Responsible management of a public facility in such a case requires that an objective assessment be made of the hazard, the risk of failure, and the loss in benefits due to any restriction imposed. A decision analysis model was used to provide a convenient format for presentation of all available information and to permit

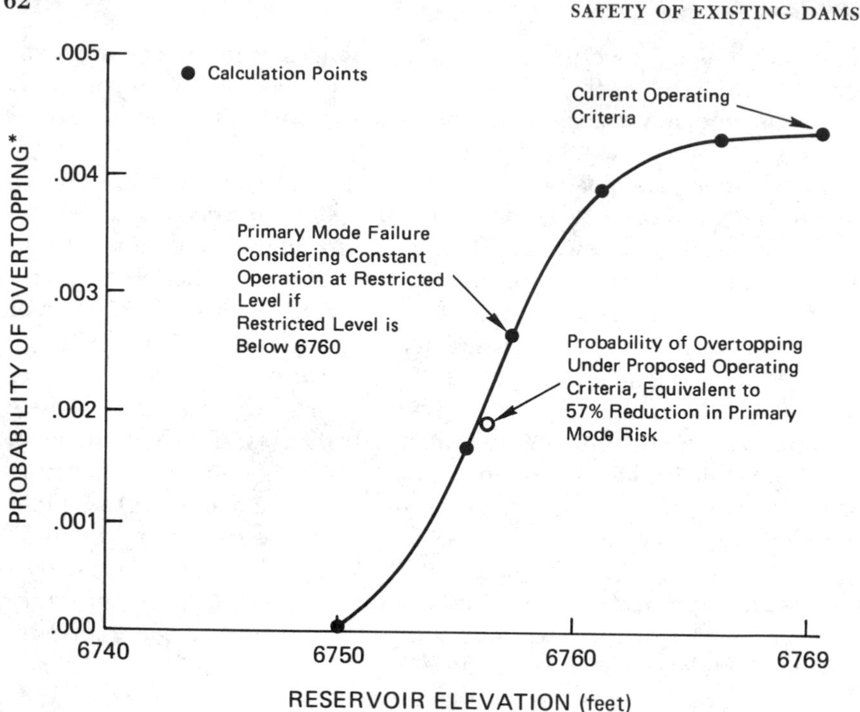


FIGURE 3-1 Risk of overtopping versus reservoir elevation of Jackson Lake. (*From liquefaction failure due to earthquake plus effect of a seiche. Primary mode: liquefaction at base level of 6750.)

an objective evaluation of the data. Although the data were limited, the risk analysis clearly showed a marked decrease in the risk of overtopping with decreasing reservoir elevation.

Analysis of the flood hydrograph for the most likely hypothetical failure mode likewise showed a significant decrease in potential hazard with decreasing reservoir elevation. Examination of these relationships and determination of a reservoir operation procedure that minimizes adverse impact from a restricted reservoir level permitted establishment of a reservoir operation plan that provides for a meaningful reduction in risk to the public while maintaining the usefulness of the reservoir.

## Case II: Island Park Dam

Island Park Dam is located on the Snake River in eastern Idaho. Its purpose is to provide storage and regulation of water for supplemental irrigation. There are some flood control benefits associated with the dam but only on an informal basis relying on forecasting techniques within the drainage basin. The reservoir is popular as a recreation area and has an abundance of campsites as well as private homes along its shores.

Island Park Dam is a zoned earthfill structure approximately 91 feet in height. The exterior slopes are 4:1 upstream and 2:1 downstream. The embankment materials consist of a 3-foot thickness of riprap on the upstream face, an impervious central core flanked by shells of semipervious material, a rockfill section on the downstream side, and a zone of selected free-draining material at the downstream toe of the semipervious section. A long dike extends to the east of the dam. The dike is a homogeneous embankment with riprap on the upstream face. The structure height of the dike is generally less than 15 feet, and exterior slopes are 3:1 upstream and

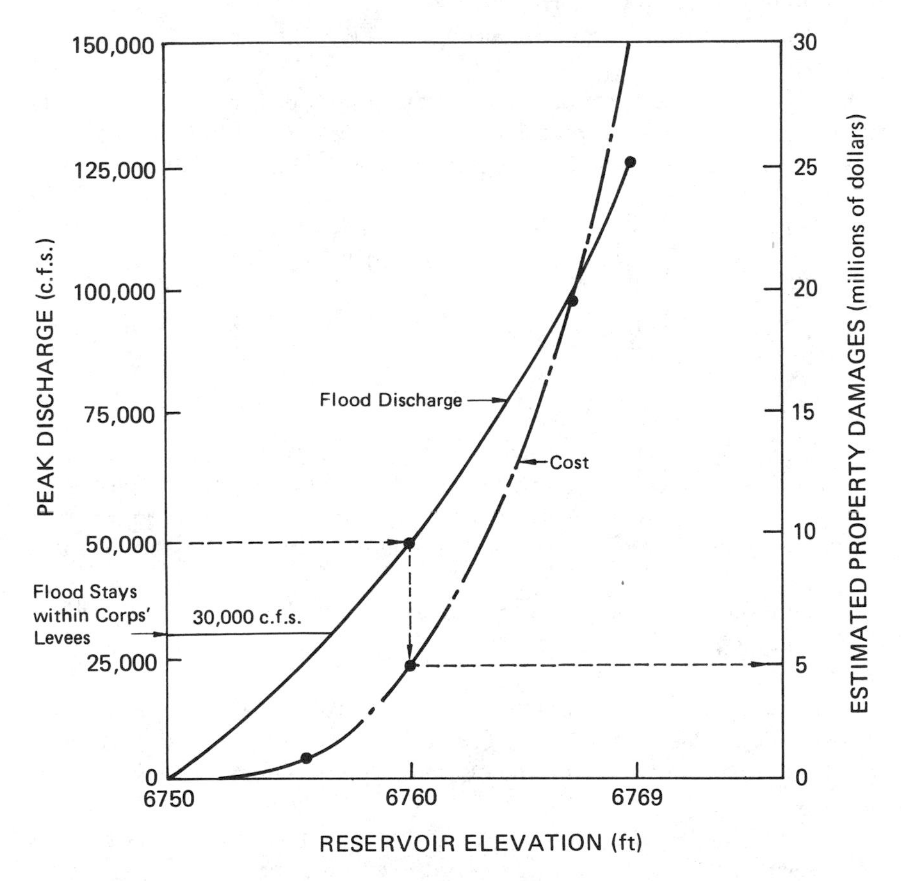


FIGURE 3-2 Jackson Lake flood and flood effects* versus reservoir restriction level. (*Given that overtopping occurs and that failure is in the primary mode with a 400-foot breach.)

2:1 downstream. Crest length of the embankment is approximately 9,500 feet. Appurtenant structures are located on the right abutment and consist of a double-side channel concrete spillway inlet transitioning to a concrete-lined circular tunnel and a concrete-lined outlet works tunnel controlled by four 5 × 6-foot high-pressure gates.

A U.S. Bureau of Reclamation technical memorandum presents the results of studies conducted to address the problem of *providing adequate freeboard* for Island Park Dam. This could be accomplished by various combinations of the actions mentioned in the next subsection. Additional factors complicated the issue considerably. These factors included the following:

- An increase in the maximum discharge for the existing service spillway results in downstream property damage.
- An increase in reservoir elevation above 6305.0 may cause upstream property damages.
- Recent insect infestation of the surrounding forests raises the possibility of significant debris accumulation in the reservoir. Concern has been expressed about the potential plugging of the existing spillway inlet structure with debris.

### *Alternatives to Be Examined*

The report examines the following alternatives to provide adequate freeboard for Island Park Dam:

1. "Do nothing" alternative.
   a. Assume inadequate freeboard for the flood events (no plugging).
   b. Assume adequate freeboard for the flood events (no plugging).
2. Continue use of the existing structures and provide adequate freeboard for the flood events and possible plugging of the existing service spillway.
3. Raise the crest of the dam and dike to store the inflow design flood (IDF) volume due to plugging of the spillway inlet and provide adequate freeboard.
4. Raise the crest of the dam and dike to control the inflow flood volume and provide adequate freeboard. Provide an emergency overflow area at the left dike to handle the volume of water due to plugging of the existing spillway.
5. Provide an auxiliary spillway to restrict the maximum water surface with or without regard for plugging of the existing service spillway.
6. Provide an auxiliary outlet works to restrict the maximum water surface with or without regard for plugging of the existing service spillway.

*Results of the Analysis*

With so many alternatives available as possible solutions to the problem of inadequate freeboard at Island Park Dam, a framework for their examination according to a common standard was required. The framework used was a risk-based decision analysis. The methodology operates on the following basic concepts:

1. Cost is the common denominator by which various alternatives may be compared.
2. Two types of costs are determined:
   a. Costs due to construction, maintenance, etc., or capital investment. These are direct costs.
   b. Costs due to damages incurred through the normal operation or failure of a structure during flood events. These are expected risk costs.

The second types of costs have a probability of being incurred that must be factored into the calculation to allow comparison with direct costs. Total expected costs are then determined for the different alternatives. The results are tabulated below, with the component costs shown in the bar chart on Figure 3-3:

| | |
|---|---:|
| Alternative No. 1A | $ 1,790,000 |
| Alternative No. 1B | 970,000 |
| Alternative No. 2 | 1,410,000 |
| Alternative No. 3 | 3,620,000 |
| Alternative No. 4 | 2,460,000 |
| Alternative No. 5 | 11,480,000 |
| Alternative No. 6 | 7,000,000 |

*Sensitivity Analysis*

Before making any conclusions on the least-cost alternative, a sensitivity study was conducted to determine the impact of different probabilities of occurrence (for the flood events and spillway inlet flow restriction) on the computed risk costs. The probabilities are a reflection of engineering judgment, which is not constant from individual to individual. Therefore, three individuals from the field of dam and spillway design were polled for their response to the following question. Responses to the question are provided along with the judgment used by this study.

What is the probability that a spillway inlet restriction (such as described in this study) will occur during the 100-year flood, 1,000-year flood, and IDFs?

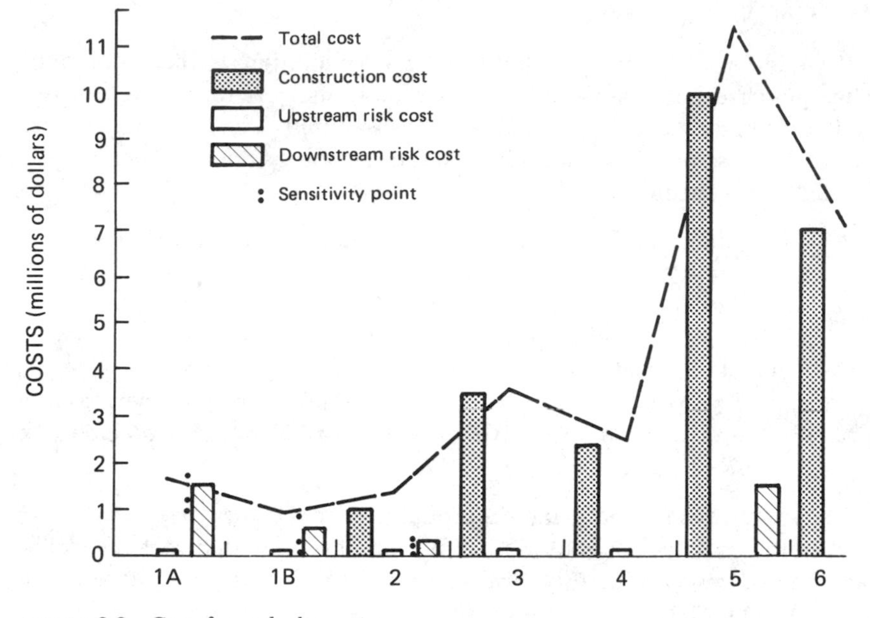


FIGURE 3-3 Costs for each alternative.

The results of using the above probabilities in the analysis are plotted on Figure 3-3 as points next to the bars representing results from the original study. Note that the only significant impact on risk costs are on the downstream risk costs. The upstream risk costs are virtually the same for each set of probability assumptions. Also, only alternatives 1A, 1B, and 2 are affected. (A second sensitivity analysis was made by assuming that the peak and volume IDFs have a recurrence interval of 1,000 years rather than 10,000 years.)

## *Conclusions*

Notwithstanding the "other factors" presented for consideration by the decision maker, the results of the risk-based decision analysis for Island Park Dam (hydrologic aspects) are as follows:

- Alternatives 5 and 6 are not justified on the basis of risk.
- Modification of reservoir operation may essentially eliminate upstream and downstream risk costs without any structural modification.
- Some combination of alternatives 1B, 2, and 4 should be required if formal modification of reservoir operation is not possible.

- A decision is required as to the effectiveness of log booms in protecting against accumulation of debris at the spillway inlet causing flow restriction.

The report makes it clear that this analysis is intended to provide input to the decision-making process and does not compromise the design process. There is no single exact solution to the problem. The effort is on presenting available data in a clear and concise manner. Conclusions and recommendations are made to give the decision maker insight to the reasoning.

## Case III: Willow Creek Dam

Willow Creek Dam is located approximately 4 miles northwest of Augusta, Montana, in Lewis and Clark County. The reservoir stores and regulates irrigation water and has an active storage capacity of approximately 32,400 acre-feet at a normal water surface elevation of 4142.0 feet. Approximately 2,600 acre-feet are available for flood storage to the crest of the existing spillway at elevation 4144.0.

The main embankment, a homogeneous earthfill structure, was constructed on Willow Creek about 1-1/2 miles upstream of the confluence with the Sun River by the U.S. Bureau of Reclamation between 1907 and 1911. The dam was raised in 1917 and again in 1941 with five dikes placed in low saddles in the northern shoreline. The existing embankment is 93 feet high and 650 feet long with a crest elevation of 4154.0 feet. A grass-lined, uncontrolled, open-channel emergency spillway is located about 3,600 feet north of the dam. The spillway crest consists of a 700-foot-long, 6-foot-deep buried concrete cutoff wall protected by riprap on both upstream and downstream sides.

A flood in 1964 caused a small flow, approximately 30 cubic feet/second, through the grass-lined natural-channel emergency spillway at Willow Creek Reservoir and resulted in an erosion phenomenon known as headcutting. This occurrence for a minimum flow caused concern that spillway flows expected from the PMFs could result in erosion cutting back to and failing the spillway crest wall, releasing the reservoir into the Sun River Valley.

### *The Decision Problem*

Examination of the situation at Willow Creek showed the basic elements to be as follows:

- The existing emergency spillway was hydraulically capable of passing the revised PMF.

- A real but uncertain potential existed for erosion to occur to an extent great enough to cause loss of the reservoir through the spillway for discharges ranging from those with a reasonable probability of occurrence to those with a highly remote probability of occurrence.
- Loss of the reservoir through the spillway appeared to constitute a low hazard considering the remoteness of the site, the time factors associated with failure development, the interrelationship of the flood on the Sun River to the discharge from Willow Creek, and the incremental nature of property damages that would result from a postulated failure.

It was determined that consideration should be given to providing for a flood less than the PMF and at the same time increase confidence in the capability of the spillway to pass the more probable lesser magnitude flows without experiencing serious headcutting problems.

To determine the cost-effectiveness of these options, the costs of providing for the PMF were compared with the costs of providing for a lesser flood plus the inherent risk costs associated with providing for less than the PMF.

The final decision between designing for the PMF and a lesser event would consider the cost comparison between the two schemes as well as other factors that are not incorporated into the cost analysis (agency credibility, funding, public acceptability, etc.).

*Risk Assessment*

The decision analysis study for Willow Creek spillway modification alternatives requires as input the risk cost associated with alternatives that provide for an IDF less than the PMF.

The risk cost is not an actual expenditure but rather the cost put "at risk" by providing for a specified design level of event. The annual risk cost is computed by multiplying the damages and losses resulting from a failure event times the annual probability of the event occurring. The total risk cost includes this product for all potential events exceeding the specified design level event integrated over the design life of the project.

Damage costs are determined by appraising the consequences of structure behavior for various conditions. Secondary damage costs such as loss of employment and water and power supplies are not addressed. The only primary cost considered is the direct cost due to inundation. This damage cost may be estimated from that experienced during previous floods, as was done for the initial assessment for the cost analysis. A preferred method employs current aerial photographs of the affected area since this includes

TABLE 3-2 Cost Analysis of Design Alternatives

| Spillway Discharge ($ft^3/s$) | Recurrence Interval, I (years) | Probability of Exceedance, P (percent) | Construction Costs, C (dollars) | Risk Cost, C (dollars) | Total Expected Cost, C (dollars) |
|---|---|---|---|---|---|
| 500 | 235 | 34.7 | — | 6,030,000 | 6,030,000 |
| 1,000 | 434 | 20.6 | — | — | — |
| 5,500 | 4,000 | 2.5 | 380,000 | 360,000 | 740,000 |
| 10,000 | 12,600 | 0.8 | — | — | — |
| PMF | 92,000 | 0.1 | 2,500,000 | 10,900 | 2,511,000 |

any construction that may have taken place since the last flood and may permit categorizing the damage as to building type or use.

Several design alternatives capable of controlling the PMF were developed, the least costly of which was estimated at a total cost of $2.5 million. It was found that a design protecting the existing emergency spillway for flows less than those of the PMF would be more cost-effective ($0.74 million versus $2.51 million). The different cost components are presented in Table 3-2 for the different alternatives.

## REFERENCES

Bohnenblust, H., and Vanmarcke, E. H. (1982) "Decision Analysis for Prioritizing Dams for Remedial Measures: A Case Study," MIT Department of Civil Engineering Research Report R82-12.

Hagen, V. K. (1982) "Re-evaluation of Design Floods and Dam Safety," *Transactions of 14th International Congress on Large Dams*, Vol. 1, Rio De Janeiro, Brazil, pp. 475-491.

Langseth, D. (1982) "Spillway Evaluation in Dam Safety Analysis," MIT Ph.D. thesis.

Lin, J. S. (1982) Probabilistic Evaluation of the Seismically Induced Permanent Displacements in Earth Dams, MIT Ph.D. thesis, Report No. R82-21.

McCann, M. W., Jr., Franzini, J. B., and Shah, H. C. (1983a) *Preliminary Safety Evaluation of Existing Dams—Volume I*, Department of Civil Engineering, Stanford University, Stanford, California.

McCann, M. W., Jr., Franzini, J. B., and Shah, H. C. (1983b), *Preliminary Safety Evaluation of Existing Dams—Volume II, A User Manual*, Department of Civil Engineering, Stanford University, Stanford, California.

U.S. Bureau of Reclamation (1980) *Safety Evaluation of Existing Dams (SEED)*, Government Printing Office, Washington, D.C.

Vanmarcke, E. H. (1974) "Decision Analysis in Dam Safety Monitoring" in *Proceedings Engineering Foundation Conference on the Safety of Dams*, Henniker, New Hampshire, published by ASCE, pp. 127-148.

Vanmarcke, E. H., and Bohnenblust, H. (1982) *Risk-Based Decision Analysis for Dam Safety*, MIT Department of Civil Engineering Research Report R82-11.

## RECOMMENDED READING

Germond, J. P. (1977) "Insuring Dam Risks," *Water Power and Dam Construction*, June, pp. 36-39.

Gruner, E. (1975) "Discussion of ICOLD's 'Lessons from Dam Incidents'," *Schweizerische Bauzeitung*, No. 5, p. 174.

Howell, J. C., Bowles, D. S., Anderson, L. R., Canfield, R. V. (1980) *Risk Analysis of Earth Dams*, Department of Civil and Environmental Engineering, Utah State University, Logan.

Mark, R. K., and Stuart-Alexander, D. E. (1977) "Disasters as a Necessary Part of Benefit-Cost Analysis," *Science*, 197 (September), pp. 1160-1162.

McCann, M. W., Jr., Shah, H. C., and Franzini, J. B. (1983), *Application of Risk Analysis to the Assessment of Dam Safety*, Department of Civil Engineering, Stanford University, Stanford, Calif.

Pate, M.-E. (1981) *Risk-Benefit Analysis for Construction of New Dams: Sensitivity Study and Real Case Applications*, MIT Department of Civil Engineering Research Report No. R81-26.

Schnitter, N. (1976) "Statistische Sicherheit der Talsperren," *Wasser, Energie, Luft*, 68 (5), pp. 126-129.

Shah, H. C., and McCann, M. W., Jr. (1982) Risk Analysis—It May Not Be Hazardous to Your Judgment, paper presented as a keynote lecture at the Dam Safety Research Coordination Conference, Denver, Colorado.

# 4
# Hydrologic and Hydraulic Considerations

## GENERAL APPROACH

### Purpose

One aspect of the investigation of the overall safety of an existing dam is an assessment of the probable performance of the project during future extreme flood conditions. Because the actual timing and magnitude of future flood events are indeterminate, such assessments cannot be based on rigorous analyses but must be made by analyzing the probable function of the project during hypothetical design floods. Generally one such hypothetical flood, called the spillway design flood (SDF), is adopted as a tool for assessing the capability of a project to withstand extreme flood conditions.

In general, there are no legal standards for the magnitude of the flood that a project must pass safely or the type of analyses to be used. Also, there are no procedures and criteria for such assessments that are universally accepted in the engineering profession. However, a number of state agencies have adopted procedures and criteria for hydrologic and hydraulic investigations that provide appropriate guidelines for assessments of dam safety where the governmental agency responsible for dam safety has not specified the bases to be used. The current procedures and criteria considered appropriate for general application are described in this chapter.

It should be recognized that hydrologic and hydraulic analyses for dam safety assessments represent a very specialized and complex branch of engineering for dams. For any project where the consequence of dam failure would be serious, these analyses should be directed by an engineer trained and experienced in this specialized field. Because this is a complex field, the coverage in this chapter is limited to some of the basic concepts involved, and the reader is directed to the references and recommended reading sections of this chapter for further development of subjects of special interest.

## Data Needs

Before beginning any hydrologic or hydraulic analysis, a review should be made of any available design records for the dam and spillway. Information such as discharge rating curves or tables, storage capacity, and pertinent dimensions of spillways and outlets should be made available. The procedures used in the design of the spillway should be compared with currently accepted techniques and criteria. If it is evident that the original design bases gave results substantially in accord with current practices, further investigation may be limited to verifying that the existing project meets the adopted design objectives. If, however, the bases used, such as design storms, are considerably different from what would be acceptable today, then a revised flood study is required.

The review of project design bases should consider data made available since project construction. For example, in recent years the National Weather Service (formerly the U.S. Weather Bureau) has updated almost all its storm estimates for the United States. Also, any available flood studies for areas within or near the watershed should be consulted. Sources of such information may be the U.S. Army Corps of Engineers, U.S. Bureau of Reclamation, Soil Conservation Service, the state dam safety agency, and other dam owners in the area.

Hydrometeorologic characteristics of the watershed of the dam, such as the mean annual precipitation and mean annual flow, should be developed, and past storage fluctuations or reservoir seasonal operation records should be evaluated. Data on unusual hydrologic events, such as peak instantaneous flood flows and operation of gates to spill excess runoff, should be obtained. Such background information permits comparisons of recorded hydrologic events and computed hypothetical events that may be helpful in assessing project safety.

Hydrologic and hydraulic analyses provide only part of the data needed for decisions on whether a dam will provide a reasonable level of safety during future floods. Data on probable effects of project operation during extreme floods or dam failure on developments around the reservoir shores and downstream from the dam are also needed. The impact of these effects on the public welfare and the dam owner's liability for damages must be considered.

## Simplified Procedures

The amount and complexity of the hydrologic and hydraulic analyses appropriate for an assessment of dam safety depend on the importance of the project, the hazards involved, and the data available. For an important project or one presenting significant hazards to downstream areas, detailed development and routing of design floods, as discussed later in this chapter, beginning with the section Spillway Capacity Criteria, would be advisable. However, in some cases the safety investigation of an existing dam will not require such sophisticated analyses. As noted below, several simple techniques are available that may provide acceptable degrees of accuracy. These approximate procedures should be used with care and under the direction of an experienced hydraulic and/or hydrologic engineer.

### *Comparisons with Historical Peak Discharges*

Evaluations of the relative magnitude and the credibility of flood peak discharge estimates can be obtained by comparison with known historical peak discharges in other watersheds. Also, such information about maximum floods of record in similar hydrologic regions can provide a basis for estimating flood potential at a given site.

Data on streamflow and flood peaks are collected and published by the U.S. Geological Survey in its water supply papers, hydrologic investigation atlases, and circulars. Figure 4-1 shows a comparison of observed flood peaks in a region based on such data.

In the past a number of empirical formulas have been advanced for describing the relationship between drainage area characteristics and maximum observed flood discharges. Figure 4-1 shows curves based on two such relationships that have been used extensively, the Myer and Creager formulas. The Creager formula is

$$Q = 46CA^{(0.894\,A^{-0.048})}$$

or its equivalent

$$q = 46CA^{(0.894\,A^{-0.048}-1)},$$

where $Q$ is the total discharge in cubic feet per second, $q$ is the unit discharge in cubic feet per second per square mile, $A$ is the drainage in square miles, and $C$ depends on the drainage basin characteristics. $C$ is a coefficient dependent on many factors, such as the following:

- *Storm rainfall.* Intensity, areal distribution, orientation, direction, trend of great storms, and effect of ocean and mountain ranges.

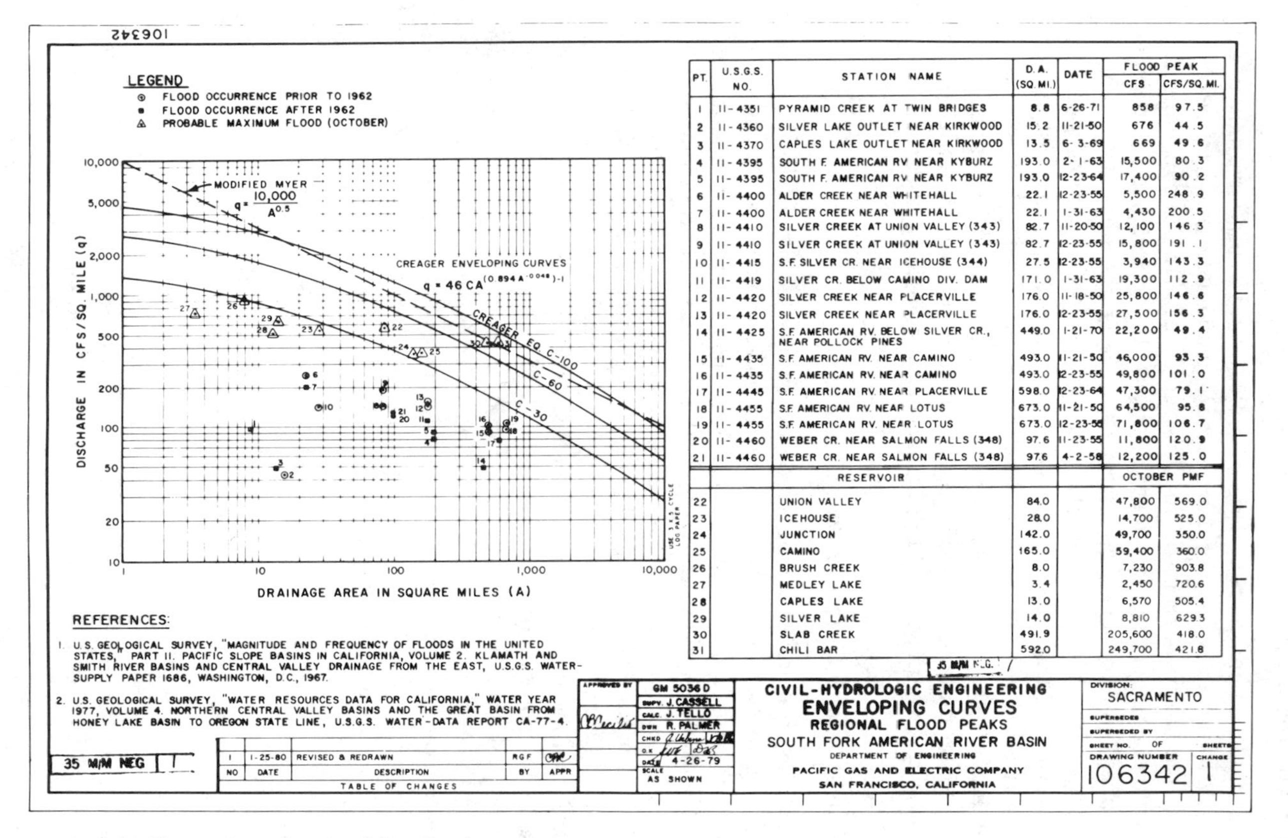


| PT. | U.S.G.S. NO. | STATION NAME | D.A. (SQ. MI.) | DATE | FLOOD PEAK CFS | FLOOD PEAK CFS/SQ. MI. |
|---|---|---|---|---|---|---|
| 1 | 11-4351 | PYRAMID CREEK AT TWIN BRIDGES | 8.8 | 6-26-71 | 858 | 97.5 |
| 2 | 11-4360 | SILVER LAKE OUTLET NEAR KIRKWOOD | 15.2 | 11-21-50 | 676 | 44.5 |
| 3 | 11-4370 | CAPLES LAKE OUTLET NEAR KIRKWOOD | 13.5 | 6-3-69 | 669 | 49.6 |
| 4 | 11-4395 | SOUTH F. AMERICAN RV NEAR KYBURZ | 193.0 | 2-1-63 | 15,500 | 80.3 |
| 5 | 11-4395 | SOUTH F. AMERICAN RV NEAR KYBURZ | 193.0 | 12-23-64 | 17,400 | 90.2 |
| 6 | 11-4400 | ALDER CREEK NEAR WHITEHALL | 22.1 | 12-23-55 | 5,500 | 248.9 |
| 7 | 11-4400 | ALDER CREEK NEAR WHITEHALL | 22.1 | 1-31-63 | 4,430 | 200.5 |
| 8 | 11-4410 | SILVER CREEK AT UNION VALLEY (343) | 82.7 | 11-20-50 | 12,100 | 146.3 |
| 9 | 11-4410 | SILVER CREEK AT UNION VALLEY (343) | 82.7 | 12-23-55 | 15,800 | 191.1 |
| 10 | 11-4415 | S.F. SILVER CR. NEAR ICEHOUSE (344) | 27.5 | 12-23-55 | 3,940 | 143.3 |
| 11 | 11-4419 | SILVER CR. BELOW CAMINO DIV. DAM | 171.0 | 1-31-63 | 19,300 | 112.9 |
| 12 | 11-4420 | SILVER CREEK NEAR PLACERVILLE | 176.0 | 11-18-50 | 25,800 | 146.6 |
| 13 | 11-4420 | SILVER CREEK NEAR PLACERVILLE | 176.0 | 12-23-55 | 27,500 | 156.3 |
| 14 | 11-4425 | S.F. AMERICAN RV. BELOW SILVER CR., NEAR POLLOCK PINES | 449.0 | 1-21-70 | 22,200 | 49.4 |
| 15 | 11-4435 | S.F. AMERICAN RV. NEAR CAMINO | 493.0 | 11-21-50 | 46,000 | 93.3 |
| 16 | 11-4435 | S.F. AMERICAN RV. NEAR CAMINO | 493.0 | 12-23-55 | 49,800 | 101.0 |
| 17 | 11-4445 | S.F. AMERICAN RV. NEAR PLACERVILLE | 598.0 | 12-23-64 | 47,300 | 79.1 |
| 18 | 11-4455 | S.F. AMERICAN RV. NEAR LOTUS | 673.0 | 11-21-50 | 64,500 | 95.8 |
| 19 | 11-4455 | S.F. AMERICAN RV. NEAR LOTUS | 673.0 | 12-23-55 | 71,800 | 106.7 |
| 20 | 11-4460 | WEBER CR. NEAR SALMON FALLS (348) | 97.6 | 11-23-55 | 11,800 | 120.9 |
| 21 | 11-4460 | WEBER CR. NEAR SALMON FALLS (348) | 97.6 | 4-2-58 | 12,200 | 125.0 |
| | | RESERVOIR | | | OCTOBER PMF | |
| 22 | | UNION VALLEY | 84.0 | | 47,800 | 569.0 |
| 23 | | ICEHOUSE | 28.0 | | 14,700 | 525.0 |
| 24 | | JUNCTION | 142.0 | | 49,700 | 350.0 |
| 25 | | CAMINO | 165.0 | | 59,400 | 360.0 |
| 26 | | BRUSH CREEK | 8.0 | | 7,230 | 903.8 |
| 27 | | MEDLEY LAKE | 3.4 | | 2,450 | 720.6 |
| 28 | | CAPLES LAKE | 13.0 | | 6,570 | 505.4 |
| 29 | | SILVER LAKE | 14.0 | | 8,810 | 629.3 |
| 30 | | SLAB CREEK | 491.9 | | 205,600 | 418.0 |
| 31 | | CHILI BAR | 592.0 | | 249,700 | 421.8 |

FIGURE 4-1 Comparison of regional flood peaks.

- *Infiltration*. Character of the soil, antecedent moisture, frozen ground.
- *Geographical characteristics*. Shape and slope of watershed.
- *Natural storage*. Valley storage, tributaries, lakes, and swamps.
- *Artificial storage*. Reservoirs, channel improvements.
- *Land coverage*. Forested, cultivated, pasture, and barren areas.
- *Sudden releases of flow*. Ice and log jams, debris jams against bridges, questionable safety of upstream dams, sudden snowmelt.

A value of 100 for the coefficient $C$ seems appropriate to compare with the most extreme event of the probable maximum flood. Intermediate values with $C$ equal to 30 and 60 are also plotted on Figure 4-1 for comparison.

The Creager enveloping curve provides an estimate of the maximum peak discharge that might be expected for drainage areas generally less than 1,000 square miles. There have been flood discharges that exceeded the limits indicated by the Creager enveloping curve in several basins greater than 1,000 square miles. The Modified Myer equation was introduced by C. S. Jarvis in 1926 and is shown in Figure 4-1 solely for the purpose of comparison with the Creager equation.

## Generalized Estimates of Probable Maximum Flood Peak Discharges

A source of probable maximum flood (PMF) peak discharges is contained in the U.S. Nuclear Regulatory Commission's Regulatory Guide 1.59 (1977). In addition, enveloping curves for areas east of the 103rd meridian were developed by the U.S. Army Corps of Engineers, the U.S. Nuclear Regulatory Commission staff, and their consultants, Nunn, Snyder and Associates. The enveloping curves were found to parallel the Creager curve. Isoline maps showing the generalized PMF estimates as well as a discussion of the way they may be used are contained in Appendix 4A. Use of such generalized data should be confined to situations in which more specific PMF discharge estimates are not available and where it is not practicable to develop estimates specifically for the project.

### *Dam Break Flood Flow Formulas*

In assessing the hazard involved in the potential failure of a dam, an estimate of the downstream flood that would be produced by such failure is needed. As discussed in the section Dam Break Analyses, rather complex techniques are available to compute the characteristics of the downstream flood wave following a dam failure. However, a number of simplified methods for making rough estimates of peak downstream flows from dam

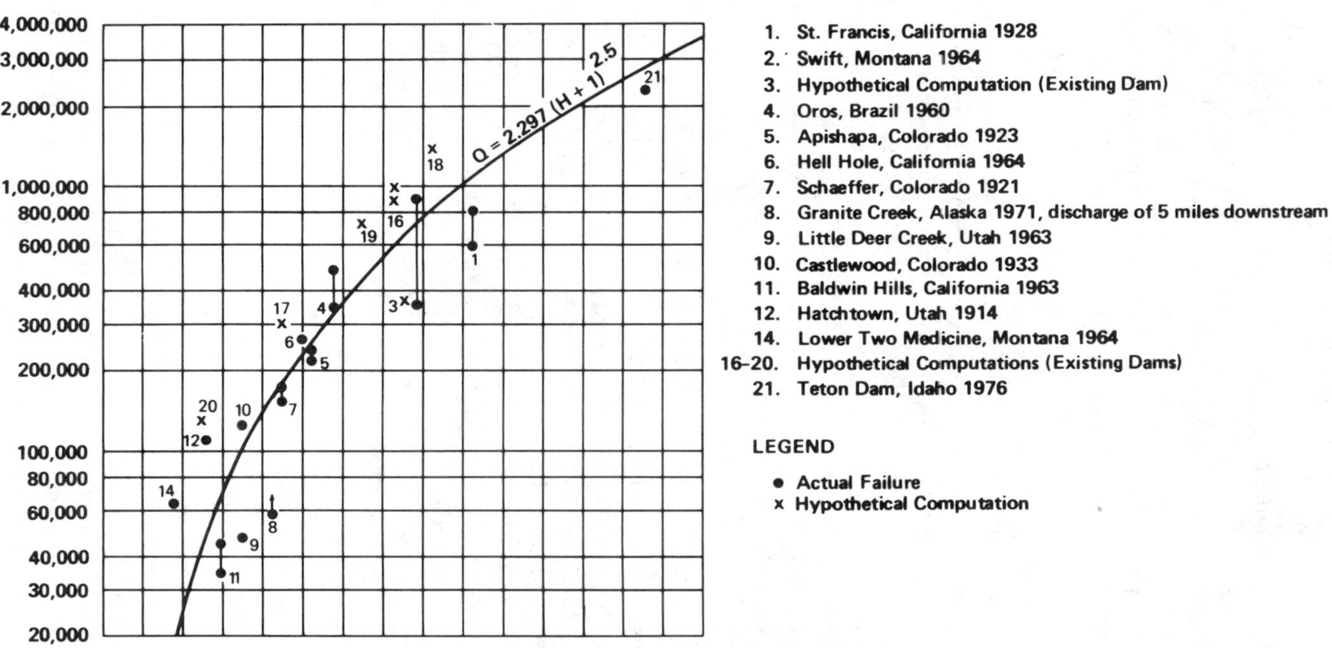

NAME OF DAM, LOCATION, YEAR OF FAILURE
1. St. Francis, California 1928
2. Swift, Montana 1964
3. Hypothetical Computation (Existing Dam)
4. Oros, Brazil 1960
5. Apishapa, Colorado 1923
6. Hell Hole, California 1964
7. Schaeffer, Colorado 1921
8. Granite Creek, Alaska 1971, discharge of 5 miles downstream
9. Little Deer Creek, Utah 1963
10. Castlewood, Colorado 1933
11. Baldwin Hills, California 1963
12. Hatchtown, Utah 1914
14. Lower Two Medicine, Montana 1964
16–20. Hypothetical Computations (Existing Dams)
21. Teton Dam, Idaho 1976
LEGEND
● Actual Failure
x Hypothetical Computation
$Q = 2.297 (H + 1)^{2.5}$
Q = PEAK FLOW IN C.F.S.
4,000,000
3,000,000
2,000,000
1,000,000
800,000
600,000
400,000
300,000
200,000
100,000
80,000
60,000
40,000
30,000
20,000
0 20 40 60 80 100 120 140 160 180 200 220 240 260 280 300
H = HEIGHT OF DAM IF OVERTOPPED, OR DEPTH OF WATER AT TIME OF FAILURE IF NOT OVERTOPPED, IN FEET

FIGURE 4-2 Estimated flood peaks from dam failures. SOURCE: U.S. Bureau of Reclamation (1977).

failures are contained in the references to this chapter. Several empirical formulas of varying degrees of accuracy have been summarized by Cecilio and Strassburger (1974). These formulas apply to instantaneous failure events only. Erosion-type failure can also utilize the same formulas with certain modifications as discussed by Cecilio and Strassburger. Other simplified approaches are presented by Sakkas (1974) and the Soil Conservation Service (1979). Figure 4-2 shows estimated downstream flood peak flows for a number of dam failures and a curve presented by Kirkpatrick (1976) as a basis for estimating such flows.

Figure 4-3 shows data developed by Vernon K. Hagen (1982) on maximum peak flows immediately downstream of dams after their failure. Hagen presents two equations based on experienced dam failures, primarily dam failures in the United States that relate peak flood flows following failure to reservoir levels and storage volumes. The equations are defined as follows (English units):

$$Q = 530\ (DF)^{0.5*}$$

and

$$Q = 370\ (DF)^{0.5},$$

where $Q$ is the peak flow in cubic feet per second immediately below the dam following its failure, and the $DF$ is a dam factor obtained by multiplying height of water in reservoir above streambed $H$ by the reservoir storage $S$ (where $H$ is vertical height measured in feet from the streambed at the downstream toe of the dam to the reservoir level at time of failure, and $S$ is storage volume of water in acre-feet in reservoir at time of failure).

As noted on Figure 4-3, the first equation envelopes all dam failures listed by Hagen, while the second equation envelopes all except the failure of the high arch Malpasset Dam in France.

## BASES FOR ASSESSING SPILLWAY CAPACITIES

In most cases assessment of the safety of an existing dam will require that the SDF be routed through the reservoir and spillway and any other outlet structures for which availability to release water from the reservoir during extreme flood events can be assured. The geometry and location of these hydraulic structures determine the quantity of water that can be discharged at any point of time during the inflow of the design flood. Desirable objectives for spillway operation and bases for assessing spillway adequacy are discussed in this section. Development of SDFs and assumptions regarding project operation for routing studies are covered in subsequent

*In the original paper the equation was written in metric units.

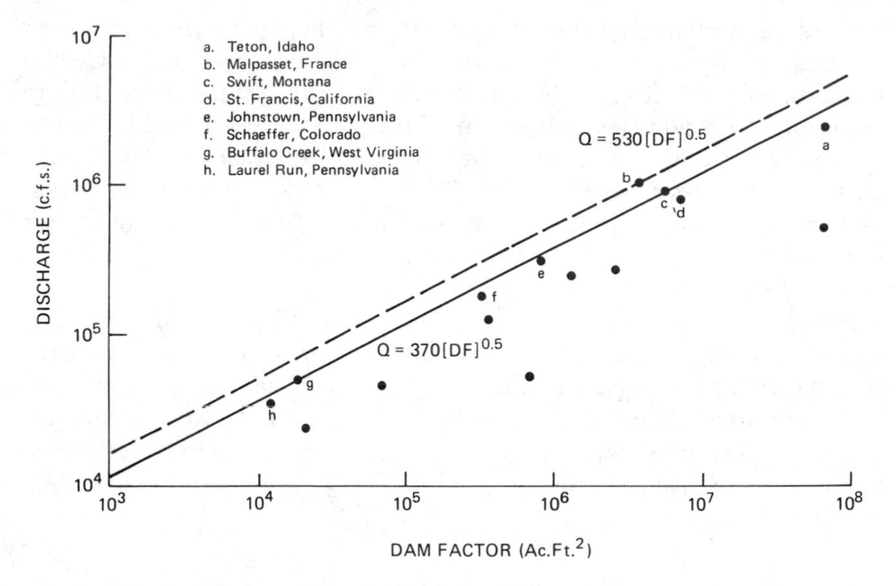


FIGURE 4-3 Peak discharge from significant dam failures.

sections. Discharge characteristics and criteria for various types of spillways and outlet works are presented in Chow (1959), Davis and Sorenson (1969), King and Brater (1963), U.S. Army Corps of Engineers (1963, 1965, 1968a), U.S. Bureau of Reclamation (1977). The bases for determining spillway capacities and freeboard allowances for dams are discussed in a report by the U.S. Army Corps of Engineers (1968b).

## Present General Situation

The design capacities of spillways for many existing dams were chosen in a subjective manner based on the magnitude of property damage and probable loss of project investment and human life in the event of dam failure during a severe flood. Also, many dams in the United States were built before data were available to assess adequately the flood-producing potentials of their watersheds. As a consequence, many existing dams have inadequate spillway capacities based on currently accepted criteria. Analyses have shown that about one-third of recorded dam failures resulted from spillway inadequacy. Also, of approximately 9,000 nonfederal dams in the United States inspected recently, almost 25% were designated as unsafe because of inadequate spillway capacity. The risk of severe consequences associated with the failure of these dams should be reduced by mitigating measures, as discussed later in this chapter.

## Currently Accepted Practices

In 1970 a working group of the United States Committee on Large Dams (designated the USCOLD Committee on Failures and Accidents to Large Dams) conducted a survey on criteria and practices used in the United States to determine required spillway capacities. A survey questionnaire was distributed, and 4 federal agencies, 2 state agencies, 2 investor-owned utility companies, and 13 private engineering firms responded. The results of the survey (USCOLD 1970), summarized below, constitute an authoritative statement of practices accepted and used by those engineers in the United States involved in the engineering of dams.

### *Design Objectives*

The USCOLD committee recognized that the costs of dams and associated facilities are influenced substantially by the degree of security to be provided against possible failure of the dam from overtopping during floods and that two basic objectives of spillway and related safety provisions are to protect the owner's investment in the project and to avoid interruptions in the services afforded by the project. But the committee noted that an additional and usually overriding requirement for security is to protect downstream interests against hazards that might be caused by the sudden failure of the dam and any ensuing flood wave.

The survey found it to be common practice that spillway capacities, in connection with other project features, should be adequate to:

- ensure that flood hazards downstream will not be dangerously increased by malfunctioning or failure of the dam during severe floods;
- ensure that services of and investment in the project will not be unduly impaired by malfunctioning, serious damage, or failure of the dam during floods;
- regulate reservoir levels as needed to avoid unacceptable inundation of properties, highways, railroads, and other properties upstream from the dam during moderate and extreme floods; and
- minimize overall project costs insofar as practicable within acceptable limits of safety.

### *Functional Design Standards*

The USCOLD committee also noted that functional design standards necessary to meet minimum security requirements for downstream areas at minimum cost usually conform with one of the following alternatives, the selection being governed by circumstances associated with specific projects and downstream developments:

*Standard 1.* Design dams and spillways large enough to ensure that the dam will not be overtopped by floods up to probable maximum categories.

*Standard 2.* Design the dam and appurtenances so that the structure can be overtopped without failing and, insofar as practicable, without suffering serious damage.

*Standard 3.* Design the dam and appurtenances in such a manner as to ensure that breaching of the structure from overtopping would occur at a relatively gradual rate, such that the rate and magnitude of increase in flood stages downstream would be within acceptable limits, and such that damage to the dam itself would be located where it could be repaired most economically.

*Standard 4.* Keep the dam low enough and storage impoundments small enough so that no serious hazard would exist downstream in the event of breaching and so that repairs to the dam would be relatively inexpensive and simple to accomplish.

### *Policies Affecting Spillway Capacity Requirements*

The following general policies (based on the accepted practices found by the USCOLD committee) provide guidance for application of the four functional design standards. These should be helpful if combined with some of the procedures in Chapter 3 under the section Flood Risk Assessments.

- When a high dam, capable of impounding large quantities of water, is constructed upstream of a populated community, a distinct hazard to that community from possible failure of the dam is created unless due care is exercised in every phase of engineering design, construction, and operation of the project to ensure complete safety. The prevention of overtopping such dams during extreme floods, including the probable maximum flood, is of such importance as to justify the additional costs for conservatively large spillways, notwithstanding the low probability of overtopping. The policy of deliberately accepting a recognizable major risk in the design of a high dam simply to reduce project cost has been generally discredited from the ethical and public welfare standpoint, if the results of a dam failure would imperil the lives and life savings of the populace of the downstream flood plain. Legal and financial capabilities to compensate for economic losses associated with major dam failures are generally considered as inadequate justification for accepting such risk, particularly when severe hazards to life are involved. Accordingly, high dams impounding large volumes of water, the sudden release of which would create major hazards to life and property damage, should be designed to conform with security Standard 1.

• Application of Standard 2 should be confined principally to the design of run-of-river hydroelectric power and/or navigation dams, diversion dams, and similar structures where relatively small differentials between headwater and tailwater elevations prevail during major floods and where overtopping would not cause either dam failure or serious damage downstream. In such cases the design capacity of the spillway and related features may be based largely on economic considerations.

• Application of Standard 3 should be limited to dams impounding a few thousand acre-feet or less, so designed as to ensure a relatively slow rate of failure if overtopped and located where hazard to life and property in the event of dam failure would clearly be within acceptable limits. The occurrence of overtopping floods must be relatively infrequent to make Standard 3 acceptable. A slow gradual rate of breaching can be accomplished by designing the dam to overtop where the breach of a large section of relatively erosion-resistant material would be involved, such as through a flat abutment section. The control may be obtained, in some cases, by permitting more rapid erosion of a short section of embankment and less rapid lateral erosion of the remaining embankment.

• Standard 4 is applicable to small recreational lakes and farm ponds. In such cases it is often preferable to keep freeboard allowances comparatively small to ensure that the volume of water impounded will never be large enough to release a damaging flood wave if the dam should fail due to overtopping. In some instances adoption of Standard 4 may be mandatory, despite the dam owner's desire to construct a higher dam, if a higher standard is not attainable. Unless appropriate safety of downstream interests can be ensured, a higher dam is not justified simply to reduce the frequency of damages to the project.

## SPILLWAY CAPACITY CRITERIA

### Recommended Spillway Capacities

Pursuant to the National Dam Inspection Act (PL 92-367), enacted August 8, 1972, the U.S. Army Corps of Engineers issued *Recommended Guidelines for Safety Inspection of Dams*, which was made an Appendix D of the *National Program of Inspection of Nonfederal Dams*, ER 1110-2-106 (1982b). This was also issued as Title 33 in the *Code of Federal Regulations*, Part 222. These documents contain guidelines for determining spillway capacity requirements for low, intermediate, and high dams with low, significant, and high hazard classifications. Similar guidelines have been used by the U.S. Bureau of Reclamation and the Soil Conservation

Service. State and local agencies, private companies, and engineering design firms have generally adopted the guidelines of one of the federal agencies in determining spillway capacity requirements, although in many cases, particularly for low and intermediate height dams, no specific guidelines were followed.

As the U.S. Bureau of Reclamation and the Soil Conservation Service guidelines for spillway capacity requirements are generally consistent with those developed by the Corps of Engineers, and the latter are more specific and inclusive, use of guidelines for spillway capacity requirements that were adopted for the National Program of Inspection of Nonfederal Dams is recommended. These guidelines are presented in Tables 4-1, 4-2, and 4-3.

### Alternative Guidelines for Spillway Capacity Requirements

Dams being reevaluated for safety or being enlarged or improved may have spillways smaller than required to pass safely some floods, as indicated in Table 4-3. A smaller capacity spillway may be acceptable if it can be shown that when a specific dam fails due to overtopping by a flood that just exceeds the routing capacity of a reservoir the resulting dam break flood would not cause additional loss of life and/or a significant increase in damage to improved properties over that which would occur prior to the dam failure. In this case, however, the minimum-sized spillway should safely pass the 100-year flood.

As previously noted, in some cases dam owners may not be willing or able to enlarge spillway capacities of existing dams in accordance with the guidelines presented in Table 4-3 because of financial constraints, or a dam may have been in existence for 25 to 50 or more years without any threat of being overtopped, which appears to support the owner's belief that the risk of dam failure and consequent damages is small. Such cases will present difficult problems to governmental agencies having regulatory responsibilities for dam safety. In each such case the agency and the dam owner should

**TABLE 4-1** Dam Size Classification

| | Impoundment | |
|---|---|---|
| Category | Storage (acre-feet) | Height (feet) |
| Small | $< 1{,}000$ and $\geq 50$ | $< 40$ and $\geq 25$ |
| Intermediate | $\geq 1{,}000$ and $< 50{,}000$ | $\geq 40$ and $< 100$ |
| Large | $\geq 50{,}000$ | $\geq 100$ |

TABLE 4-2 Hazard Potential Classification

| Category | Urban Development | Economic Loss |
|---|---|---|
| Low | No permanent structure for human habitation | Minimal (undeveloped to occasional structures or agriculture) |
| Significant | No urban development and no more than a small number of habitable structures[a] | Appreciable (notable agriculture, industry, or structures) |
| High | Urban development with more than a small number of habitable structures[a] | Excessive (extensive community, industry, or agriculture) |

[a]Because this definition does not cite a specific number of lives that could be lost, difficulty was experienced in determining whether dams should be categorized as having "significant or high hazard potential." The issue was clarified by emphasizing that the hazard potential classification should be based on the density of downstream development containing habitable structures. For example, dams located upstream of isolated farmhouses would be classified as having significant hazard potential, and those located upstream of several houses or a residential development would be classified as having high hazard potential.

SOURCE: U.S. Army Corps of Engineers (1982b).

seek to provide as a minimum those corrective and mitigating improvements that could be made to increase the spillway capacity within the owner's financial means. Any increase in spillway capacity, even though less than indicated by Table 4-3, would decrease the risk of dam failure. However, all parties involved should recognize that such partial steps will not meet the design objectives to protect their interests.

## DESIGN FLOODS

By definition, the SDF is the reservoir inflow-discharge hydrograph used to estimate the spillway discharge capacity requirements and corresponding maximum surcharge elevation in the reservoir. The surcharge elevation is obtained by routing the SDF through the reservoir and spillway.

Practices in establishing SDFs in the United States have undergone continuous evolution through several periods. At the present time there are

**TABLE 4-3** Hydrologic Evaluation Guidelines: Recommended Spillway Design Floods

| Hazard | Size | Spillway Design Flood (SDF)[a] |
|---|---|---|
| Low | Small | 50- to 100-year frequency |
| | Intermediate | 100-year to 1/2 PMF |
| | Large | 1/2 PMF to PMF |
| Significant | Small | 100-year to 1/2 PMF |
| | Intermediate | 1/2 PMF to PMF |
| | Large | PMF |
| High | Small | 1/2 PMF to PMF |
| | Intermediate | PMF |
| | Large | PMF |

[a]The recommended design floods in this column represent the magnitude of the spillway design flood (SDF), which is intended to represent the largest flood that need be considered in the evaluation of a given project, regardless of whether a spillway is provided; i.e., a given project should be capable of safely passing the appropriate SDF. Where a range of SDF is indicated, the magnitude that most closely relates to the involved risk should be selected.

100-year = *100-year exceedance interval.* The flood magnitude expected to be exceeded, on the average, of once in 100 years. It may also be expressed as an exceedance frequency with a 1% chance of being exceeded in any given year.

PMF = *probable maximum flood.* The flood that may be expected from the most severe combination of critical meteorologic and hydrologic conditions that are reasonably possible in the region. The PMF is derived from the probable maximum precipitation (PMP), which information is generally available from the National Weather Service, NOAA. Most federal agencies apply reduction factors to the PMP when appropriate. Reductions may be applied because rainfall isohyetals are unlikely to conform to the exact shape of the drainage basin. In some cases local topography will cause changes from the generalized PMP values; therefore, it may be advisable to contact federal construction agencies to obtain the prevailing practice in specific areas.

SOURCE: U.S. Army Corps of Engineers (1982b).

four methods currently in significant use for the derivation of the SDF: (1) an envelope curve method that uses the actual recorded peak flows for the river of concern or for the general region; (2) frequency-based floods using standard statistical techniques that convert the historic peaks into a probability of occurrence curve; (3) the hydrometeorologic approach that maximizes the combination of all of the appropriate physical parameters involved in flood development on the particular drainage area in question;

and (4) a site-specific determination of the maximum flood for which failure of the dam would significantly affect downstream losses.

**Envelope Method**

The envelope process has probably been in use the longest of the three methods. It consists of plotting all of the floods for the river in question or for the hydrologically similar region, against a physical parameter, such as drainage area. An envelope curve is then drawn, and the appropriate peak discharge can be picked off for use as a design parameter. Obviously, the confidence in the flood value thus selected will vary with the length of the data base used in the envelope curve delineation. This procedure is usually acceptable only for rough estimates. It should not be considered as an accurate method for establishing the full flood-producing potential of a basin for determining an SDF. As suggested in the section Simplified Procedures, this method is preferably used only for comparison.

**Frequency-Based Floods**

The second method, statistical manipulation of recorded flood peaks, either for the river in question or for hydrologically similar areas, to produce a return period curve gained great favor in the United States in the early 1940s. Much time and energy have been devoted to determine the statistical distribution that provides the "best fit" for the existing data and that can be extrapolated beyond the period of record with the most confidence. This basic method has severe limitations in that the data base, i.e., usually annual flood peaks, in the United States is of relatively short duration. Very few records go back 100 years, and the majority are less than 50, so that only limited confidence can be placed in use of an extrapolated curve to predict flood events expected to occur only once in one or two centuries on the average. This is not to say that the method is not a useful tool for the hydrologist, but it does have severe drawbacks for use in deriving the SDF.

To achieve some consistency of approach within the federal agencies, the Hydrology Committee of the Water Resources Council issued Bulletin 15, *A Uniform Technique for Determining Flood Flow Frequencies*, in December 1967. This bulletin recommends use of the Pearson Type III distribution with log transformation of the data (log-Pearson Type III distribution) as a base method for flood-flow frequency studies.

Bulletin No. 15 was subsequently extended and updated by Bulletin No. 17 (Bulletin No. 17B, U.S. Department of Interior, Interagency Committee on Water Data, 1982, is the current edition) and its subsequent revisions. This update provides a more complete guide for flood-flow frequency analysis, incorporating currently accepted technical methods with sufficient detail to promote uniform application. This guide is limited to defining flood potentials in terms of peak discharge and exceedance probability at locations where a systematic record of peak flows is available.

The U.S. Geological Survey publishes information on a regional basis for estimating discharge frequencies for ungaged areas. Such data are published in Water Supply Papers with the general title *Magnitude and Frequency of Floods in the United States*, subtitled with the name of the basin. These are presented in the form of regression equations or by use of the "index flood" method.

Most applications of flood frequency analysis methods deal only with peak flood flows and do not produce hydrographs of flow that can be used for routing floods through a reservoir. If such routing is required, an auxiliary method of developing the hydrograph is necessary. Such a method is described in the U.S. Army Corps of Engineers' (1982a) document *Hydrologic Analysis of Ungaged Watersheds Using HEC-1*.

Frequency-based flood analysis should be used only for small-sized and low-hazard or low-risk dams, where accepted practice allows use of design floods with average return periods of 50 to 100 years. Small dams affected by flows from upstream reservoirs should not be evaluated by the frequency type of analysis. For such situations sequential flood routing of flood discharges from the upstream reservoirs and intermediate areas should be used.

### Hydrometeorologic Approach

In the 1940s the U.S. Army Corps of Engineers and the U.S. Weather Bureau (now the National Weather Service) embarked on a study to determine probable maximum precipitation (PMP) magnitudes based on the synoptic processes that generate such floods. In the present report, careful consideration is given to the meteorology of storms that produced major floods in various areas of the United States. The synoptic features of the storm, such as dew point temperatures and rainfall amounts, were cataloged, as were the depth-area-duration (D-A-D) values produced by these storms. It was then possible hypothetically to maximize these D-A-D rainfall amounts by increasing the storm dew point temperature and other factors affecting rainfall to the maximum appropriate values. Adjustments were made for the natural barrier effects on the D-A-D amounts for different areas in the appropriate storm trajectories. The end result of these studies was a series of generalized D-A-D isopleths for use in selecting the PMP meteorologically appropriate for an area. These generalized data are commonly used for design studies in broad level regions, such as the Central states.

In mountainous regions precipitation due to lifting of air by ground slopes, called orographic precipitation, is complemented by precipitation due to meteorological processes that would be present were the mountains not there, called convergence precipitation. In estimating maximum rainfalls these two rain-producing processes are maximized and summed. To facilitate estimate of the PMP, separate isohyetal maps showing orographic and convergence indices are presented in National Weather Service reports.

The National Weather Service has published data for estimating hypothetical storms ranging from the frequency-based storm to the PMP event. There are two major hydrometeorological reports (HMRs) that are applicable for areas east of the 105th meridian: U.S. Weather Bureau (1956) and U.S. Department of Commerce (1978). Before using either of these reports the user should consult with the regulatory agency or the National Weather Service on the appropriate report to use. A U.S. Department of Commerce (1978) report has an auxiliary report: National Weather Service (1982), which describes in detail the application of the report. The U.S. Department of Commerce (1980) HMR Report No. 53 provides estimates of the PMP for 10-square-mile areas. As indicated by the entries in the references for this chapter, the National Weather Service also has issued a considerable number of special reports on individual watersheds. Several hydrometeorological reports of varying degrees of application are available for areas west of the 105th meridian. These areas are shown in Figure 4-4. Ordinarily, an estimate of PMP developed specifically for an area in which a reservoir is situated should be used instead of generalized estimates. As an example, in the Tennessee area, specific hydrometeorological reports that address the orographic effects of the Appalachian Mountains should be used over generalized estimates that neglect such effects.

In areas not covered by specific National Weather Service reports or where doubtful estimates are available, transposition and maximization of major historical storms, D-A-D studies, or storm sequence studies may be necessary. Procedures in determining the PMP are outlined in many of the hydrometeorological reports mentioned above.

In some geographic areas the SDF may result from snowmelt runoff or from the combination of extreme rainfall and snowmelt. Pure snowmelt hydrographs tend to have characteristic shapes and usually result in floods of large volume. Methods have been developed for determining snowmelt hydrographs based on rates of solar radiation. However, historical snowmelt flood hydrographs from the study basin or similar basins are often the best guide to hydrograph shapes and can be adjusted on the basis of water equivalent at the beginning of the melt season.

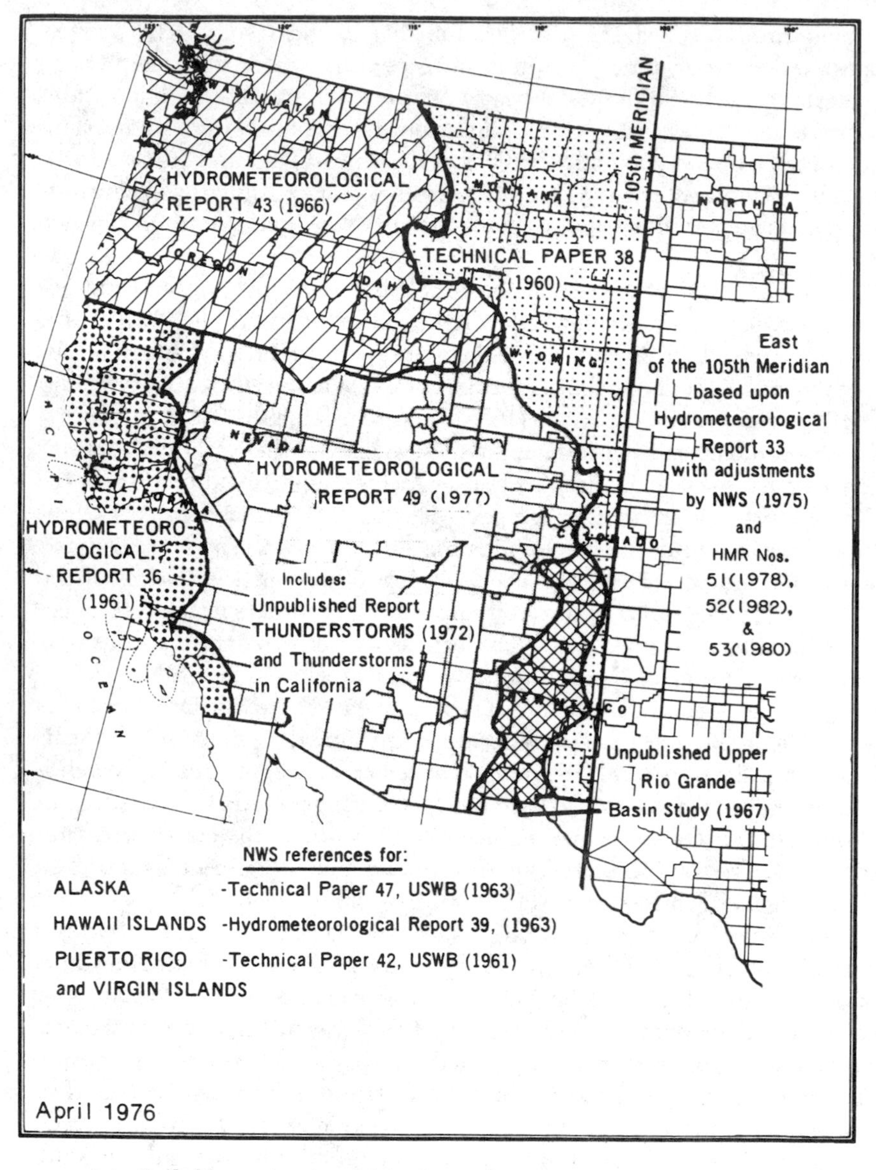


FIGURE 4-4 Probable maximum precipitation study regions.

### Site-Specific Determinations

For many small reservoirs the failure of the dam during a flood of the magnitude of the PMF would not significantly affect damages in downstream areas because of the devastation that would be produced by the natural flood. Where the dam already exists, the available spillway capacity is not adequate to pass the PMF, and it would be difficult or excessively costly to provide such spillway capacity, a lesser spillway capacity may be justified. One procedure that has been used is to consider a number of major floods, less severe than the PMF, that would cause overtopping and failure of the dam and to determine, by analyses of flood waves that would result from the respective failures, the maximum general flood in which the dam failure produces significant additional losses downstream. Such a flood is then adopted as the design flood for spillway improvements. Of course, this selection of design flood is very much dependent on the level of existing developments in the area downstream that would be affected by dam failure. With further development downstream, a larger design flood may be indicated. For this reason, use of this method for selection of an SDF is not well adapted to the design of a new project.

### Credibility of Maximum Flood Estimates

Usually, estimates of probable maximum rainfalls and the attendant PMFs far exceed any rainfalls and floods experienced in the areas involved. Such rainfalls and floods are in the class of natural events that are rarely experienced in a person's lifetime. Perhaps naturally, the adoption of such a standard for design or evaluation of a project can raise doubts as to the possibility that such rainfalls can occur or that the criteria are reasonable. Of course, there is no way to ascertain that probable maximum rainfalls will ever be experienced over a given area. However, as identified in Appendix 4B, records of major storm rainfalls in the United States show that literally dozens of storms have produced rainfalls exceeding 50% of PMP estimates and a considerable number have almost reached PMP magnitudes. These data (U.S. Army Corps of Engineers 1982b) show that enough near-PMP events occur each decade to make consideration of such rainfalls clearly reasonable.

## ANALYSIS TECHNIQUES

Following selection of the basis for the SDF, a number of decisions must be made relative to the conditions of the reservoir and the drainage basin to be assumed to exist antecedent to the design flood and to the methods to be used in developing the reservoir inflow hydrograph, routing the inflow hydrograph through the reservoir storage, and determining the capability of the project to withstand safely such a hydrologic event. As noted before, if the PMF is adopted as a design flood, such analyses usually must deal with phenomena well beyond the range of past experiences in the basin. Often, hydrologic relationships derived from past experience should be adjusted to reflect the rainfall and runoff conditions visualized in the PMF. The objective of the analyses should be to reflect reasonably the probable maximum flood-producing potential of the basin without illogical pyramiding of improbabilities. Guidance for some aspects of those analyses is presented below.

## Providing for Seasonal Variations

Seasonal variations in such basin characteristics as dominant storm types, vegetative cover, ground moisture, and snowpack can often significantly affect the runoff-producing capability of a drainage area. Also, seasonal changes in reservoir operations plans may affect the ability of the project to withstand major floods. To determine the most critical PMF estimate for such conditions, the PMP estimate for the dominant storm type of each season should be used with the most critical basin and project conditions characteristic of the respective season. This may require consideration of small-area thunderstorm-type rainfalls as well as large-area cyclonic-type disturbances.

## Placement of PMP Storm Over a Basin

An isohyetal map of PMP amounts may be shifted to any position over the basin not inconsistent with the meteorologic conditions on which it is based. Several storm centerings should be considered in order to determine the most critical condition. In areas where a specific oval-shaped isohyetal pattern is used to estimate the design storm, placing the storm center near the downstream end of the drainage area produces the most critical condition for peak inflow but not for maximum volume.

## Distribution of PMP to Basin Subareas

For a number of reasons it is often desirable to subdivide large drainage areas in computing PMF hydrographs. Such subdivisions may be required because a number of reservoirs are operational in the basin or to secure subareas adaptable to unit hydrograph application. Nonuniformity of physical or hydrologic conditions in the area also may make subdivision into areas of fairly uniform characteristics desirable. One method of distributing PMP values to subareas is to transpose a selected PMP isohyetal pattern to the drainage area and determine the depth-duration curves for each area. Another method is to place the most intense rainfall over the most critical portion of the basin and derive depth-duration curves for successively large areas. Precipitation curves are then calculated corresponding to adjacent and mutually exclusive subareas.

**Rainfall Time Distribution**

To compute the flood hydrograph for the PMP, it is necessary to specify the time sequence of the precipitation. Usually, the estimate of PMP is derived by 6-hour increments. These increments should be arranged in a sequence that will result in a reasonably critical flood hydrograph. Ideally, the PMP time sequence should be modeled after historically observed storms if such storms show that major storm rainfalls have a predominant pattern.

When no predominant rainfall pattern is evident from past records or has not been developed, a number of guides are available for arranging the rainfall into patterns consistent with the meteorologic processes involved and that will produce reasonably critical hydrographs. One such guide is as follows:

1. Group the four heaviest 6-hour increments of the PMP in a 24-hour sequence, the next highest four increments in a 24-hour sequence, etc.
2. For the maximum 24-hour sequence, arrange the four 6-hour increments ranked 1, 2, 3, and 4 (maximum to minimum) in the order 4, 2, 1, 3. Other days may be arranged in similar order.
3. Arrange the 24-hour sequences such that the highest period is near the center of the storm and the second, third, etc., are distributed in a manner similar to step 2 above.
4. The 6-hour increments may be further subdivided into 1-hour increments by determining the incremental differences from a depth-duration plot (mass curve) of the total PMP storm. The six 1-hour increments ranked 1, 2, 3, 4, 5, and 6 (maximum to minimum) should be arranged in the order of 6, 4, 2, 1, 3, 5.

Typical rainfall distribution patterns applicable in California and the northwest states are shown in Figure 4-5. In some other parts of the country the time distribution pattern suggested in the *Standard Project Flood Determinations* (U.S. Army Corps of Engineers, 1965a) is adopted.

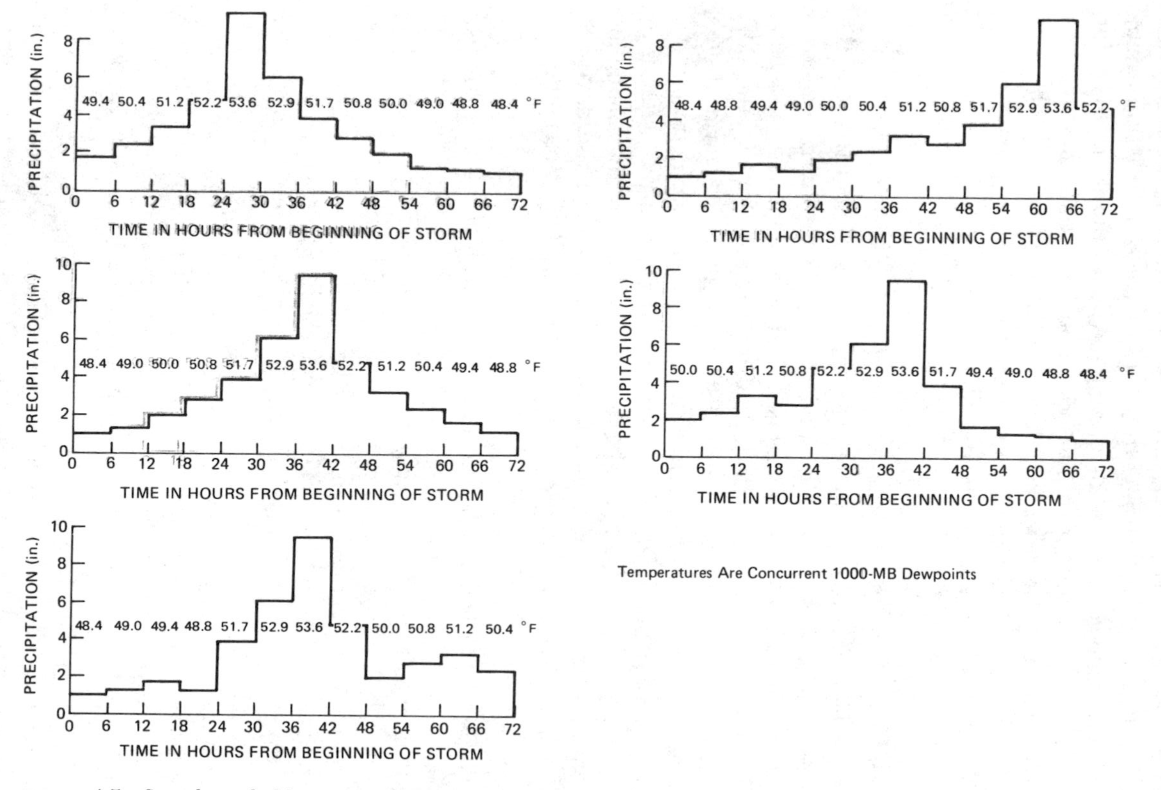


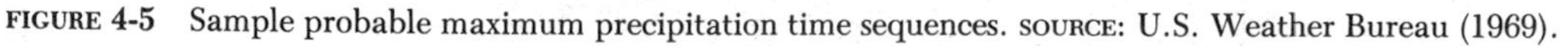

FIGURE 4-5 Sample probable maximum precipitation time sequences. SOURCE: U.S. Weather Bureau (1969).

### Antecedent Storms

Studies of historical storms in certain regions indicate that it is possible to have significant rainfalls occurring before or after the major flood-producing storms. Such meteorological sequences should be considered in studies involving a PMF estimate in an area where several reservoirs exist. As a general rule, the critical PMP in a small basin results primarily from extremely intense small-area storms, whereas in large basins the critical PMP usually results from a series of less intense large-area storms.

A number of National Weather Service studies cover the time sequencing of storms. For areas not covered by these studies, it is often considered that the PMF is preceded 3 to 5 days earlier by a flood that is 40 to 60% of the principal flood. Assumed antecedent conditions generally provide wet and saturated ground conditions prior to the occurrence of a PMP.

### Antecedent Snowpack and Snowmelt Floods

Flood flows in many parts of the United States frequently depend on the rate and volume of melting from snow that has accumulated during the winter months. The volume of water available in such form for flood runoff depends on the depth, density, and area of snow accumulation. The rate of melt depends on meteorologic factors, such as temperature, cloudiness, wind movement, humidity, and on basin physiographic factors, such as elevation, shape, orientation, and type of vegetation. The months in which the greatest snowmelt occurs will vary from one locality to another and from one year to another, depending on the geographic location, the prevailing climate, and meteorologic variations of the season. However, in most western states, major floods resulting from snowmelt occur between April 1 and June 30.

Estimates of PMF in areas subject to snow accumulation require the evaluation of snowmelt as added contributions to runoff. The initial snowpack condition is important both from the consideration of snowmelt and for the storage and delay of liquid water in the snowpack. For the PMP-plus-snow flood conditions, it may be assumed that sufficient water equivalent exists to provide snowmelt continuously through the storm period throughout the entire range of elevation. Also, it may be assumed that the preceding melt and rainfall have provided drainage channels through the snowpack and have conditioned it to provide runoff without significant delay; thus, water excesses from rain and snowmelt during the storm period are immediately available for runoff. To estimate the initial snowpack condition at the beginning of the PMP, records of snow accumulation and water content prior to historical floods should be investigated.

Methods of estimating the probable maximum snowmelt flood, whether the flood is entirely from thermal action on snow or from a combination of snowmelt and rain, have been developed by the U.S. Army Corps of Engineers. Snowmelt evaluation for basins may be accomplished either through the use of simplified generalized equations, sometimes known as the energy-budget method, or indirectly through use of snowmelt indexes, also called the degree-day method. In basin applications for design floods, the former are more appropriately used because of the requirement for direct rational evaluation of all factors affecting snowmelt and extending them to the given design condition. This involves detailed computations of major scope, but they are justified for the design of major water control projects. For daily streamflow forecasting uses, however, a simple snowmelt index is usually adequate when considering the overall accuracy of forecasts and time limitations in their preparation. The detailed applications of these methods are described in U.S. Army Corps of Engineers (1956, 1960, 1962), U.S. Bureau of Reclamation (1966), and U.S. Weather Bureau (1966a).

### Loss Rates

The absorption capability of a watershed depends on many factors, and its determination is subject to several uncertainties. However, reasonable estimates of loss rates can be based on detailed seasonal rainfall-runoff studies performed for past floods within the watershed. These studies should cover antecedent precipitation conditions, base flow, and the type of soils in the watershed.

The selection of loss rates is a major consideration in deriving hypothetical floods because a major portion of the rainfall is lost to interception or localized ponding and infiltration. The degree of conservatism in the estimate of the peak design flow is subject to the degree of conservatism by which the loss rates are assumed or applied. The loss rate is often not as significant in determining the PMF as it is in determining the 100-year flood. Seasonal variation in minimum loss rates, which should be considered as representative of the most extreme conditions for the season for the hypothetical flood, should be applied. Typical values used throughout the United States are in the range of 0.10 to 1.0 inch initial loss followed by a uniform rate of 0.05 to 0.15 inch per hour.

### Runoff Models

The hydrologic response characteristics of the watershed to precipitation are embodied in a mathematical model termed a runoff model. A runoff model translates precipitation excess over a watershed to its resulting flood hydrograph. A number of different types of runoff models have been used, including computer-based mathematical models (as discussed under Dam Break Analyses later in this chapter). A model may be derived from rainfall and runoff data for historical floods or by use of so-called synthetic procedures based on generalized empirical relationships between a basin's physical characteristics and its hydrologic characteristics. When practicable, a runoff model should be verified by using it to reproduce one or more historical floods.

It is often necessary to divide a watershed into subareas on the basis of size, drainage pattern, installed and proposed regulation facilities, vegetation, soil and cover type, and precipitation characteristics. Runoff models are often derived for each subarea, and these subarea models are connected and combined by channel routing.

The unit hydrograph has been found to be a very powerful tool in watershed modeling. The unit hydrograph is defined as the hydrograph that represents a unit volume of runoff (customarily 1 inch) from the study basin generated during a finite rainfall excess period. Thus, a 1-hour unit graph is the hydrograph that would be produced by 1 inch of rainfall excess (runoff) generated in a 1-hour period. Unit hydrograph derivation is discussed in several hydrology textbooks and publications. Generally, its use should be limited to drainage areas not exceeding 2,000 square miles.

Unit hydrographs are usually derived in one of two ways depending on the extent of the data available. In the case of gaged drainage basins, historical flood events are analyzed by separating out the base flow component and calculating the depicted surface runoff volume. The event ordinates are then proportionally reduced or increased as appropriate to give a hydrograph that represents 1 inch of runoff from the basin under the particular temporal and spatial distribution of the actual storm precipitation. In the ideal case, several floods resulting from the type of synoptic situation that has been adjudged the most critical for that particular basin, i.e., frontal rainfall, convective rainfall, snowmelt, or a mixed snow/rain event, are available for separation analysis. These separated hydrographs and the resulting *derived* unit hydrographs are then compared with the actual rainfall event to establish the time parameter of the appropriate unit hydrographs. An infiltration analysis of the rainfalls and runoff volumes for the historical floods will often aid in determining the lengths of periods of rainfall excess represented by the derived unit hydrographs. After selection of a unit hydrograph and its time parameters, methods are available to correct the unit hydrograph to a time basis that is relevant to the PMP increments and to the basin characteristics.

In the case of ungaged basins the unit hydrograph can be derived from climatically and geomorphologically similar basins for which data are available or from various mathematical models. For similar basins the procedure is the same as that described above. For the situation where there are no streamflow data and no appropriately similar basin available, the mathematical approach permits derivation of a synthetic unit hydrograph based on the physical aspects of the basin (U.S. Army Corps of Engineers 1959).

Various adjustments have been made to runoff models derived as above in attempts to make the models more nearly fit the runoff conditions anticipated during a PMF. Such refinements have attempted to account for the expected increase in overland and channel flow velocities during periods of very high runoff rates, for rain falling directly on a large reservoir surface, and for the shortened flow distances in the watershed because of a large reservoir expanse above the dam. These adjustments result in making the unit hydrograph peak earlier and higher.

By combining critically arranged PMP increments with the adopted runoff model (with appropriate allowances for base flow and the recession limbs of antecedent storm hydrographs), a PMF hydrograph for the basin is obtained. If the runoff model represents the entire watershed, this PMF hydrograph is considered the inflow hydrograph for the reservoir. If the watershed has been subdivided, subbasin PMF hydrographs are appropriately routed and combined to obtain such an inflow hydrograph.

## Base Flow

The base flow in a river at the beginning of the main hypothetical flood should be equivalent to the receded flow of any antecedent flood assumed or considered in the study. In the absence of an assured antecedent flood, as may be the case for small drainage basins, a reasonable base flow such as the mean annual flow should be added to the principal flood.

## Channel Routing

Outflow hydrographs from each subarea in a watershed above a dam under investigation should be routed through the river channel up to a point where they can be combined with another subarea hydrograph. This routing and combining should proceed from the uppermost subarea to the most downstream area adjacent to the reservoir of the dam under investigation.

This procedure is generally called channel or streamflow routing. For rainflood or snowmelt flood studies where the rate of flow is not rapid, hydrologic storage-routing techniques can be used. There are several generic names for channel routing, but the most commonly used techniques are the Muskingum method and the Modified-Puls storage-indication method (U.S. Army Corps of Engineers 1969), or the progressive average-lagg method (Straddle-Stagger). Both of these methods assume that the outflow at the downstream end of a channel reach is a function of the storage between successive water surface profiles in the reach. To allow for the steeper water surface slopes in rapidly rising streams, the Muskingum method uses the concept of wedge storage in the reach.

In applying the hydrologic method of routing, care should be taken in making assumptions for the coefficients used in a specific routing procedure. These assumptions can make a significant degree of change in the attenuation of the flood peak. If in doubt about the coefficients, it is suggested that these should be derived from historical floods recorded at streamflow gaging stations with proper allowance for extrapolating the data to greater floods.

### Antecedent Reservoir Level

It is difficult in most cases to estimate the initial reservoir level that is likely to prevail at the beginning of a SDF, except when the storage space is so small as to assure frequent filling. If a long period of streamflow records is available, hypothetical routing studies will provide some index to reservoir elevation probabilities, but even these computed relations may be greatly altered in the future if changing conditions result in substantial alterations in the river or regulation plan.

For projects where the flood control storage space is appreciable, it may be appropriate to select starting water surface elevations below the top of flood control storage for routings. The U.S. Army Corps of Engineers' practice is to assume that 50% of the available flood control storage is filled at the beginning of the SDF. Conservatively high starting levels should be estimated on the basis of hydrometeorological conditions reasonably characteristic for the region and flood release capability of the project. Necessary adjustments of reservoir storage capacity due to existing or future sediment or other encroachment may be approximated when accurate determination of deposition is not practical.

In view of the uncertainties involved in estimating initial reservoir levels that might reasonably be expected to prevail at the beginning of an SDF, it is common practice, particularly for small and intermediate height dams with a single low-level outlet, to assume that the reservoir is initially filled to the "normal full pool level." This reservoir level should be in accordance with the operational practice for the season of occurrence of the SDF.

### Use of Gated Spillways

Spillway crest gates are frequently used to provide required spillway capacity and to control the release of spillway discharges. The gates are designed to operate under controlled conditions during the occurrence of major floods. Proper operation of spillway gates for flood control can prevent downstream flooding that might otherwise occur with a fixed crest spillway. Also, the gates must be operated properly to prevent the release of reservoir discharges larger than those of the natural flood. The gates should be maintained and tested on a regular basis to assure that they will be fully operational during the flood season.

For some projects it may be unsafe to assume that regulating gates would be attended during the occurrence of the SDF. The lag time between intense rainfall and the occurrence of the peak reservoir level could be too short for the gate operator to operate the gates properly, especially if the storm occurs at night. In this case gates should be assumed to be in the open position if they are normally open during the flood season, or closed if they are normally closed. No credit should be taken for any gate operation in the SDF routing unless it can be assured that the gates would be operated properly. Some regulatory agencies require that spillway gates at remote dam sites be locked in open position during usual heavy precipitation seasons, particularly if heavy snowfalls may prohibit access.

### Use of Low-Level Outlets

Current practice is to assume no credit for water releases through any low-level outlet in routing the SDF. It is normal practice to assume that these structures are either clogged or inoperable. Many existing dams, particularly those of low and intermediate heights, have single, relatively small, low-level outlets that are controlled in some cases by a single gate or valve that is not operated regularly. These outlets cannot be relied on to be fully operational during an SDF. Although in some cases flow releases through power penstocks and turbines have been assumed to occur in the spillway flood routing, in most cases the power plant is assumed to be inoperational. For existing dams, consideration may be given to taking credit for power releases within certain limitations and to full assurance that the power plant could be operated safely during an SDF.

### Reservoir Routing

The computation by which the interrelated effects of the inflow hydrograph, reservoir storage, and discharge from the reservoir are evaluated is called reservoir routing. Generally, such routings assume a level pool within the reservoir. However, for long, narrow reservoirs some adjustment for the so-called wedge storage during rapidly rising pool levels may be in order. The basic tools for such a routing are the reservoir elevation-storage curve and the spillway rating curve. The elevation-storage curve is a plot or table showing accumulated storage volumes versus reservoir stages or water surface elevations. The spillway rating curve is a plot or table showing the discharge capability of the spillway facilities (i.e., outlets, uncontrolled spillway weirs, spillway gates, etc.) versus reservoir stages or water surface elevations.

All reservoir routing procedures use variations of the volumetric conservation equation:

$$I - O = \Delta S,$$

where $I$ is reservoir inflow, $O$ is the outflow or discharge, and $\Delta S$ is the change of storage, all for the same time interval. Usually it is assumed that outflow will vary linearly through a short routing interval and the equation is written:

$$I - \frac{(O_1 + O_2)}{2} = S_2 - S_1,$$

where the subscripts designate instantaneous values of $O$ and $S$ at the beginning and end of the routing interval. At any point in a routing computation, the value of $I$ will be available from the inflow hydrograph and values of $S_1$, and the corresponding $O_1$ will have been determined by the routing for the previous time interval. Thus the routing for an interval consists of finding the value of $S_2$ and the corresponding $O_2$ that will satisfy the above equation. This can be done by trial and error, but the solution is more direct if the equation is written:

$$I + \left(S_1 - \frac{O_1}{2}\right) = \left(S_2 + \frac{O_2}{2}\right).$$

By developing curves or tabulations of values of the terms in parentheses, a routing can be speedily accomplished by either graphical or arithmetic means.

The results of the routing procedure are a reservoir stage hydrograph and an outflow hydrograph representing the attenuation of the PMF inflow hydrograph by reservoir storage, spillway, and outlet facilities.

## Freeboard Allowance

It is common practice to provide an extra height of dam over the computed maximum reservoir level for the design PMF. This added height, termed freeboard, is an allowance for waves, wave runup, and wind surge or pileup of water that could be caused by strong winds over the full reservoir surface. The type of dam, its geographical location and directional orientation, and the synoptic situation producing the PMP are all factors to be considered in the determination of freeboard allowance.

A major consideration, the synoptic situation, should deal with the timing of the maximum winds, their orientation and duration that accompany the study storm. For example, some types of storm systems, such as hurricanes, usually have the heavy rain-producing mechanism in the forefront, followed by the maximum winds. This scenario on certain watersheds could result in the maximum winds occurring after the PMF has filled the reservoir.

For assessing freeboard requirements, estimates are needed for the velocities, directions, and durations of winds that reasonably could occur with the reservoir at or near full pool. The "fetch," or maximum over-water distance adjacent to the dam in the direction of the wind, and depths of the reservoir also are needed in estimating wind effects. A number of approaches to computing these wind effects are available in U.S. Army Corps of Engineers' reports (1968b, 1976). The results of such computations will give the wave height and setup and runup elevations for the selected conditions. Common practice for major dams is to add from 3 to 5 feet to the maximum computed water surface level, including wind effects, to establish top of dam.

## DAM BREAK ANALYSES

Knowledge of the nature and extent of catastrophic flooding and resulting risk to downstream life and property following collapse of a dam is a critical step in assessing and improving the safety of existing dams. Disasters caused by past dam failures and results from the recent Corps of Engineers' dam inspection program, have focused the attention of the public and federal and state officials on finding mitigatory measures for unsafe dams, on emergency action planning and preparedness, and on means for assessing public safety and predicting probable damage in the event of failure of existing dams.

To better assess the hazard potential of a dam in a systematic and equitable manner, an analytical approach called dam break analysis should be used. Dam break analysis serves two primary goals. First, it provides information to the engineer about classifying the potential hazard of a dam for determining recommended spillway capacity. Second, it predicts flood depths and wave arrival times and identifies areas that could be affected by flood water should a failure occur. Estimation of downstream flooding times and identification of flood inundation areas permit rational development and implementation of emergency preparedness, warning, and evacuation plans. Such plans, coupled with frequent inspection and conscientious maintenance work, could minimize the loss of life resulting from a failure.

Many types of dam break models exist. The objective of each is to simulate the failure of a dam (i.e., produce a dam failure hydrograph and/or route the hydrograph downstream). Some modeling procedures can be performed readily by hand, such as the Soil Conservation Service (SCS) TR #66 Simplified Dam Breach Model and the Uniform Dam Failure Hydrograph procedure described by Hagen (1982), while others such as the National Weather Service Dam Break Flood Forecasting Computer Model and the Corps of Engineers HEC-1 Flood Hydrograph Computer Package are very complex and require computer analysis. In general, the handworked methods represent simplified approaches to a complex phenomenon. Generally, simplified methods should be applied only to small dams and in instances where other dams are not involved or where damage to downstream developments is not significant.

### Inventory of Dam Break Models

Several methods and computer models for analyzing the likelihood of dam failure and its potential downstream flooding effect are detailed in the literature. The decision as to which method to use for a given situation ultimately depends on the judgment of a qualified engineer. However, the choice usually involves the following considerations:

- General availability, acceptance, and documentation of a model or method.
- Capability of the model or method to simulate the conditions, assumptions, and uniqueness of a given dam situation.
- Resources available to the user in the way of data, finances, and computer facilities for applying the method or model.
- Purpose to which the results of the dam break analysis would be applied. (For example, where approximate downstream impacts resulting from postulated single dam failures are needed for preliminary planning, simple handworked methods can be considered first. Conversely, in situations where detailed flood wave arrival times, depths, and accurate inundation mapping are required, or where complex situations arise from multiple dams, the user should consider the computer model approach.)

Several factors usually have to be evaluated or assumed whenever dam failures are postulated. The type of dam failure and mechanism causing failure require careful consideration, if a realistic breach is to be assumed. Size and shape of breach, reservoir storage, height of overtopping, and timing of breach formation are critical factors in the determination of the dam failure hydrograph. Although considerable investigation has been conducted on historical dam failures, there is not enough information to predict all of the critical parameters with accuracy and consistency. Appendix 4C sets out guidelines used by one organization for assumptions relating to the breaching of dams. For the areas downstream from small dams the length of time assumed for breach development after erosion action is under way will significantly affect the estimated flood heights. However, little data are available on which to base such assumptions. The material of which the dam is structured will influence the time for breach development and estimates of such times should take into account conditions specific to the site.

The following list of dam break models and modeling procedures are available, are documented in the literature, and have been widely used and accepted by both federal and private sector users:

- *Soil Conservation Service (1979) TR #66 Simplified Dam Breach Model.* A quick, handworked method for (1) estimating maximum dam breach discharge from an empirical curve or equation based on historical dam failure data for height of dam versus maximum discharge and (2) estimating maximum discharge and stage at selected downstream floodplain points, utilizing a simplified version of the simultaneous storage routing—Kinematic routing method (e.g., Attenuation-Kinematic, or Att-Kin model).
- *Uniform Procedure for Dam Failure Hydrographs* (Hagen 1982). A simple, quick, handworked procedure developed by V. K. Hagen of the U.S. Army Corps of Engineers for computing the critical failure hydrograph for all dams regardless of type and condition. This method uses an envelope curve or equation, based on historical dam failure data, to relate a "dam factor," the product of dam height and reservoir storage, to peak discharge from a postulated dam failure. Since this procedure produces only a dam failure discharge hydrograph, the user must apply his own channel routing techniques for determining downstream consequences.
- *National Weather Service (NWS) Dam Break Flood Forecasting Model* (Fread 1982). A computer model analysis consisting of two parts: (1) simulation of outflow hydrograph due to instantaneous or time-dependent erosion-type of dam break through an assumed hydraulic weir opening or orifice breach, while simultaneously considering the effects of the reservoir storage depletion and the inflow hydrograph via either a storage or hydraulic routing technique and (2) routing the generated outflow hydrograph through the downstream valley by dynamic hydraulic method to give flow time and stage at user-specified cross-section points. This model allows the

user to specify an inflow design flood hydrograph, select a method of reservoir routing, specify geometry and time of breach, specify depth of reservoir at time of breach, and specify hydraulic characteristics of downstream channel. The National Weather Service model has the capability of handling reservoir rim slides, multiple dams (series), bridge effects, and other complex downstream channel geometry conditions.

• *U.S. Army Corps of Engineers HEC-1 Flood Hydrograph Computer Package.* This computer model analysis consists of two parts: (1) evaluation of the overtopping potential of the dam, by simulation of outflow hydrograph due to time-dependent, erosion-type of dam breach through an assumed hydraulic weir opening, while simultaneously considering the effects of the reservoir storage depletion and the inflow hydrograph via storage routing technique and (2) estimation of downstream hydraulic-hydrologic consequences resulting from the assumed failure of the dam.

While the HEC-1 allows the user to specify similar breach and downstream conditions as in the NWS model, a major feature of this model allows the user either to specify an inflow design flood hydrograph or to input design precipitation and watershed characteristics for automatically generating an inflow flood hydrograph.

• *Some Other Methods and Models.* There are several other methods available for estimating peak discharge from a postulated dam failure, ranging from simple handworked methods to sophisticated two-dimensional computer models. Although a brief listing of some of these methods is provided below, the reader is encouraged to consult with federal agency experts, consulting engineers, state dam safety engineers, and the technical literature on the uniqueness, advantages, and limitations of each method before using any of them:

(a) TAMS Model. Developed by Balloffet for describing the propagation of a dam collapse wave in a natural channel (1974).

(b) TVA Dam Breaching Program developed by the TVA Flood Control Branch in 1973. Predicts if a flood overtopped earth embankment will fail and, if so, the time and rate of failure.

(c) Two-dimensional dam breach wave model developed by Strelkoff in 1978.

(d) U.S. Army Corps of Engineers' Hydrologic Engineering Center dimensionless depth and distance curves, based on Sakkas's work, in 1980. Estimates maximum flood depths at downstream points from an instantaneous and completely failed dam.

(e) Classic dam break equations for instantaneous ("pull the plug") and complete or partial dam failures, developed by Keulegan in 1961. Uses simple equations and cross-sectional depth-discharge rating curves. User must

TABLE 4-4 Summary of Seven Selected Dam Break Model Capabilities

| Dam Breach | Assumed Type of Failure | | | | | | | | | | Routing | | | | | | | | Losses | Comp. CPU Time | | Rel. Manpwr. Needs | |
|---|---|---|---|---|---|---|---|---|---|---|---|---|---|---|---|---|---|---|---|---|---|---|---|
| | Complete | | | | | Partial | | | | | Reservoir | | | | Channel | | | | | | | | |
| Model or Method | Sudden | Triangular | Rectangular | Trapezoidal | Trap./Rec. | Gradual | Triangular | Rectangular | Trapezoidal | Trap./Rec. | Hydrologic | Hydraulic | Implicit | Explicit | Hydrologic | Hydraulic | Implicit | Explicit | Expan./Contr. | Low | High | Low | High |
| SCS | Peak Failure Discharge Based on Historical Events Curve | | | | | | | | | | | | | | ● | | | | | | | ● | |
| Uniform Procedure (Hagen) | Peak Failure Discharge Based on Historical Events Curve | | | | | | | | | | | | | | | | | | | | | ● | |
| NWS | ● | ● | ● | ● | | ● | ● | ● | ● | | ● | ● | ● | | | ● | | ● | ● | | ● | | ● |
| HEC-1 | ● | ● | ● | ● | | ● | ● | ● | ● | | ● | | | | ● | | | | ● | ● | | ● | |
| Classic Dam - Breach Equations | ● | ● | ● | ● | ● | | ● | ● | ● | ● | | | | | | | | | | | | ● | |
| TAMS | ● | ● | | | ● | ● | ● | | | ● | | ● | | ● | | ● | | ● | ● | | ● | | ● |
| TVA (SOCH) | ● | ● | ● | ● | ● | | ● | ● | ● | ● | | ● | | ● | | ● | | ● | ● | | ● | | ● |

furnish channel routing system to provide data for assessing downstream flood consequences.

### Assessment of Available Dam Break Models

The selection of a method or model for analyzing a postulated dam break should depend on the type and quality of information that the user desires for a given dam failure situation. The user should consider the following three major parts of analyses associated with dam failure for comparing the relative merits and capabilities of the available procedures:

1. Type of reservoir routing desired for design flood hydrograph (may also include generation of inflow flood hydrograph).
2. Assumed mode of failure for generation of dam break hydrograph: instantaneous versus gradual failure; partial versus complete cross-section failure; orifice versus weir-type failure for embankment dams; type of breach geometry assumed (rectangular, triangular, trapezoidal); type of time-dependent erosion model used (for embankment dams) (When does erosion begin? How long does breach erosion take?).
3. Whether peak failure discharge or estimated failure hydrograph is needed and the need for and type of downstream channel routing of dam break hydrograph.

Few comparative analyses of different available procedures and models have been made. One study conducted at the University of Tennessee compared peak dam failure discharges for two simple handworked methods (SCS and Classic Equations-Keulegan) and two computer models (NWS and HEC-1) for a single 36-foot-high embankment subjected to a PMF-level flood (Tschantz and Majib 1981). The computed peak flows from this study compared as follows:

| Procedure | Peak Flow at Dam (cfs) |
|---|---|
| SCS | 71,335 |
| Classic equations | 76,000 |
| NWS | 85,950 |
| HEC-1 | 87,000 |

It should be noted in comparing these results that the first two handworked methods assume instantaneous failure, while the two computer model applications assume gradual dam breach by erosion. The study also compared peak flows, stages, and time of peak flows at selected points downstream from the failed dams. The four methods demonstrated little (±8%) difference among flood profile depths.

Table 4-4 compares important features and capabilities for seven selected dam break analysis methods.

## MITIGATING INADEQUATE HYDRAULIC CAPACITIES

Earlier sections of this chapter are concerned primarily with one question: Do a dam and reservoir acting together have the hydraulic capacity to withstand the adopted design flood? This section considers some of the alternative actions available when the answer to that question is *no*. In Chapter 3 the general types of measures available to mitigate deficiencies of unsafe dams are discussed, and the impacts that may be expected from use of each type are set out. In other chapters remedial measures for various types of structural deficiencies are discussed.

It is often said that a dam is unsafe because its spillway has inadequate capacity. What is usually meant is that the development must be considered unsafe because the reservoir and spillway acting together are not capable of controlling the adopted design flood to the extent needed to ensure the structural integrity of the dam, thus creating potential for dam failure with uncontrolled release of stored water and serious damages to persons and properties downstream. It is well to keep in mind that the spillway, the dam, the reservoir, and the uses being made of the downstream areas are all part of the danger scenario we usually contemplate when we speak of an inadequate spillway, for such an overall view of the situation will be helpful in deciding on the most feasible mitigation plan.

Each dam with inadequate hydraulic capacity is very much a unique case. Hence, it is not feasible to set out a list of mitigating measures and state precisely under what circumstance each should be used. In a specific situation the feasibility of each type of measure would very much depend on such aspects as the nature of the site, the type and condition of the development, the benefits it produces, the mode of operation, and the technical and financial resources available to the dam owner. In addition, the policies of the state dam safety regulatory agency and the legal and moral responsibilities of the dam owner to those endangered by the project must be considered.

### Types of Mitigating Measures

Since most projects with inadequate hydraulic capacities would be endangered only during very rare flood events, the most feasible mitigation measures may involve operations or losses that could not be tolerated on a more frequent basis. The approaches that have been used to meet such problems can be classed as follows:

- Increases in discharge capacity.
- Increases in reservoir storage capacity.

- Diversions of runoff from the reservoir.
- Modification of dams to permit overflow.
- Modification of project operations.
- Modifications of use of downstream areas.
- Do nothing or perform minimal repairs. (The dam owner may choose to do a minimal amount to remedy an "unsafe" structure or "inadequate" spillway. The reasons for this could be lack of funds, lack of a practical method for correction due to space limitation, or a combination of many site-specific circumstances. In arriving at a decision the dam owner must consider the potential for liability from property damage upstream as well as downstream, the potential for loss of life, and the moral obligation to avoid unnecessary hazard to those who might be affected by dam failure. To make the "do nothing" or "minimal repairs" approach a possible viable alternative, it is very important that an effective plan be formulated for an early warning to all who may be affected by a sudden increase in water level to protect against or at least to minimize the possibility of loss of life (see the section Emergency Action Planning). The above could also be a deliberate or designed procedure once the dam owner has determined that the "failure" of a structure or portion of a structure at a development under a low flood level may have a minimal incremental effect downstream. On the other hand, should the owner partially correct a deficiency, he could compound potential problems downstream.

## Increasing Discharge Capacity

This is the most direct approach to solving the problem of inadequate spillway capacity, if it is found feasible. Some approaches to providing more spillway capacity are as follows:

- Increasing crest length of ungated spillways.
- Lengthening and adding gates to gated spillways.
- Lowering crest of existing spillway. If needed to maintain pools for project operation, flashboards or gates can be used on the lowered crest.
- Constructing a new spillway. This could involve reconstruction of a section of dam to serve as an auxiliary or emergency spillway or locating a new spillway in an abutment area or remote from the dam in a low saddle at the reservoir rim. A new remote saddle spillway could introduce new problems if it would discharge flood flows into areas or into a stream of another drainage basin that would not have been affected by such flows without the new spillway.
- Improving the hydraulic efficiency of the existing spillway. At some projects such measures as removal of land masses projecting into approach

or retreat channels, removal of debris and deposits from such channels, or provision of guide walls or improved pier noses will add significantly to the discharge capacity of the spillway. Often the last two or three spillway bays on each end of a spillway are less efficient or have less flow capacity than the other bays. Severe flow angle of approach can cause flow separations and flow drawdown along the piers of these spillway bays. In some cases the flow reductions are as much as 20% in each bay. With proper guide walls or with modified pier noses to guide the incoming flow, uniform flow distributions across the bay could be achieved. Another benefit could be the reduction of structural vibration and cavitation erosion.

- Providing a fuse plug levee or dike with crest lower than top of main dam designed to wash out and provide emergency spillway capacity. Figure 4-6 shows a system proposed by Harza Engineering Company for providing a fuse plug in an existing dam embankment. Only locations where the emergency discharge would not endanger the main dam or other important facilities should be considered for fuse plugs. Stoplogs, flashboards, or needle beam closures for new spillway structures can provide the same type of emergency spillway capacity but with better control of the spillway operation.

### Increasing Reservoir Storage Capacities

In some cases it may be feasible to alleviate a problem of inadequate spillway capacity at an existing dam by making substantially more reservoir

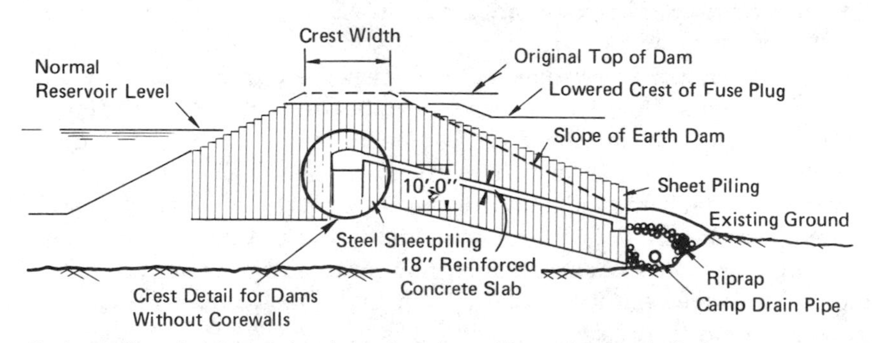


*Section. Spillway chute for typical earth dam. Fuse plug not illustrated to show detail of structure.*

FIGURE 4-6 Harza Engineering Co. scheme for constructing fuse plug spillways.

storage available by raising the top of the nonoverflow section of the dam, thus providing storage above operating pool levels for a major part of the SDF and reducing spillway capacity requirements. Also, increasing the height of a dam can effectively increase the flow capacity not only for the spillway but also for the other flow conveyance structures. Before such measures can be adopted, it is necessary to check whether the structures and geology downstream can withstand the additional energy of discharge without detrimental erosion and whether increase in reservoir stages would give problems with stability of structures. Also, it is necessary to check whether raising the structure would increase or create a problem in upstream channels or around the reservoir due to the increased water level. Some means of increasing the effective height of dams are as follows:

- Adding or increasing height of parapet walls on the upstream side of the dam. On a masonry dam the upstream handrail may be removed and replaced by a solid concrete parapet of the same height or slightly higher. On a fill dam a concrete or masonry parapet could be used.
- For larger height increases, concrete mass may be required to be placed on top and on the downstream face of a masonry dam. For embankment sections fill may be required to be placed on top and downstream face of fill. To place new mass concrete against old mass concrete, very special care in design and in construction must be exercised.
- The use of flashboards on top of concrete dam. Flashboards can be designed to fail when reservoir height rises to a certain level.
- Install inflatable dam similar to the "Faber" type on top of dam. The Faber is to be inflated by pumping water to fill the Faber tubing under pressure. This was used effectively for Mangla diversion in Pakistan to temporarily increase diversion water level.
- Increasing height of a concrete dam by adding a concrete cap on top and installing posttensioned cables, tying old and new structures to the foundation.
- Steel sheet piles can be driven in certain types of fill dams to increase height.

### Diversions Upstream from Reservoir

At some projects it may be feasible to direct runoff from the reservoir to mitigate a spillway capacity problem. Usually such opportunities will be found only at impoundments in relatively flat terrain or at relatively small impoundments. The possible dangers to others and legal implications of diverting runoff from natural channels should be considered before such a plan is adopted.

## Modification of Dams to Permit Overflow

As discussed in Chapters 6 and 7, dams can be constructed to withstand overflow. A masonry dam generally can withstand overflow if the added hydraulic loading does not endanger the stability of the structure and if the overflow will not erode the foundation at the toe of the dam or damage other downstream facilities, such as outlet valves and controls. For an earth or rockfill dam there is the added problem of erosion of the top and downstream face of the dam by the overflow. The ability of masonry dams to resist overturning and sliding forces resulting from high reservoir stages and overflow can be improved by installation of high-capacity, posttensioned ties through the dam into the foundation. Armoring the foundation area at the toe can improve erosion resistance at this important location.

Considerable guidance is available on designing for overtopping of rockfill dams (Curtis and Lawson 1967, Gerodetti 1981, Leps 1973, Olivier 1967, Parlin et al. 1966, Sarkaria 1968, and Wilkins 1963). Generally, designing an earth dam to permit overtopping should be considered only where the dam is low and the depth of overtopping would be small and the duration short. The rate of erosion and subsequent breaching of an earth dam would very much depend on the depth of overtopping, the geometry of the top of the dam, and the erosive characteristics of the soil of which the dam is composed. Thus, the indicated shallow overtopping of an existing small dam composed of erosion-resistant materials during a short interval in the passage of an SDF hydrograph may not require remedial action.

## Modification of Project Operations

In some instances it may be feasible to modify project operations to substantially increase the reservoir storage space that would be available to regulate the design flood. In considering such a plan the effect of the modified operations on project benefits, the effectiveness of the increased storage in mitigating spillway capacity problems, and the assurance that the storage would, in fact, be available when needed should be appraised.

## Modification of Uses of Downstream Areas

Seldom will it be feasible for the owner of an unsafe dam to reduce the hazard of dam failure by changing the uses of downstream areas, but this situation might arise if the dam and the downstream areas have the same owner. Downstream damage potentials can often be greatly reduced by just a little forethought in developing the areas. A classic example of the lack of such forethought involved the building of a new hospital in a narrow mountain valley subject to natural flooding from the river that ran through the valley. Areas flat enough and extensive enough for either the hospital or its accompanying parking lot were scarce, but two areas were available, each of which could accommodate either the hospital or the parking lot but not both. One was in the low, narrow flood plain of the river. The other was on a low bluff well above the flood plain. Unfortunately, the developers placed the hospital on the low area and the parking lot on the bluff. The folly of this selection has been amply demonstrated during subsequent floods.

## REFERENCES

Balloffet, A. (1979) *Dam Break Flood Routing, Two Cases*, Columbia University, New York, N.Y.

Balloffet, A., et al. (1974) "Dam Collapse Wave in a River," *Hydraulics Journal*, ASCE, Vol. 100, No. HY5 (May), Proceedings Paper 10523, pp. 645–665.

Cecilio, C. B., and Strassburger, A. G. (1974) *Downstream Hydrograph from Dam Failures*, Proceedings, The Evaluation of Dam Safety, Engineering Foundation Conference, November 28–December 3, 1974, published by ASCE, 1977.

Chow, V. T. (1959) *Open Channel Hydraulics*, McGraw-Hill, New York.

Creager, W. P., and Justin, J. D. (1963) *Hydroelectric Handbook*, John Wiley & Sons, New York.

Curtis, R. P., and Lawson, J. D. (1967) "Flow Over and Through Rockfill Dams," *Journal of the Hydraulics Division*, Proceedings of the ASCE, Vol. 93, No. HY5 (September), pp. 1–21.

Davis, C. V., and Sorenson, K. E. (1969) *Handbook of Applied Hydraulics*, 3d ed., McGraw-Hill, New York.

Fread, D. L. (1982) *DAMBRK: The NWS Dam-Break Flood Forecasting Model*, Hydrologic Research Laboratory, Office of Hydrology, National Weather Service, NOAA, Silver Spring, Md.

Gerodetti, M. (1981) "Model Studies of an Overtopped Rockfill Dam," *Water Power and Dam Construction*, (September), pp. 25–31.

Hagen, V. K. (1982) Re-evaluation of Design Floods and Dam Safety, paper presented at 14th ICOLD Congress, Rio de Janeiro.

Kevlegan, G. H. (1950) "Wave Motion," pp. 711–768 in *Engineering Hydraulics*, H. Rouse, ed., John Wiley & Sons, New York.

King, H. W., and Brater, E. F. (1963) *Handbook of Hydraulics*, 5th ed., McGraw-Hill, New York.

Kirkpatrick, G. W. (1976) *Evaluation Guidelines for Spillway Adequacy*, Proceedings of the Engineering Foundation Conference, The Evaluation of Dam Safety, Asilomar Conference Ground, Pacific Grove, Calif., November 28–December 3, 1976, published by ASCE, pp. 395–414.

Leps, T. M. (1973) "Flow Through Rockfill," offprint from *Embankment Engineering*, Ronald Hirshfeld, ed., John Wiley & Sons, New York, pp. 87–107.

National Weather Service (1975) *Hydrometeorological Report No. 33*, Washington, D.C.

National Weather Service (1977) *Hydrometeorological Report No. 49*, Washington, D.C.

National Weather Service (1978) *Hydrometeorological Report No. 51*, Washington, D.C.

National Weather Service (1980) *Hydrometeorological Report No. 53*, Washington, D.C.

National Weather Service (1982) "Application of Probable Maximum Precipitation Estimates—United States East of the 105th Meridian," *Hydrometeorological Report No. 52*, NOAA, Washington, D.C., August.

Newton, D. W., and Cripe, M. W. (1973) *Flood Studies for Safety of TVA Nuclear Plants—Hydrologic and Embankment Breaching Analysis*, Flood Hydrology Section, TVA, Knoxville, Tenn.

Olivier, H. (1967) *Through and Overflow Rockfill Dams—New Design Techniques*, Proceedings, Institution of Civil Engineers, Paper No. 7012, pp. 433–471.

Parlin, A. K., Trollope, D. H., and Lawson, J. D. (1966) "Rockfill Structures Subject to Water Flow," *Journal of the Soil Mechanics and Foundations Division*, Proceedings of the ASCE, Vol. 92, No. SM6 (November), pp. 135–151.

Sakkas, J. G. (1974) *Dimensionless Graphs of Floods from Ruptured Dams*, Report prepared for the Hydrologic Engineering Center, Davis, Calif.

Sarkaria, G. S., and Dworsky, B. H. (1968) "Model Studies of an Armoured Rockfill Overflow Dam," *Water Power*, November, pp. 445–462.

Soil Conservation Service (1979) "Simplified Dam-Breach Routing Procedure," Technical Release No. 66, March.

Strelkoff, A. K. (1978) "Computing Two-Dimensional Dam-Break Flood Waves," *Hydraulics Journal*, ASCE, September.

Tennessee Valley Authority (1973) *Dam Breaching Program*, Knoxville, Tenn., November.

Tschantz, B. A., and Mojib, R. M. (1981) *Application of and Guidelines for Using Available Dam Break Models*, Tennessee Water Resources Research Center, Report No. 83, University of Tennessee, Knoxville.

U.S. Army Corps of Engineers (1956) "Snow Hydrology," *Summary Report of Snow Investigations*, North Pacific Division, Portland, Oreg., June.

U.S. Army Corps of Engineers (1959) *Flood Hydrograph Analyses and Computations*, EM1110-2-1405, August 31.

U.S. Army Corps of Engineers (1960) *Runoff from Snowmelt*, Engineering and Design Manual, EM1110-2-1406, Washington, D.C.

U.S. Army Corps of Engineers (1962) *Generalized Snowmelt Runoff Frequencies*, Technical Bulletin No. 8, Sacramento District, September.

U.S. Army Corps of Engineers (1963) *Hydraulic Design of Reservoir Outlet Structures*, EM1110 2-1602, Office of the Chief of Engineers, Washington, D.C., August 1.

U.S. Army Corps of Engineers (1965a) *Standard Project Flood Determinations*, EM 1110-2-1411, Washington, D.C.

U.S. Army Corps of Engineers (1965b) *Hydraulic Design of Spillways*, EMl110-2-1603, Office of the Chief of Engineers, Washington, D.C., March 31.

U.S. Army Corps of Engineers (1968a) *Policies and Procedures Pertaining to Determination of Spillway Capacities and Freeboard Allowances for Dams*, EC 1110-2-27, Change 1, February 19.

U.S. Army Corps of Engineers (1968b) *Hydraulic Design Criteria*.

U.S. Army Corps of Engineers (1969) *Routing of Floods Through River Channels*, EM 1110-2-1408, March 1.

U.S. Army Corps of Engineers (1976) *Wave Runup and Wind Setup on Reservoir Embankments*, ETL 1110-2-221, November 29.

U.S. Army Corps of Engineers (1980) Hydrologic Engineering Center Report No. 8, *Dimensional Graphs of Floods From Ruptured Dams*, Davis, Calif. (based on work developed by Sakkas 1974-1980).

U.S. Army Corps of Engineers (1982a) *Hydrologic Analysis of Ungaged Watersheds Using HEC-1*, Training Document No. 15, The Hydrologic Engineering Center, Davis, Calif.

U.S. Army Corps of Engineers (1982b) *National Program of Inspection of Nonfederal Dams, Final Report to Congress* (contains ER 1110-2-106, September 26, 1979).

U.S. Bureau of Reclamation (1966) *Effect of Snow Compaction From Rain on Snow* Engineering Monograph No. 35.

U.S. Bureau of Reclamation (1977) *Design of Small Dams*, Denver, Colo.

U.S. Committee on Large Dams (1970) Survey: *Criteria and Practices Utilized in Determining the Required Capacity of Spillways*, USCOLD, Boston, Mass.

U.S. Department of Commerce (1977) "Probable Maximum Precipitation Estimates, Colorado River and Great Basin Drainages," *Hydrometeorological Report No. 49*, NOAA, National Weather Service, Silver Spring, Md., September.

U.S. Department of Commerce (1978) "Probable Maximum Precipitation Estimates, United States East of the 105th Meridian," *Hydrometeorological Report No. 51*, NOAA, National Weather Service, Washington, D.C., June.

U.S. Department of Commerce (1980) "Seasonal Variation of 10-Square-Mile Probable Maximum Precipitation Estimates, United States East of the 105th Meridian," *Hydrometeorological Report No. 53*, NOAA, National Weather Service, Silver Spring, Md., April.

U.S. Department of Interior, Bureau of Reclamation (1952) "Discharge Coefficients for Irregular Overfall Spillways," *Engineering Monograph No. 9*.

U.S. Department of Interior (1982) "Guidelines for Determining Flood Flow Frequency," *Bulletin No. 17B*, Hydrology Subcommittee, Interagency Advisory Committee on Water Data.

U.S. Nuclear Regulatory Commission (1977) "Design Basis Floods for Nuclear Power Plants," Regulatory Guide 1.59, Revision 2, August.

U.S. Weather Bureau (1956) "Seasonal Variation of the Probable Maximum Precipitation East of the 105th Meridian," *Hydrometeorological Report No. 33*, Washington, D.C.

U.S. Weather Bureau (1960) "Generalized Estimates of Probable Maximum Precipitation West of the 105th Meridian," *Technical Paper No. 38*, Washington, D.C.

U.S. Weather Bureau (1961) "Generalized Estimates of Probable Maximum Precipitation and Rainfall-Frequency Data for Puerto Rico and Virgin Islands," *Technical Paper No. 42*, Washington, D.C.

U.S. Weather Bureau (1961/1969) "Interim Report—Probable Maximum Precipitation in California," *Hydrometeorological Report No. 36*, Washington, D.C., October 1961, with revisions in October 1969.

U.S. Weather Bureau (1963a) "Probable Maximum Precipitation in the Hawaiian Islands," *Hydrometeorological Report No. 39*, Washington, D.C.

U.S. Weather Bureau (1963b) "Probable Maximum Precipitation Rainfall-Frequency Data for Alaska," *Technical Paper No. 47*, Washington, D.C.

U.S. Weather Bureau (1966a) "Meteorological Conditions for the Probable Maximum Flood on the Yukon River above Rampart, Alaska," *Hydrometeorological Report No. 42*, Environmental Science Services Administration, Washington, D.C., May.

U.S. Weather Bureau (1966b) "Probable Maximum Precipitation, Northwest States," *Hydrometeorological Report No. 43*, Washington, D.C.

Wilkins, J. K. (1963) "The Stability of Overtopped Rockfill Dams," Proceedings Fourth Australia-New Zealand Conference on Soil Mechanics and Foundation Engineering.

## APPENDIX 4A GENERALIZED ESTIMATES OF PROBABLE MAXIMUM FLOOD PEAKS

Figures 4A-1 through 4A-6 present generalized estimates of probable maximum flood (PMF)peak discharges. The maps may be used to determine PMF peak discharge at a given site with a known drainage area as follows:

1. Locate the site on the 100-square-mile map, Figure 4A-1.
2. Read and record the 100-square-mile PMF peak discharge by straight-line interpolation between the isolines.
3. Repeat Steps 1 and 2 for 500, 1,000, 5,000, 10,000, and 20,000 square miles from Figures 4A-2 through 4A-6.
4. Plot the six PMF peak discharges so obtained on logarithmic paper against drainage area, as shown on Figure 4A-7.
5. Draw a curve through the points. Reasonable extrapolations above and below the defined curve may be made.
6. Read the PMF peak discharge at the site from the curve at the appropriate drainage area.

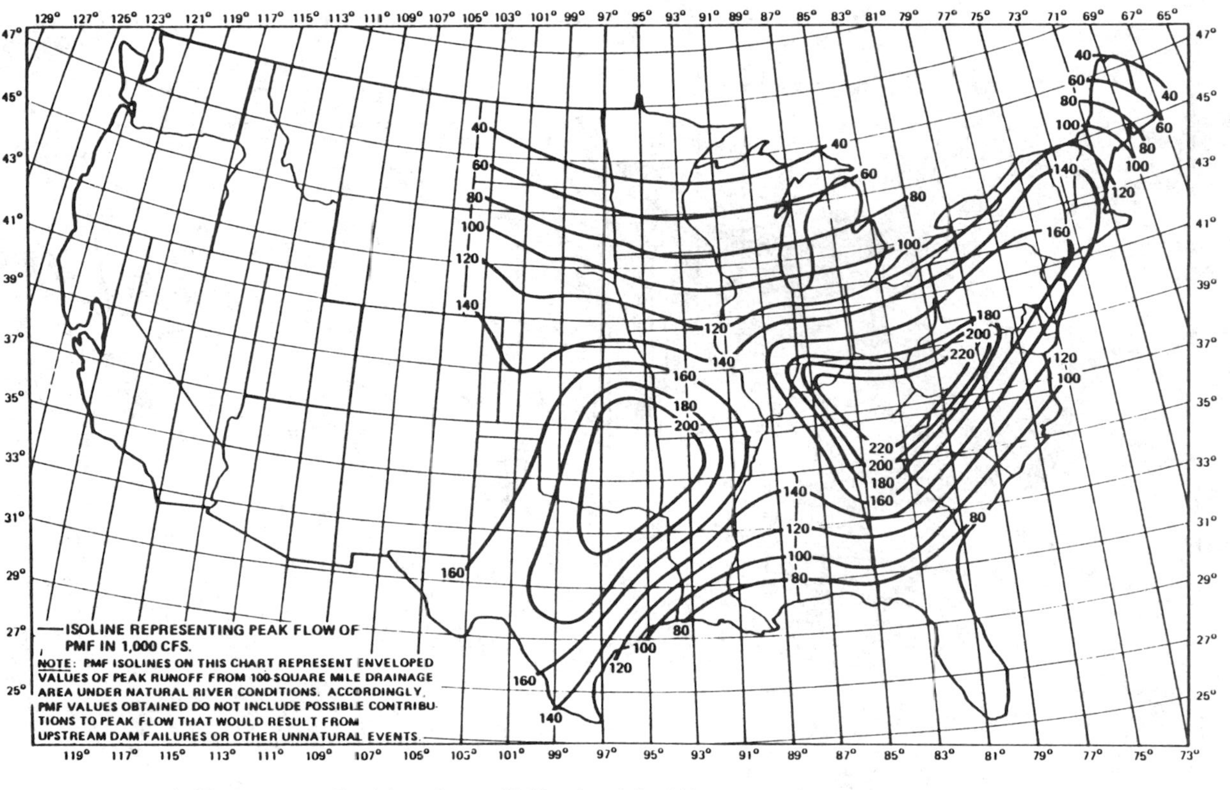
ISOLINE REPRESENTING PEAK FLOW OF PMF IN 1,000 CFS.
NOTE: PMF ISOLINES ON THIS CHART REPRESENT ENVELOPED VALUES OF PEAK RUNOFF FROM 100-SQUARE MILE DRAINAGE AREA UNDER NATURAL RIVER CONDITIONS. ACCORDINGLY, PMF VALUES OBTAINED DO NOT INCLUDE POSSIBLE CONTRIBUTIONS TO PEAK FLOW THAT WOULD RESULT FROM UPSTREAM DAM FAILURES OR OTHER UNNATURAL EVENTS.

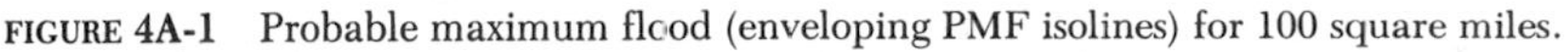

FIGURE 4A-1 Probable maximum flood (enveloping PMF isolines) for 100 square miles.

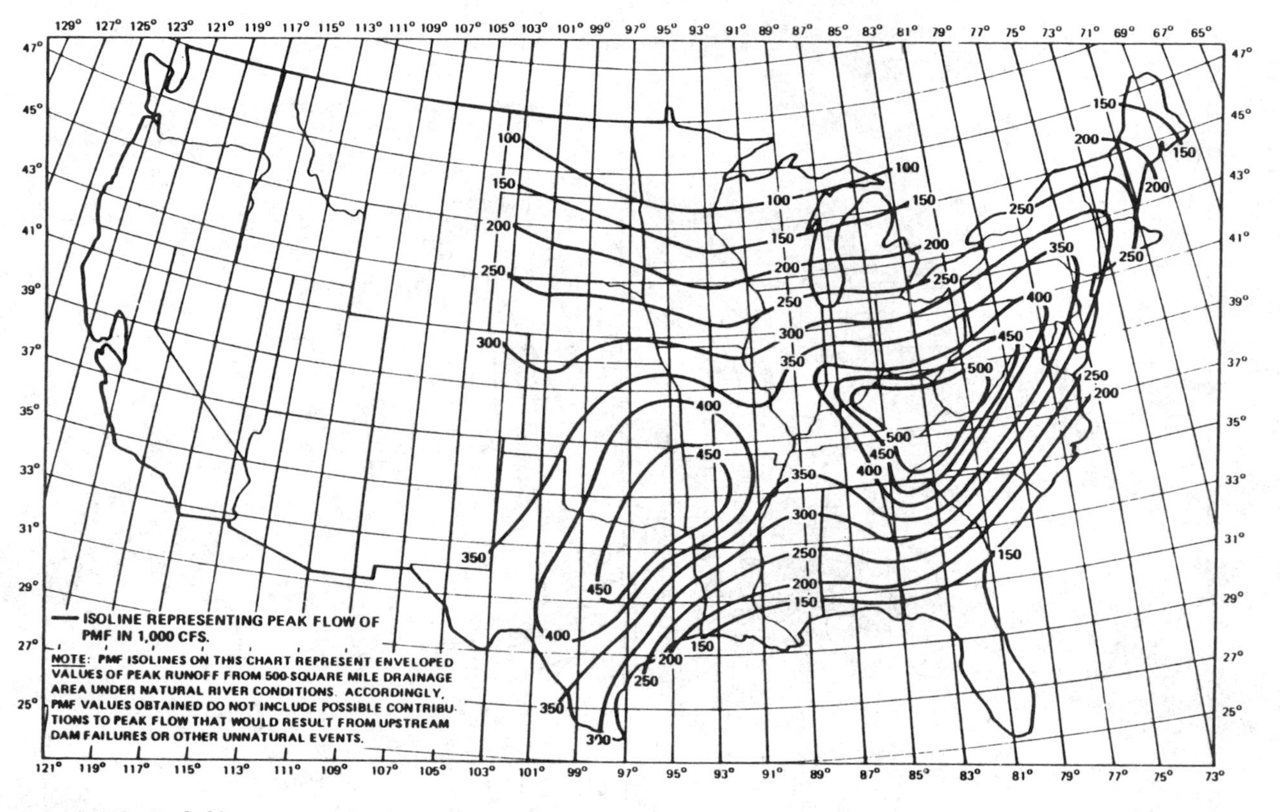
ISOLINE REPRESENTING PEAK FLOW OF PMF IN 1,000 CFS.
NOTE: PMF ISOLINES ON THIS CHART REPRESENT ENVELOPED VALUES OF PEAK RUNOFF FROM 500-SQUARE MILE DRAINAGE AREA UNDER NATURAL RIVER CONDITIONS. ACCORDINGLY, PMF VALUES OBTAINED DO NOT INCLUDE POSSIBLE CONTRIBUTIONS TO PEAK FLOW THAT WOULD RESULT FROM UPSTREAM DAM FAILURES OR OTHER UNNATURAL EVENTS.

FIGURE 4A-2 Probable maximum flood (enveloping PMF isolines) for 500 square miles.

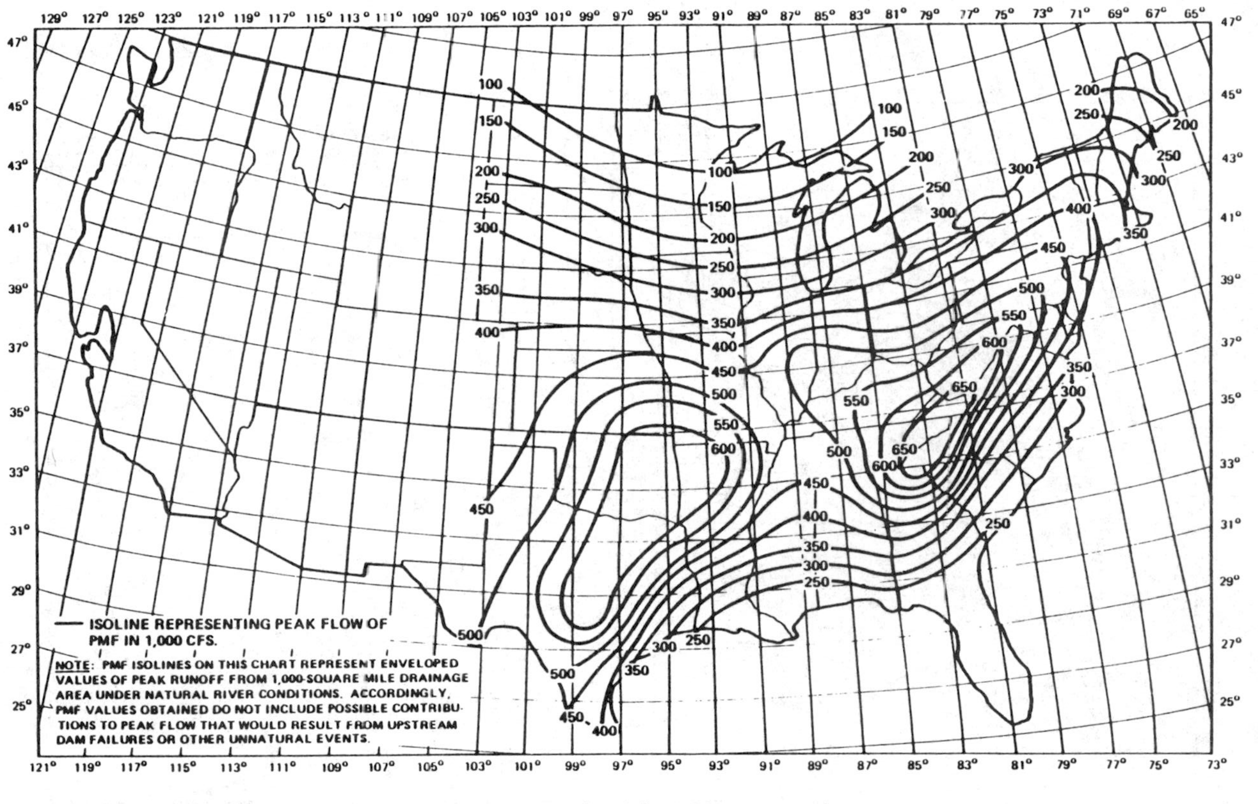
ISOLINE REPRESENTING PEAK FLOW OF PMF IN 1,000 CFS.
NOTE: PMF ISOLINES ON THIS CHART REPRESENT ENVELOPED VALUES OF PEAK RUNOFF FROM 1,000-SQUARE MILE DRAINAGE AREA UNDER NATURAL RIVER CONDITIONS. ACCORDINGLY, PMF VALUES OBTAINED DO NOT INCLUDE POSSIBLE CONTRIBUTIONS TO PEAK FLOW THAT WOULD RESULT FROM UPSTREAM DAM FAILURES OR OTHER UNNATURAL EVENTS.

FIGURE 4A-3 Probable maximum flcod (enveloping PMF isolines) for 1,000 square miles.

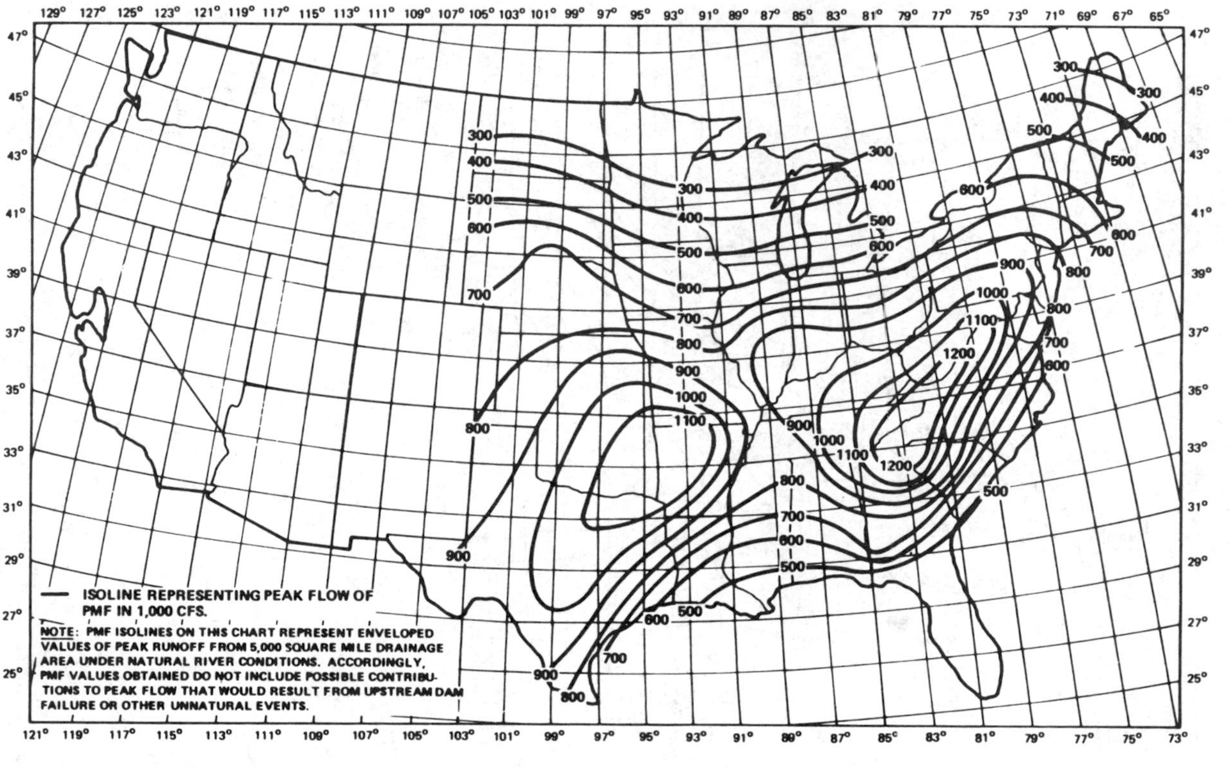
ISOLINE REPRESENTING PEAK FLOW OF PMF IN 1,000 CFS.
NOTE: PMF ISOLINES ON THIS CHART REPRESENT ENVELOPED VALUES OF PEAK RUNOFF FROM 5,000 SQUARE MILE DRAINAGE AREA UNDER NATURAL RIVER CONDITIONS. ACCORDINGLY, PMF VALUES OBTAINED DO NOT INCLUDE POSSIBLE CONTRIBUTIONS TO PEAK FLOW THAT WOULD RESULT FROM UPSTREAM DAM FAILURE OR OTHER UNNATURAL EVENTS.

FIGURE 4A-4 Probable maximum flood (enveloping PMF isolines) for 5,000 square miles.

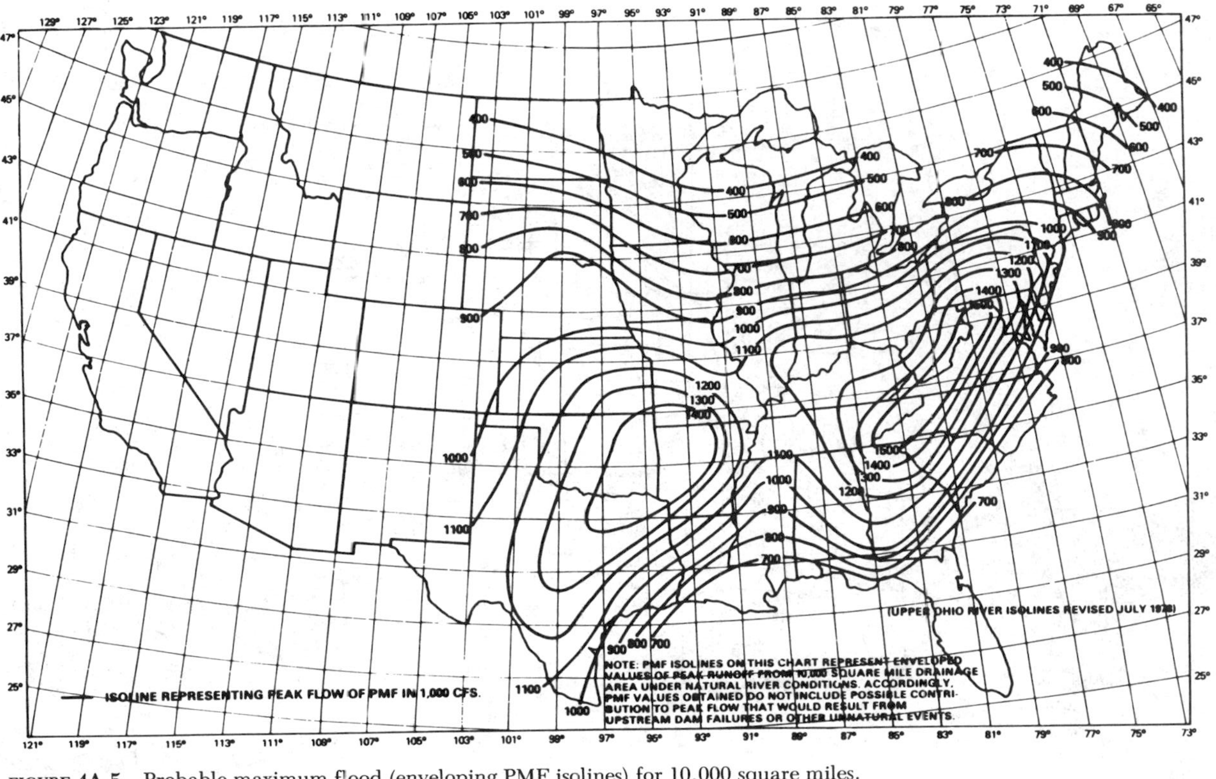
ISOLINE REPRESENTING PEAK FLOW OF PMF IN 1,000 CFS.
NOTE: PMF ISOLINES ON THIS CHART REPRESENT ENVELOPED VALUES OF PEAK RUNOFF FROM 10,000 SQUARE MILE DRAINAGE AREA UNDER NATURAL RIVER CONDITIONS. ACCORDINGLY, PMF VALUES OBTAINED DO NOT INCLUDE POSSIBLE CONTRIBUTION TO PEAK FLOW THAT WOULD RESULT FROM UPSTREAM DAM FAILURES OR OTHER UNNATURAL EVENTS.
(UPPER OHIO RIVER ISOLINES REVISED JULY

FIGURE 4A-5 Probable maximum flood (enveloping PMF isolines) for 10,000 square miles.

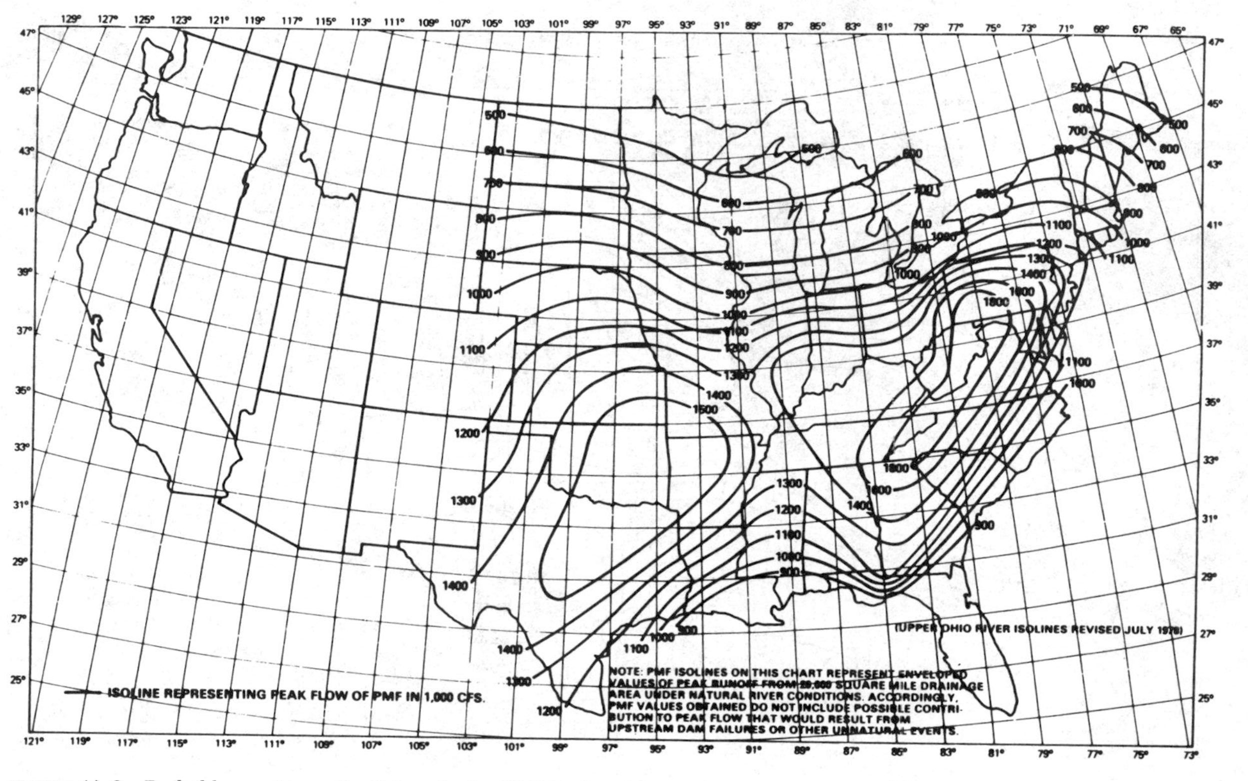
NOTE: PMF ISOLINES ON THIS CHART REPRESENT ENVELOPED VALUES OF PEAK RUNOFF FROM 20,000 SQUARE MILE DRAINAGE AREA UNDER NATURAL RIVER CONDITIONS. ACCORDINGLY, PMF VALUES OBTAINED DO NOT INCLUDE POSSIBLE CONTRIBUTION TO PEAK FLOW THAT WOULD RESULT FROM UPSTREAM DAM FAILURES OR OTHER UNNATURAL EVENTS.
(UPPER OHIO RIVER ISOLINES REVISED JULY 1978)
ISOLINE REPRESENTING PEAK FLOW OF PMF IN 1,000 CFS.

FIGURE 4A-6 Probable maximum flood (enveloping PMF isolines) for 20,000 square miles.

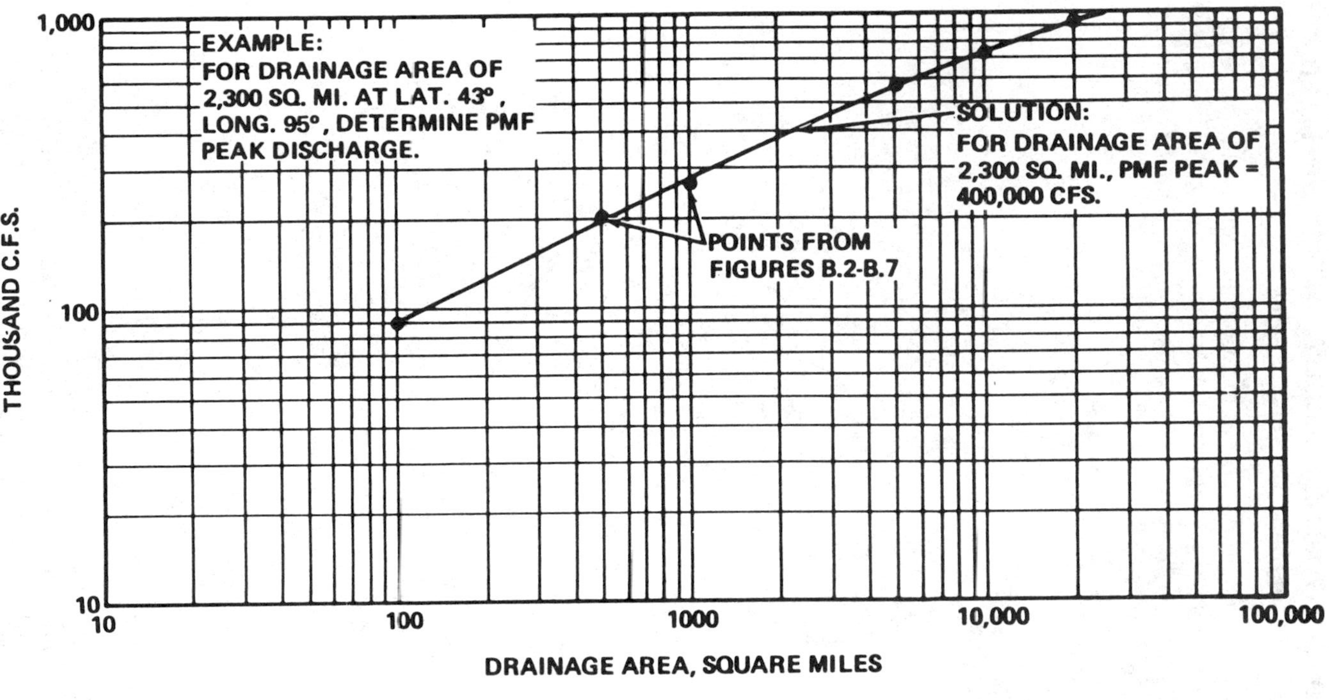


FIGURE 4A-7 Examples of use of enveloping isolines.

# APPENDIX 4B STORMS EXCEEDING 50% OF ESTIMATED PROBABLE MAXIMUM PRECIPITATION

Figures 4B-1 through 4B-5 show locations of storms that have exceeded 50% of the estimated probable maximum precipitation (PMP) value for the indicated durations and sizes of drainage areas. The specific storms are identified in Tables 4B-1 and 4B-2.

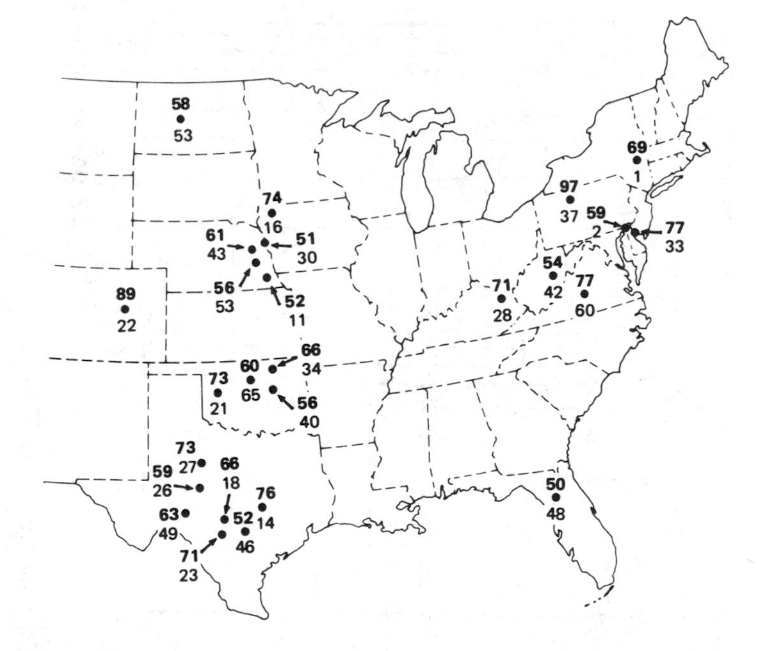


FIGURE 4B-1 Observed point rainfalls exceeding 50% of all-season PMP, United States east of 105th meridian for 10 square miles, 6 hours. (Large number is the percentage of the PMP, small number is storm index; see Table 4B-1.) SOURCE: U.S. Army Corps of Engineers (1982b).

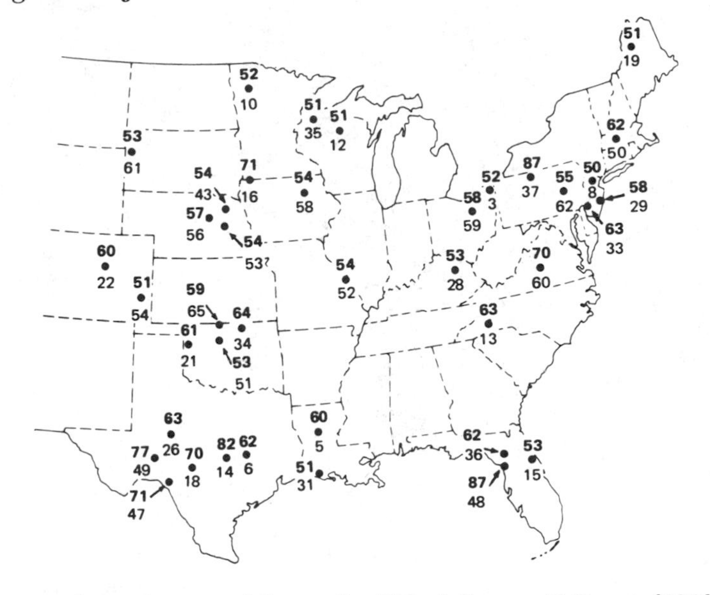


FIGURE 4B-2 Observed point rainfalls exceeding 50% of all-season PMP, east of 105th meridian for 200 square miles, 24 hours. (Large number is the percentage of the PMP, small number is storm index.) SOURCE: U.S. Army Corps of Engineers (1982b).

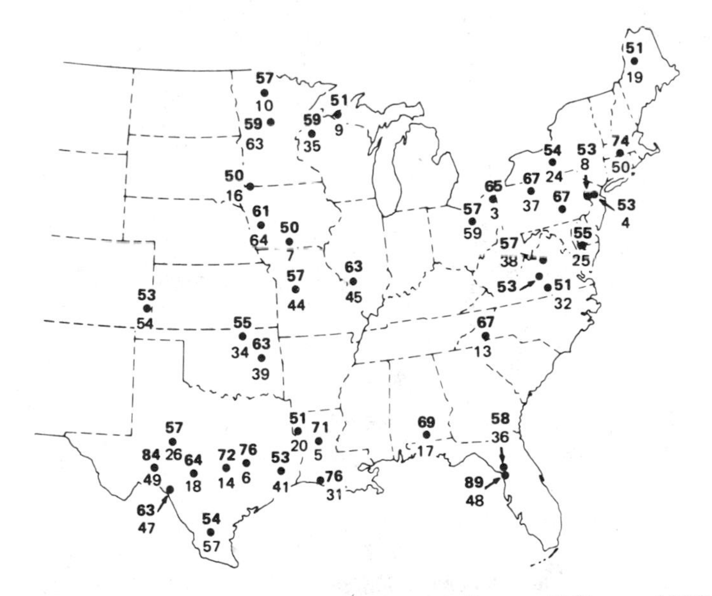


FIGURE 4B-3 Observed point rainfalls exceeding 50% of all-season PMP, east of 105th meridian for 1,000 square miles, 48 hours. (Large number is the percentage of the PMP, small number is storm index.) SOURCE: U.S. Army Corps of Engineers (1982b).

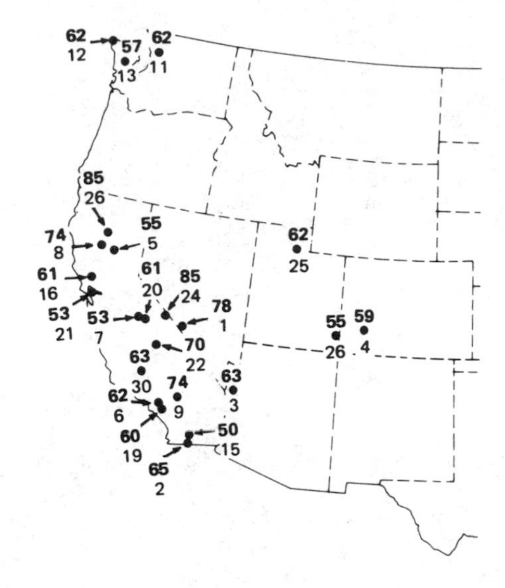


FIGURE 4B-4 Observed point rainfalls exceeding 50% of all-season PMP, west of continental divide for 10 square miles, 6 hours. (Large number is the percentage of the PMP, small number is storm index.) SOURCE: U.S. Army Corps of Engineers (1982b).

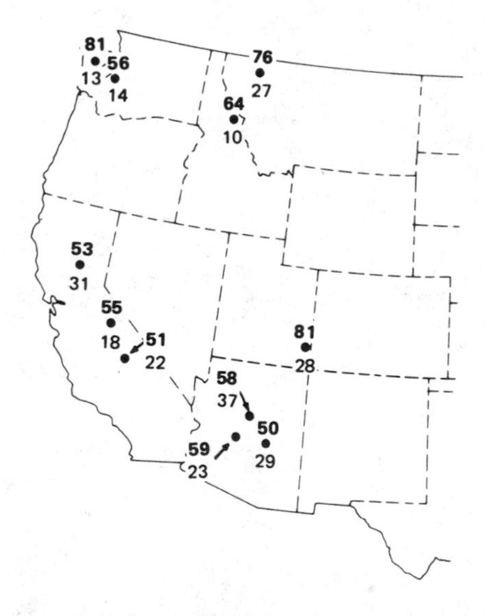


FIGURE 4B-5 Observed point rainfalls exceeding 50% of all-season PMP, west of continental divide for 1,000 square miles and duration between 6 and 72 hours. (Large number is the percentage of the PMP, small number is storm index.) SOURCE: U.S. Army Corps of Engineers (1982b).

**TABLE 4B-1** Identification of Storms Exceeding 50% PMP, East of 105th Meridian

| Storm Date | Index No. | Corps Assignment No. (if available) | Storm Center | | | |
|---|---|---|---|---|---|---|
| | | | Town | State | Lat. | Long. |
| 7/26/1819 | 1 | — | Catskill | NY | 42°12′ | 73°53′ |
| 8/5/1843 | 2 | — | Concordville | PA | 39°53′ | 75°32′ |
| 9/10-13/1878 | 3 | OR 9-19 | Jefferson | OH | 41°45′ | 80°46′ |
| 9/20-24/1882 | 4 | NA 1-3 | Paterson | NJ | 40°55′ | 74°10′ |
| 6/13-17/1886 | 5 | LMV 4-27 | Alexandria | LA | 31°19′ | 92°33′ |
| 6/27-7/11/1899 | 6 | GM 3-4 | Turnersville | TX | 30°52′ | 96°32′ |
| 8/24-28/1903 | 7 | MR 1-10 | Woodburn | IA | 40°57′ | 93°35′ |
| 10/7-11/1903 | 8 | GL 4-9 | Paterson | NJ | 40°55′ | 74°10′ |
| 7/18-23/1909 | 9 | UMV 1-11B | Ironwood | MI | 46°27′ | 90°11′ |
| 7/18-23/1909 | 10 | UMV 1-11A | Beaulieu | MN | 47°21′ | 95°48′ |
| 7/22-23/1911 | 11 | — | Swede Home | NB | 40°22′ | 96°54′ |
| 7/19-24/1912 | 12 | GL 2-29 | Merrill | WI | 45°11′ | 89°41′ |
| 7/13-17/1916 | 13 | SA 2-9 | Altapass | NC | 35°33′ | 82°01′ |
| 9/8-10/1921 | 14 | GM 4-12 | Taylor | TX | 30°35′ | 97°18′ |
| 10/4-11/1924 | 15 | SA 4-20 | New Smyrna | FL | 29°07′ | 80°55′ |
| 9/17-19/1926 | 16 | MR 4-24 | Boyden | IA | 43°12′ | 96°00′ |
| 3/11-16/1929 | 17 | UMV 2-20 | Elba | AL | 31°25′ | 86°04′ |
| 6/30-7/2/1932 | 18 | GM 5-1 | State Fish Hatchery | TX | 30°01′ | 99°07′ |
| 9/16-17/1932 | 19 | — | Ripogenus Dam | ME | 45°53′ | 69°09′ |
| 7/22-27/1933 | 20 | LMV 2-26 | Logansport | LA | 31°58′ | 94°00′ |
| 4/3-4/1934 | 21 | SW 2-11 | Cheyenne | OK | 35°37′ | 99°40′ |
| 5/30-31/1935 | 22 | MR 3-28A | Cherry Creek | CO | 39°13′ | 104°32′ |
| 5/31/1935 | 23 | GM 5-20 | Woodward | TX | 29°20′ | 99°28′ |
| 7/6-10/1935 | 24 | NA 1-27 | Hector | NY | 42°30′ | 76°53′ |
| 9/2-6/1935 | 25 | SA 1-26 | Easton | MD | 38°46′ | 76°01′ |

**TABLE 4B-1** Identification of Storms Exceeding 50% PMP, East of 105th Meridian (*continued*)

| Storm Date | Index No. | Corps Assignment No. (if available) | Storm Center | | | |
|---|---|---|---|---|---|---|
| | | | Town | State | Lat. | Long. |
| 9/14-18/1936 | 26 | GM 5-7 | Broome | TX | 31°47′ | 100°50′ |
| 6/19-20/1939 | 27 | — | Snyder | TX | 32°44′ | 100°55′ |
| 7/4-5/1939 | 28 | — | Simpson | KY | 38°13′ | 83°22′ |
| 8/19/1939 | 29 | NA 2-3 | Manahawkin | NJ | 39°42′ | 74°16′ |
| 6/3-4/1940 | 30 | MR 4-5 | Grant Township | NB | 42°01′ | 96°53′ |
| 8/6-9/1940 | 31 | LMV 4-24 | Miller Island | LA | 29°45′ | 92°10′ |
| 8/10-17/1940 | 32 | SA 5-19A | Keysville | VA | 37°03′ | 78°30′ |
| 9/1/1940 | 33 | NA 2-4 | Ewan | NJ | 39°42′ | 75°12′ |
| 9/2-6/1940 | 34 | SW 2-18 | Hallet | OK | 36°15′ | 96°36′ |
| 8/28-31/1941 | 35 | UMV 1-22 | Haywood | WI | 46°00′ | 91°28′ |
| 10/17-22/1941 | 36 | SA 5-6 | Trenton | FL | 29°48′ | 82°57′ |
| 7/17-18/1942 | 37 | OR 9-23 | Smethport | PA | 41°50′ | 78°25′ |
| 10/11-17/1942 | 38 | SA 1-28A | Big Meadows | VA | 38°31′ | 78°26′ |
| 5/6-12/1943 | 39 | SW 2-20 | Warner | OK | 35°29′ | 95°18′ |
| 5/12-20/1943 | 40 | SW 2-21 | Nr. Mounds | OK | 35°52′ | 96°04′ |
| 7/27-29/1943 | 41 | GM 5-21 | Devers | TX | 30°02′ | 94°35′ |
| 8/4-5/1943 | 42 | OR 3-30 | Nr. Glenville | WV | 38°56′ | 80°50′ |
| 6/10-13/1944 | 43 | MR 6-15 | Nr. Stanton | NB | 41°52′ | 97°03′ |

| | | | | | | |
|---|---|---|---|---|---|---|
| 8/12-15/1946 | 44 | MR 7-2A | Cole Camp | MO | 38°40′ | 93°13′ |
| 8/12-16/1946 | 45 | MR 7-2B | Nr. Collinsville | IL | 38°40′ | 89°59′ |
| 9/26-27/1946 | 46 | GM 5-24 | Nr. San Antonio | TX | 29°20′ | 98°29′ |
| 6/23-24/1948 | 47 | — | Nr. Del Rio | TX | 29°22′ | 100°37′ |
| 9/3-7/1950 | 48 | SA 5-8 | Yankeetown | FL | 29°03′ | 82°42′ |
| 6/23-28/1954 | 49 | SW 3-22 | Vic Pierce | TX | 30°22′ | 101°23′ |
| 8/17-20/1955 | 50 | NA 2-22A | Westfield | MA | 42°07′ | 72°45′ |
| 5/15-16/1957 | 51 | — | Hennessey | OK | 36°02′ | 97°56′ |
| 6/14-15/1957 | 52 | — | Nr. E. St. Louis | IL | 38°37′ | 90°24′ |
| 6/23-24/1963 | 53 | — | David City | NB | 41°14′ | 97°05′ |
| 6/13-20/1965 | 54 | — | Holly | CO | 37°43′ | 102°23′ |
| 6/24/1966 | 55 | — | Glenullin | ND | 47°21′ | 101°19′ |
| 8/12-13/1966 | 56 | — | Nr. Greely | NB | 41°33′ | 98°32′ |
| 9/19-24/1967 | 57 | SW 3-24 | Falfurrias | TX | 27°16′ | 98°12′ |
| 7/16-17/1968 | 58 | — | Waterloo | IA | 42°30′ | 92°19′ |
| 7/4-5/1969 | 59 | — | Nr. Wooster | OH | 40°50′ | 82°00′ |
| 8/19-20/1969 | 60 | NA 2-3 | Nr. Tyro | VA | 37°49′ | 79°00′ |
| 6/9/1972 | 61 | — | Rapid City | SD | 44°12′ | 103°31′ |
| 6/19-23/1972 | 62 | — | Zerbe | PA | 40°37′ | 76°31′ |
| 7/21-22/1972 | 63 | — | Nr. Cushing | MN | 46°10′ | 94°30′ |
| 9/10-12/1972 | 64 | — | Harlan | IA | 41°43′ | 95°15′ |
| 10/10-11/1973 | 65 | — | Enid | OK | 36°25′ | 97°52′ |

SOURCE: U.S. Army Corps of Engineers (1982b).

**TABLE 4B-2** Storms with Rainfall Exceeding 50% of PMP, West of Continental Divide

| Storm Date | Index No. | Storm Center | | | | Duration for 1000 mi² |
|---|---|---|---|---|---|---|
| | | Town | State | Lat. | Long. | |
| 8/11/1890 | 1 | Palmetto | NV | 37°27 | 117°42 | |
| 8/12/1891 | 2 | Campo | CA | 32°36 | 116°28 | |
| 8/28/1898 | 3 | Ft. Mohave | AZ | 35°03 | 114°36 | |
| 10/4–6/1911 | 4 | Gladstone | CO | 37°53 | 107°39 | |
| 12/29/1913–1/3/1914 | 5 | — | CA | 39°55 | 121°25 | |
| 2/17–22/1914 | 6 | Colby Ranch | CA | 34°18 | 118°07 | |
| 2/20–25/1917 | 7 | — | CA | 37°35 | 119°36 | |
| 9/13/1918 | 8 | Red Bluff | CA | 40°10 | 122°14 | |
| 2/26–3/4/1938 | 9 | | CA | 34°14 | 117°11 | |
| 3/30–4/2/1931 | 10 | — | ID | 46°30 | 114°50 | 24 |
| 2/26/1932 | 11 | Big Four | WA | 48°05 | 121°30 | |
| 11/21/1933 | 12 | Tatoosh Island | WA | 48°23 | 124°44 | |
| 1/20–25/1935 | 13 | — | WA | 47°30 | 123°30 | 6 |
| 1/20–25/1935 | 14 | — | WA | 47°00 | 122°00 | 72 |
| 2/4–8/1937 | 15 | Cyamaca Dam | CA | 33°00 | 116°35 | |
| 12/9–12/1937 | 16 | — | CA | 38°51 | 122°43 | |
| 2/27–3/4/1938 | 17 | — | AZ | 34°57 | 111°44 | 12 |
| 1/19–24/1943 | 18 | — | CA | 37°35 | 119°25 | 18 |
| 1/19–24/1943 | 19 | Hoogee's Camp | CA | 34°13 | 118°02 | |
| 1/30–2/3/1945 | 20 | — | CA | 37°35 | 119°30 | |
| 12/27/1945 | 21 | Mt. Tamalpias | CA | 37°54 | 122°34 | |
| 11/13–21/1950 | 22 | — | CA | 36°30 | 118°30 | 24 |
| 8/25–30/1951 | 23 | — | AZ | 34°07 | 112°21 | 72 |
| 7/19/1955 | 24 | Chiatovich Flat | CA | 37°44 | 118°15 | |
| 8/16/1958 | 25 | Morgan | UT | 41°03 | 111°38 | |
| 9/18/1959 | 26 | Newton | CA | 40°22 | 122°12 | |
| 6/7–8/1964 | 27 | Nyack Ck. | MT | 48°30 | 113°38 | 12 |
| 9/3–7/1970 | 28 | — | UT | 37°38 | 109°04 | 6 |
| 9/3–7/1970 | 29 | — | AZ | 33°49 | 110°56 | 6 |
| 6/7/1972 | 30 | Bakersfield | CA | 35°25 | 119°03 | |
| 12/9–12/1973 | 31 | — | CA | 39°45 | 121°30 | 48 |

SOURCE: U.S. Army Corps of Engineers (1982b).

## APPENDIX 4C GUIDELINES FOR BREACH ASSUMPTION

Presented here are guidelines for formulating emergency action plans with no admission, written or implied, that the structures are a hazard. Additional criteria for dam break analysis are given for use in a specific computer model. These guidelines are representative of criteria developed by an investor-owned utility, which were submitted and accepted by a regulatory agency. Prior to any use of these criteria, however, they should be discussed with the specific regulatory agency with which the dam owner has to deal.

1. All dams upstream or downstream from a given dam under investigation should be evaluated.
2. If failure of an upstream dam is found to cause or compound the failure of the dam under consideration, an evaluation shall be made.
3. If the upstream dam belongs to a different owner, efforts should be made to obtain the necessary information to perform a dam break evaluation.
4. If the downstream dam is owned and operated by another regulatory agency, the upstream owner should inform the downstream owner of the result of the upstream dam break evaluation.
5. Evaluation of flood-prone areas resulting from dam breaks should be made using available U.S. Geological Survey 7.5-minute quadrangle topographic maps. Cross-section intervals should be limited to 1 mile and a maximum of 10 miles except within reservoirs. Cross sections should be taken at or near a development or populated area.
6. Inundation maps shall be prepared for areas where potential flooding from the hypothetical dam break could cause significant hazard to human habitation.
7. No stability analysis or any type of geologic investigation shall be performed in connection with the dam break studies. However, results of earlier studies to satisfy regulatory requirements as to safety of the dam or dams in question may be used as references.
8. For all dam break analyses where a storm is not in progress, all reservoirs shall be assumed to be at normal maximum operating water levels unless otherwise noted. In so doing, all flashboards shall be assumed installed and gates in position.
9. For river channel base flow, use mean annual flow. When using the National Weather Service (NWS) Dam Break Model, it is often necessary to increase the base flow to an exceedingly high value before the computer program can run. However, it has been shown that base flows are insignificant compared to the failure flood waves.

10. The most likely mode of dam failure need not be the most severe; only the most severe shall be assumed. Initial failure of gates or other appurtenant features other than the dam shall not be considered because they will not produce the most critical flows.

11. For dam break analysis with no concurrent storm in progress, no cause of failure shall be assumed. Dams are considered to fail only to prepare an emergency action plan. Results of the study should not reflect in any way on the structural integrity of the dam or dams and are not to be construed as such. All reports and/or inundation maps shall contain such a qualifying statement.

12. Simultaneous failure of two or more adjacent or in-tandem dams shall not be considered. However, successive or domino-type failures shall be considered.

13. When the upper dam fails, and the lower dam is earth or rock, consider the lower dam to fail if it overtops. When the lower dam is concrete and stability analyses indicate a low factor of safety for overturning or sliding, failure should be assumed, if significant overtopping would occur.

14. When a lower dam is overtopped from an upstream dam failure, all flashboards and gates shall be assumed to fail.

15. For concrete dams (gravity or arch), assume instantaneous failure. When using the NWS Dam Break Model, a time of failure equal to 0.15 hour would be equivalent to instantaneous failure.

16. The shape for an instantaneous failure is similar to the geometry of the channel at the center line of the dam.

17. Any partial width but full depth failure for concrete gravity dams may be considered.

18. Complete failure for concrete arch dams is considered. When using the NWS Dam Break Model, the parabolic shape of the failure breach should be transformed into an equivalent trapezoidal section.

19. For earth and rockfill dams, assume failure by erosion. The word *erosion* should not be construed as a cause of failure but as a rate of failure dependent on time.

20. The shape of failure breach may either be trapezoidal, parabolic, or rectangular. For partial instantaneous breach of concrete gravity dams, assume a rectangular shape. For erosion failure, assume a parabolic or trapezoidal shape. For complete instantaneous failure, assume a parabolic or trapezoidal shape.

21. For mixed dams (concrete and embankment), assume failure to produce the largest but reasonable flow.

22. Final breach shape of embankment dams should be trapezoidal when the maximum base is equal to the height of the dam. A proportion-

ately smaller base may be assumed if the geometry of the original river channel is small compared to the height of the dam.

23. The maximum width of breach may be estimated by dividing the area of the dam (measured along the axis) by the maximum height of the dam. It could also be considered equivalent to the most hydraulically efficient section. Dimensions for this efficient section are given in most textbooks on hydraulics (e.g., Chow's *Open Channel Hydraulics*).

24. The failure outflow hydrograph may be estimated by flood routing (hydraulic or hydrologic method) or by the approximate triangle method supplemented by empirical formulas.

25. To arrive at reasonable dam break mechanics, postulated failures shall be compared with historical dam failures.

26. All flood routing shall be done with the most practical and economical procedures. Computer models that simulate the dynamic passage of a dam break flood through river channels should be used when practical.

27. All flood routing shall be continued downstream until attenuated to a peak equivalent to the recorded precipitation-caused flood peak. In certain cases where the historical precipitation-caused flood peaks have caused damage in the area, continued routing may be necessary until attenuation to a mean annual flood peak value.

28. All bridges along the pathway of a dam break flood may be assumed to have failed prior to the arrival of the flood peak if it is determined that they will be overtopped. The NWS Dam Break Model can be used to route through bridges.

29. Flood wave arrival times shall be indicated at points on inundation maps with significant population.

30. A report documenting all pertinent assumptions, evaluations, and scenarios in the dam failure analysis shall be prepared for each study.

# 5
# Geologic and Seismological Considerations

## GENERAL GEOLOGIC CONSIDERATIONS

Defective foundation and reservoir conditions have perhaps been responsible for a majority of dam failures and accidents (see Chapter 2). This chapter will discuss the major geologic features and rock types that may contribute to the development of serious conditions at a dam site.

In the 5 billion or so years of the earth's life, many changes have taken place in the rocks forming the earth's crust. These changes are continuing. The result is a great variety in the type of rock from place to place and also great differences in the quality of the rock. Engineers and geologists, after much experience with dam site geology, frequently associate certain defects with each class of rock. While rocks differ greatly in the kind and size of their mineral constituents, the broadest general classification of rock is based on the way in which they are formed: igneous, sedimentary, and metamorphic rocks. Coincidentally, certain types of defects in dam foundations often seem to have some correlation with each of these broad classifications.

### Rock Classification

#### *Igneous Rocks*

Igneous rocks are formed by the solidification of molten rock or lava. If solidification takes place slowly at great depths in the earth, the rocks are called plutonic or intrusive rocks and are composed of masses of crystalline particles. Depending on the relative amounts and sizes of these constituents, these rocks will have various names, such as granite, diabase, or gabbro. When the molten rock is expelled from the interior to the surface of the earth, it cools quickly, the crystal sizes are generally very small, and the dissolved gases in the rock expand. These rocks are called volcanic or extrusive rocks. These include basalt, rhyolite, obsidian, and pumice.

A common characteristic of plutonic rocks is their large crystal sizes and the interlocking of grains of different minerals. As the rock cools, it shrinks. When the internal tensile stresses resulting from this shrinkage become greater than its strength, the rock cracks and develops a regular pattern of joints. At the surface, weather processes break the rock down almost completely into residual soil.

Volcanic rocks have their own special problems. The material may be ejected explosively in the form of rocks, molten bombs, or small, dust-like particles or may flow like a dense liquid, such as lava. Deposits of the ejected material, such as pumice, are apt to be porous, with easy permeability and erodibility to flow of water. A reservoir rim or dam foundation containing such ejecta might require extensive (and often very difficult) grouting before a satisfactory reduction of seepage can be achieved. The rock mass, with large or small open bubbles from expanded gases, tends to be weak and needs careful study to determine if it has sufficient strength for heavily loaded structures. Volcanic rock often tends to break down easily by weathering to leave a weak residue. Each lava flow lasts only a relatively short time. Therefore, a deep deposit of lava may be made up of many individual flows. Since the surface of each flow tends to deteriorate, the mass is frequently characterized by interfaces of altered material and sometimes volcanic ash; these latter materials may be mechanically weak and very permeable to water flow. Washing and grouting often improve such foundation materials. Sometimes the flowing mass cools and hardens on the surface, and the included gases may escape while the liquid interiors simply run out and, in either case, can leave a hollow pipe of considerable size and length. Careful exploration is needed in regions of volcanic deposits for assurance that reservoir leakage can be held to an acceptable minimum.

### *Sedimentary Rock*

Rock at the earth's surface is continually being broken down not only by tectonic forces but also by the process of weathering. Weathering processes include the freezing of water in cracks accompanied by expansion of the ice and subsequent fracturing of the rock. Another weathering process is that of extreme temperature changes that will cause fracturing. The flow of water through rock can weaken some minerals and this will leave the remainder unsupported, so they are removed easily by wind or water erosion.

Large-scale weathering processes, such as river erosion or glaciation, not only will remove large masses of rock by abrasion but may also remove support from large masses and leave the balance of the rock in a state of stress that is conducive to fracture and further breakdown. The broken particles resulting from the weathering process are often transported by air, ice, water, or gravity and then redeposited. As the modes of deposition can be extremely variable, there may be considerable differences in the successive layers of the deposited or sedimentary material. Hence, sedimentary rocks are often characterized by the great variety in the material in the successive layers. Much time can elapse between depositional modes, and this provides opportunity for decomposition of the top surface of the layer. The result can be a very weak interface between the layers, and this will require washing and grouting to increase the strength and impermeability. If the sedimentary deposits are subjected to heat and pressure from overlying material or to cementation from dissolved minerals in water, the layers may greatly increase in strength.

In addition to the weathering processes, tectonic activities such as earth movements may cause joints or faults to form in the sedimentary layers. These discontinuities may be tightly closed, open, or filled with other materials or minerals that have been transported into the openings.

Many sedimentary rocks take their names from the size of the rock particles in the deposit. Thus, a rock made up of gravel or larger particles is called conglomerate (the contained particles are usually rounded or subrounded) or breccia (the contained particles are angular). If sand-size particles are compressed or cemented into a coherent mass, the result is a sandstone; similarly, silt-size particles constitute a siltstone and clay-size particles result in a claystone or shale. Calcareous mud or sand may become limestone. Deposits of highly carboniferous materials, such as vegetation, become coal. Any or all of these may be present in successive layers of a sedimentary rock formation.

In addition to the mud or weak zones between successive layers, sedimentary rocks can have other problems. For example, if the cementing material in sandstone or conglomerate is water soluble, it will go into solution or become very weak when saturated by a reservoir, and the rock may revert to its original form, i.e., a sand or gravel. This action is considered the cause of the collapse of the St. Francis Dam in California. The sandy shaley conglomerate foundation disintegrated under the water action during the first filling of the reservoir. Some shales that are wet in situ tend to fall apart to a powder on drying out. These are called air-slaking shales. Limestones and dolomites can be dissolved by the weak natural acid formed by the combination of water and carbon dioxide in the water. Hence, over a long time period, water moving through these types of carbonate rocks can dissolve out large portions of the rock and result in natural pipes or caverns (Figure 5-1). The overlying surface may then develop sinkholes and is called karst. Great care must be taken to grout and otherwise seal off cavernous limestone; otherwise a reservoir founded on this rock will not hold water. A similar action can occur in rock formations primarily composed of gypsum.

### *Metamorphic Rocks*

These are formed from other rocks when they are subjected to great heat and pressure. In igneous rocks, like granite, the mineral grains will be reoriented to planar sheetlike forms called gneiss or schist. These may have a decided plane of weakness parallel to the mineral planes. In sandstone most of the minerals (except quartz) may disappear, and the quartz grains will fuse together into a glassy solid called quartzite. Under metamorphic processes, shale goes through transitional stages to ultimately form a slate. The latter will have decided cleavage planes, but these often are at angles to the original bedding of the shale. Limestone and dolomite when metamorphosed become marble.

FIGURE 5-1 Caverns in dolomite foundation of a gravity concrete dam. (Later filled with concrete to ensure adequate bearing for the dam.)

Since the process of metamorphism is not uniform, metamorphic rocks have weak zones due to pressure, heat, or chemical action. These can result in sheeted areas, gouge seams, or even faults. Weakened areas, when found, must be removed and replaced with concrete so that the dam will have a uniformly strong foundation. Jointing in metamorphic rocks will be formed during the metamorphic processes and will often form irregular-shaped or wedge-shaped blocks.

## ROCK TYPES

There are hundreds, probably thousands, of different varieties of rock that could be encountered in dam construction. Of course at any one site the likelihood is that there would be only a few to several rock types. However, because there are so many possibilities, it is not feasible to list here all the problems that might develop from every rock type that might be encountered at a dam. Therefore, only listed are those rock types that most commonly might be encountered or those that have been reported as causing problems at a dam.

### Rock Types and Their Performance

*Amphibolite*: Metamorphic. Dark-colored with little or no quartz. May have poor weathering resistance. Tends to break along foliations.

*Anhydrite*: Sedimentary. $CaSO_4$. Usually white or slightly tinted. Unstable in the presence of water and tends to expand when wet, which causes rapid disintegration of the rock.

*Ash*: Igneous. Usually noncoherent but may be somewhat cemented or welded. Light colored to gray. Deteriorates rapidly under water action and may be subject to considerable settlement because of its low density. High content of silica. (see *Tuff*)

*Basalt*: Igneous. Dark colored, fine grained. Little or no quartz. Complex chemical formula including Na, Al, Si, O, Ca, Mg, Fe, and K in varying amounts. May contain small, highly visible pores called vugs, which may or may not be filled with clay-like material or be interconnected. However, some basalts do contain continuous tunnels or tubes as a result of gas flowing through the material during the time of its formation. Basalts tend to crack into well-defined chunks or blocks during the cooling process. High cohesion. If fractured, a plucking action can occur when high-velocity water flows over its surface.

*Claystone*: Sedimentary. Clay constitutes greater than 25%. May be massive or stratified. Generally high in quartz. Some shales are classified as claystones and vice versa. There is no agreement as to the distinction except that shale would be expected to have considerable fissility (easy cleavage or laminations). Coherent but subject to erosion under water and other forms of weathering action. May disintegrate into fine particles or clay under severe weathering conditions. Where stratified, may be a weakened zone that would be conducive to slides.

*Conglomerate*: Sedimentary. Composed of easily visible, generally rounded fragments or pebbles in a matrix of granular material. Can be cemented with calcium carbonate, iron oxide, silica, or clay. Depending on the type of cement, it may be disintegrated easily by weathering agents.

*Diabase or Dolerite*: Igneous. Dark colored, intrusive. Composed mainly of labradorite and pyroxene minerals. Little or no quartz. Fine to medium grained. Visible particles appear to be angular. Can be associated with cavities or numerous open fissures. Generally high strength. Sometimes called trap rock.

*Dolomite*: Sedimentary. Composed chiefly of Ca, Mg, and $CO_3$. White or tinted. May be crystalline or noncrystalline. Effervesces very slowly in HCl. Can be associated with a cavernous structure. Relatively high strength.

*Gneiss*: Metamorphic. Foliated. Light to dark gray. Less than half of the minerals may show preferred parallel orientation. Commonly rich in quartz and feldspar. May contain considerable mica. Those high in mica may rapidly slake or have easy cleavage.

*Granite*: Igneous. Primarily composed of quartz and feldspar. May contain some mica. Texture usually from medium to coarse grained. White to dark gray with occasional red or pink. May be block jointed or sheeted. The feldspar may disintegrate under weathering and leave a rather granular soil with some clay admix. Grains may be strongly or poorly interlocked. Frequently becomes a catch-all term for many types of feldspathic, quartzitic intrusive igneous rocks; thus, test values can have a considerable range.

*Gypsum*: Sedimentary. $CaSO_4 . 2H_2O$. Very soft. White or colorless but can be tinted. Easily disintegrated by normal weathering processes. Water action may cause solution channels.

*Limestone*: Sedimentary. $CaCO_3$. May have impurities of other minerals. White to tinted. Rock composed solely or almost entirely of $CaCO_3$ includes chalk, coquina, and travertine. All effervesce freely with any common acid. The name always carries a warning that there may be minor or extensive cavern systems in the formation. Slowly soluble in water with a low pH. Can deteriorate under high temperatures.

*Marl*: Sedimentary. Catch-all term describing soft, loose, earthy deposits that may be coherent or noncoherent. Chiefly clay and $CaCO_3$. Gray, but other colors are frequently present. Texture may be extremely fine or granular. Often easily eroded by water.

*Micaceous Rock*: Generally igneous or metamorphic. Any rock containing easily visible quantities of the sheetlike material called mica. Latter has a highly complex formula that contains Ca, Mg, Fe, Li, Al, Si, O, H, and F. Easily split into thin plates. Color ranges from colorless to black. In a rock mass under shear stresses, it can be expected to be a weak member owing to a relatively low coefficient of sliding friction.

*Pegmatite*: Igneous. Very coarse grained with interlocking crystals. Composition similar to that of granite. High in quartz. Same color variations as granite. Larger constituents may weather loose from the binding matrix. More frequently tight interlocking is present and represents a highly durable rock.

*Peridotite*: Igneous. Coarse grained. Chiefly olivine with other iron-magnesium minerals. Dark colored. Mentioned here primarily because it commonly alters to serpentine, often a highly undesirable foundation material.

*Phyllite*: Metamorphic. Intermediate between slate and micaceous schist. Well-defined thin laminations. Usually black or dark brown. Can be split along bedding planes with some difficulty, and split surfaces may be slick. Resistant to weathering but tends to split or slab when original crustal stresses on it are relieved by excavation.

*Pumice*: Igneous. Light colored. Highly porous or vesicular. Generally composed primarily of silica that has been produced by volcanic eruption. Very lightweight and abrasive. Stony or earthy texture. Very erodible and very low strength.

*Quartzite*: Sedimentary or metamorphic. Resembles a very hard sandstone where the quartz grains are tightly cemented with silica. Also may be a metamorphic rock formed from the recrystallization of sandstone. Usually colorless. Breaks with irregular fractures across the grains. Very hard. Highly resistant to weathering forces but on occasion can disintegrate into granular material.

*Sandstone*: Sedimentary. Medium grained, composed of rounded or angular fragments. Cementing material may be silt, clay, iron oxide, silica, or $CaCO_3$. White, red, yellow, brown, or gray. May be friable, i.e., when rubbed by the fingers grains easily detach themselves. Can be well-defined bedding or very massive. Rate of disintegration depends on type of cementing material and weathering forces. May have a relatively high porosity and be considered a reservoir rock for water or oil.

*Schist*: Metamorphic. Well-developed foliation generally in thin parallel plates that may show considerable distortion. Usually high in mica. Can be very competent as an engineering material but is frequently separated easily along the foliations or planes of schistosity. Can be subject to plucking action under high-velocity water. Because of the strong forces that develop the inherent foliation, it is possible there will be shear planes between the laminations; if so, it will be a dangerous rock in slopes and tunnels.

*Serpentine*: Igneous or metamorphic. Complex formula with Mg, Fe, Si, O, and H. Usually easily recognized because of its very greasy or soapy appearance or feel. Can be granular or fibrous. Often green or greenish yellow or greenish gray. May have a vein-like appearance. Generally a secondary mineral but can be found in thick beds. Always to be regarded with caution because of its tendency to disintegrate into very incompetent material under normal weathering action. Also, can have very low shear strength.

*Shale*: Sedimentary. Extremely fine grained, composed mainly of clay-size particles but may occasionally contain silt or sand sizes. Characterized by its laminar or fissile structure. Easily cleavable. May be soft and easily scratched with a fingernail but can also be quite hard. Tends to slake rapidly when dried and then put into water. Can range from relatively light colors to black. Usually has a low shear strength and is to be regarded with caution when encountered in engineering works. Frequently regarded as a borderline material between soil and rock. Can disintegrate to a clayey mass. (See also the section Residual Soils.)

*Siltstone*: Sedimentary. Generally hard and coherent. Tends to be massive rather than laminated. Gritty feel. At least two-thirds of the constituents will be silt size. Depending on the type of cement, may disintegrate rapidly into silty deposits. Not expected to have high durability when used as a rock fill.

*Slate*: Metamorphic. No visible grains. Composed of clay-size particles. Will cleave into very hard, relatively thin plates. However, fissility occurs along planes that may not be parallel to the original bedding that is visible in the material. Very dark colored. Very durable but can break into large slabs in open slopes.

*Tuff*: Igneous or sedimentary. Usually a product of volcanic eruption. May have relatively low density or, if the individual silica grains are welded together, a high density. Depending on the cement, may be easily eroded or very resistant to erosion. Gray to yellow in color. Usually low density.

## General Comments on Rock Types

The physical properties of rock are extremely variable, even for one type of rock. This is illustrated in Table 5-1. For example, it can be seen in this table that the unconfined compressive strength of a "granite" can vary between 2,600 psi and 48,200 psi. One reason for these wide variations is the

**TABLE 5-1** Strengths of Rocks

| | Unconfined Compressive Strength (× 100 psi) | | | | Modulus of Elasticity (× 1,000,000 psi) | | | | Specific Gravity | | | |
|---|---|---|---|---|---|---|---|---|---|---|---|---|
| Rock Type | Average | Minimum | Maximum | No. Tests | Average | Minimum | Maximum | No. Tests | Average | Minimum | Maximum | No. Tests |
| Amphibolite | 22.08 | 3.60 | 40.7 | 14 | 13.93 | 9.80 | 16.30 | 9 | 2.91 | 2.71 | 3.08 | 14 |
| Anhydrite | — | — | — | — | 4.43 | 1.86 | 9.32 | 8 | 2.90 | 1.86 | 3.48 | 20 |
| Basalt | 16.3 | 0.6 | 55.6 | 195 | 6.07 | 1.58 | 13.39 | 75 | 2.59 | 1.91 | 2.99 | 195 |
| Claystone | — | — | — | — | — | — | — | — | 2.67 | 2.52 | 2.78 | 13 |
| Conglomerate | 32.8 | 15.3 | 47.8 | 7 | 1.89 | 6.75 | 13.83 | 6 | 2.72 | 2.66 | 2.81 | 9 |
| Diabase | 33.0 | 6.0 | 51.8 | 15 | 8.35 | 6.20 | 10.5 | 2 | 2.73 | 2.49 | 2.88 | 6 |
| Dolomite | 14.0 | 4.2 | 52.0 | 62 | 6.72 | 2.76 | 14.74 | 12 | 2.62 | 2.12 | 2.93 | 59 |
| Gneiss | 19.4 | 5.2 | 42.4 | 103 | 9.76 | 6.60 | 13.8 | 35 | 2.76 | 2.54 | 3.09 | 107 |
| Granite | 23.28 | 2.6 | 48.2 | 140 | 9.23 | 3.00 | 11.77 | 49 | 2.65 | 2.59 | 2.77 | 141 |
| Limestone | 10.9 | 0.2 | 37.8 | 211 | 5.03 | 0.51 | 11.05 | 62 | 2.40 | 1.21 | 4.41 | 246 |
| Marl | 7.5 | 5.5 | 9.6 | 2 | 5.90 | — | — | 1 | 2.79 | 2.78 | 2.80 | 2 |
| Pegmatite | 31.09 | 21.1 | 41.2 | 9 | 11.48 | 11.00 | 12.00 | 5 | 2.68 | 2.66 | 2.70 | 9 |
| Phyllite | — | — | — | — | — | 1.09* | 1.40* | ? | 2.74 | 2.18 | 3.34 | 7 |
| Pumice | 4.44 ± 1.97* and 3.29 ± 1.39* | | | 41 | — | — | — | — | — | — | — | — |
| Quartzite | 42.4 | 3.7 | 91.2 | 25 | 9.13 | 1.00 | 13.60 | 8 | 2.82 | 2.53 | 4.07 | 22 |
| Sandstone | 9.1 | 0.3 | 47.6 | 255 | 0.93 | 0.12 | 9.74 | 199 | 2.32 | 1.86 | 3.26 | 292 |
| Schist | 7.3 | 1.0 | 23.5 | 16 | 5.25 | 1.20 | 9.40 | 15 | 2.80 | 2.47 | 3.20 | 66 |
| Shale | 9.8 | 0.1 | 33.5 | 67 | 2.56 | 0.01 | 7.50 | 30 | 2.59 | 1.38 | 2.86 | 76 |
| Siltstone | 15.7 | 0.5 | 45.8 | 14 | 5.49 | 0.10 | 12.59 | 10 | 2.67 | 2.21 | 2.77 | 27 |
| Slate | 22.3 | 14.2 | 30.4 | 6 | — | — | — | — | 2.78 | 2.71 | 2.93 | 6 |
| Tuff | 20.7 | 3.0 | 45.5 | 6 | 6.54 | 0.58 | 12.50 | 2 | 1.85 | 1.37 | 2.78 | 10 |

*From Touloukian et al. (1981).

SOURCE: Judd (1969, 1971).

inaccuracy of the description or "naming" of the rock, which points to the lack of an acceptable engineering classification system for rock. For example, if accurate petrographic descriptions are not obtained, the rock may be inaccurately categorized as basalt, trap rock, or granite. For this reason it is desirable to determine the mode of deposition of the rock; this may provide clues to the expected performance in engineering works. For example, if the rock is a member of a flow of molten rock on the surface or at shallow depths, attention should be directed at the possibility of gas caverns or extensive cooling fractures being present. Similarly, if the rock has been deposited in water, such as many sedimentary rocks, the possibility exists that such rocks will be susceptible to severe erosion under the weathering action of wind, temperature changes, or water.

Another clue to predicting rock performance is to determine the geologic age of the material. This certainly is not a precise predictor but can assist when the age along with other information on the rock origin is known. For example, if the rock is relatively young, i.e., formed in the Cenozoic Era, it might be expected that the material would have a relatively poor coherence and tend to be highly erodible. Of course this is not always true, but it is one possible indicator. On the other hand, if the rock is extremely old, for example of the Precambrian Age, the rock might well be very durable and hard and a good-performing structural material. Obviously age by itself is not a good criterion because of the long period of time in which rocks have developed, e.g., between the Precambrian and the Cenozoic there is a period of over 500 million years.

The properties noted in the above descriptions of various rock types are summarized in Tables 5-2 and 5-3. Table 5-2 correlates the easily visible surface defects or evidence of rock behavior to the possible cause for this behavior. Table 5-3 compares the rock type with the defects that may occur and that are easily visible by surface examination. Table 5-3 can be used in two different ways: (1) if the rock name or type is known, the expected defects can be identified from the table and (2) if certain defects are observed, it may be possible to identify the rock type associated with the defect. It must be recognized, however, that the defects noted are only those that more commonly occur, because almost any of the so-called surface defects may develop from any rock type.

## GEOLOGIC STRUCTURE

### Rock Foundation Defects

Some of the potential defects in a rock foundation that will result from geologic structure are faults, joints, shear zones, and bedding and foliation

**TABLE 5-2** Major Cause of Defect

| Defect | Freeze-Thaw and Other Temperature Changes | Stress Relief | Solubility | Consolidation | Water Pressure | Water Lubrication | Piping |
|---|---|---|---|---|---|---|---|
| Block loosening | X | X | | | X | X | |
| Cracking | X | X | | | X | | |
| Disintegration (granular) | X | | | | | | X |
| Seepage (clear) | | | X | | | | |
| Seepage (muddy) | | | | | | | X |
| Settlement | | | | X | | | X |
| Slabbing | X | X | | | X | | |
| Slides | | X | | | X | X | |
| Softening | | | X | | | X | |

**TABLE 5-3** Surface Defects

| Rock Type | Block Loosening | Cracking | Disintegration (granular) | Leakage (clear) | Leakage (muddy) | Settlement | Softening | Slabbing | Slides | Foliation |
|---|---|---|---|---|---|---|---|---|---|---|
| Anhydrite | | | | X | | | X | | | |
| Ash, volcanic | | | X | | X | X | | | | |
| Basalt | X | X | | X | | | | | | |
| Claystone | | | X | | X | | X | | | X |
| Conglomerate | | | | X | | | | | | |
| Diabase | X | X | | | | | | | | |
| Dolomite | | | | X | | | | | | |
| Gneiss | X | | | | | | | | | X |
| Granite | X | X | X | | | | | X | | X |
| Gypsum | | | X | X | | X | X | | | X |
| Limestone | X | | X | X | | | | | | |
| Marl | | | | X | | X | X | | X | |
| Micaceous rock | | X | X | | | | X | X | X | X |
| Pegmatite | X | | | | | | | | | |
| Peridotite | | | | | | | X | | | |
| Pumice | | | X | | X | X | | | | |
| Quartzite | X | | | | | | | | | |
| Sandstone | X | | X | X | X | | | | | |
| Schist | | X | X | | | | | X | X | X |
| Serpentine | | | | | | X | X | | X | |
| Shale | | X | X | | | X | X | | X | X |
| Siltstone | | | X | | X | | X | | | |
| Slate | X | X | | | | | | X | | X |
| Tuff | | | X | | X | X | | | | |

planes. Faults result from a rupture in rock formations and are caused by high-magnitude tectonic forces. Joints can be formed as a result of tensile forces that develop from cooling of the liquid rock or from a weathering process, as described in the section General Geologic Considerations. The differentiation between faults and joints is that when portions of a formation move with respect to each other the discontinuity so produced is termed a fault. If there is a discontinuity but no movement has occurred, the break would be called a joint or fracture. Shear zones merely are another term for a faulted area in the crust. They generally have appreciable thickness or width and contain considerable ground-up rock derived from the parent rocks on both sides of the shear zone. Occasionally they also contain secondary material that has been transported into the zone by means of water. Foliation is a result of parallel arrangement of platy minerals and is common in metamorphic rocks such as schist and slate. Separations between bedding planes are a type of joint primarily associated with sedimentary rock. The term *fracture* can refer to a joint or a fault but always denotes a discontinuity in the rock mass. Water percolating through fractures can alter the mineral of the adjacent rock and in some cases actually dissolve portions of it. The result can be a weak, altered material. As previously noted, the fracture may be open, closed, or filled with some type of secondary material. Usually the secondary material is weaker than the surrounding rock, although occasionally there may be a high percentage of silica that actually welds the fractures together to form a rock mass that may have strength equivalent to the original unbroken mass. Joints generally tend to be more continuous and more open near the surface and to close with increasing depth. Regardless of whether the fractures are filled, it is necessary to reduce their permeability by washing and then grouting with a cementitious or resinous material. Continuous joints and fractures should always be of concern as they can result in instability in slopes adjacent to a dam or in the reservoir or downstream from the dam.

Faulting, jointing, and shear zones in carbonate rocks may contribute to the development of karstic conditions. Clay material along such discontinuities may be washed away with increased head and/or the surging action common to a hydroelectric project. One such example is the Logan Martin Dam on the Coose River in Alabama. The rock foundation is dolomite with isolated beds of limestone and scattered masses of chert. (Chert is an amorphous or cryptocrystalline sedimentary rock comprised primarily of silica, with lesser amounts of quartz.) The rock is highly jointed and cavernous. Although the bedrock was extensively grouted during construction, underseepage developed soon after reservoir impoundment. Upstream sinkholes and downstream boils developed and persisted through periods of remedial grouting during the 18 years of operation. Seepage has been monitored by stream flow measurements downstream of the dam every spring and fall of this operating period. The reservoir was filled in 1964 and by the spring of 1977, stream flow measurements downstream of the dam had increased to 675 cfs. The seepage has been slightly reduced in recent years by multirow grouting of the dam foundation and by filling upstream sinkholes with cherty residuum. As a second level of defense, a trench drain and rock bolster have been constructed along a critical section downstream of the toe of the dam.

## Failures Due to Geologic Defects

### *Waco Dam*

Structural defects have been responsible for numerous dam failures and accidents. For example, in 1961 a 1,500-foot slide occurred in the Waco Dam in Texas during construction. The earth embankment was as much as 13 feet below finished grade when the slide occurred. The dam was founded on three formations of clay shales that had varying strengths. This complexity in the foundation was due in part to faults. According to Beene (1967), the foundation failure resulted from a combination of depositional sequence and geologic structure disturbance. The weaker clay shale that failed was sandwiched between two stronger shales. Movement along closely spaced nonparallel faults caused shearing stresses in the weaker shale. The presence of a relatively pervious contact along a fault between the weak shale and the stronger shale permitted widespread distribution of uplift pressure. Figure 5-2 shows profile and embankment sections. Figure 5-3 shows pore-pressure contours in the weak shale after the slide. Beene concluded that influence of a joint system on the development and distribution of pore pressure in a clay shale cannot be predicted by laboratory tests; therefore, the embankment must be instrumented for movement and pore pressure.

### *Baldwin Hills Dam*

Another failure resulting from foundation faults was that of Baldwin Hills Reservoir in California in 1963. This occurred some 12 years after the dam had gone into operation. The asphalt lining for the reservoir was founded on thinly bedded and poorly consolidated sands and silty sands. There were

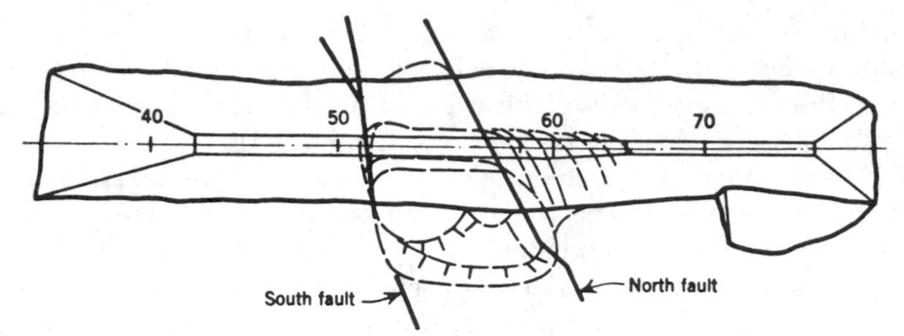


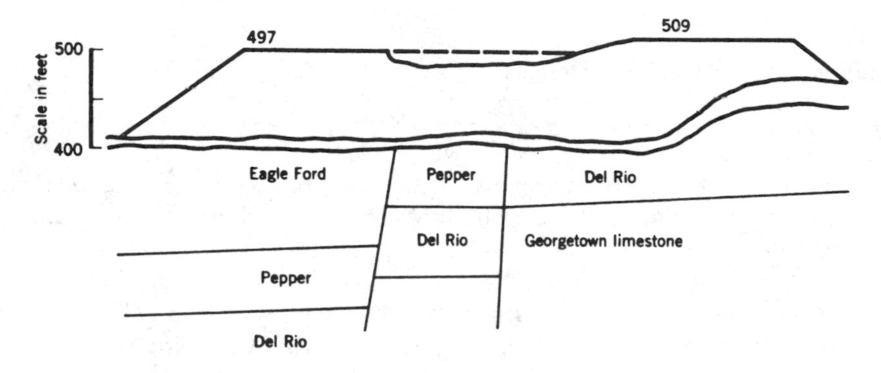


PLAN AND PROFILE

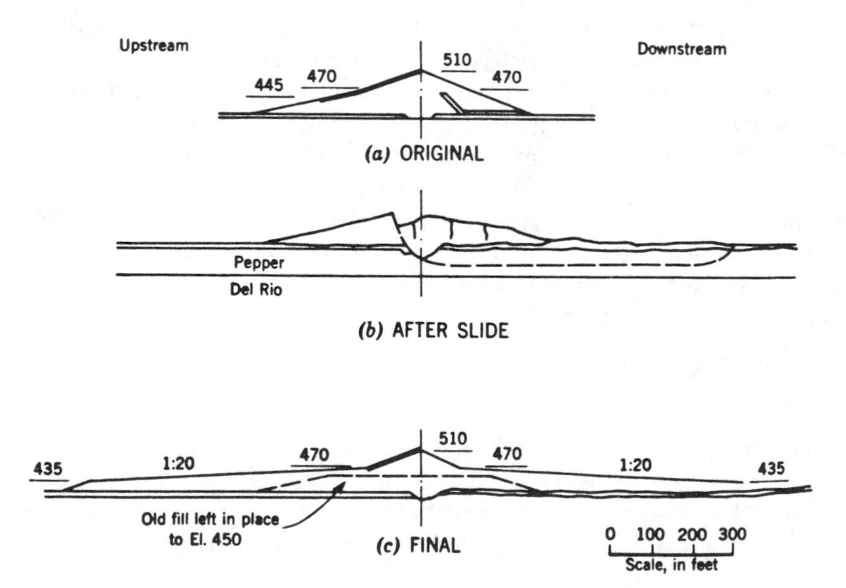


FIGURE 5-2 Embankment sections of Waco Dam. SOURCE: Beene (1967).

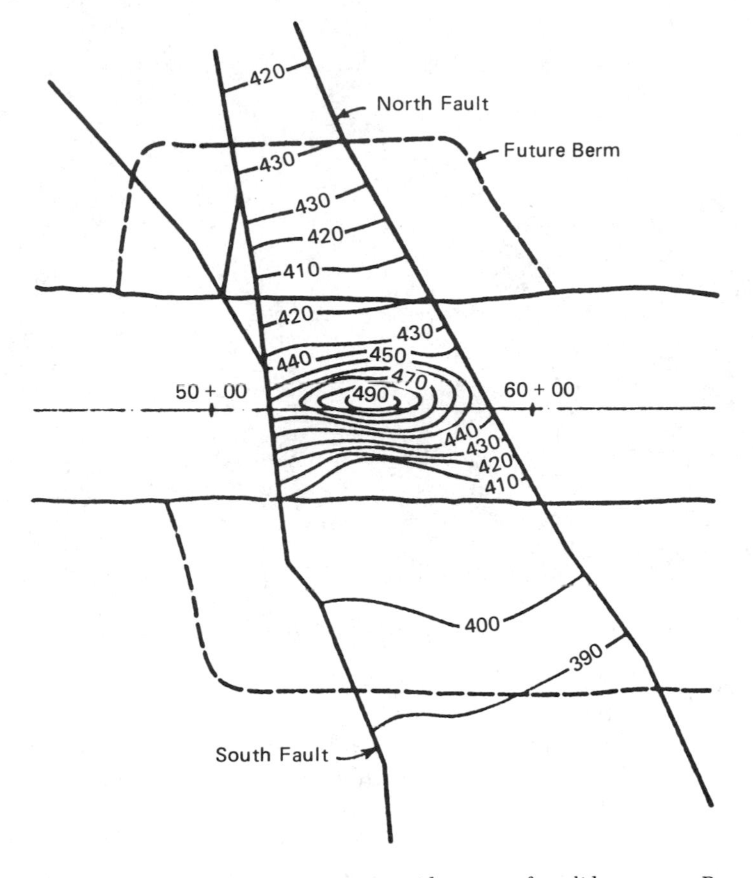


FIGURE 5-3 Pore pressure contours at mid-pepper after slide. SOURCE: Beene (1967).

numerous known faults in the foundation. Because of this, special attention was paid to the design of the dam and the reservoir was highly instrumented. According to the State Engineering Board of Inquiry (Jansen et al. 1967), slow movements occurred on these faults with progressively increasing displacement. This ruptured the asphalt lining and allowed water under pressure to enter the faults and pipe out (remove) the filling material in the faults. This erosion proceeded very rapidly and undermined the dam with its subsequent complete failure. The long-term movement and development of these stresses appear to have been caused primarily by subsidence, and the latter had been observed for many years in this area (Leps 1972). Figure 5-4 shows a view of the dam after failure.

FIGURE 5-4 Baldwin Hills Reservoir after failure.

*Teton Dam*

Open joints in the foundation need careful consideration in design. The failure at Teton Dam in 1976 in Idaho may in part have been caused by the existence of open joints in the foundation. According to the Independent Panel to Review the Cause of Teton Dam Failure (1976), the volcanic rock in the foundation was highly permeable and moderately to intensely jointed. The foundation was grouted during construction, but the grout curtain was not sufficiently effective, and there were open joints in the upstream and downstream faces of the right abutment key trench; these provided conduits for ingress and egress of water during reservoir filling. The independent panel considered the placement of highly erodible soil (the core of the dam) adjacent to the heavily jointed rock a major factor contributing to the failure.

*Malpasset Dam*

Often it may be the combination of deficiencies in the foundation that causes a failure or an incident. One such case was the Malpasset Dam failure in France in 1959. According to Bellier (1976), this was the first total failure in the history of arch dams. The failure was a result of foliation dipping downstream, arch stress parallel with the foliation, and a fault plunging beneath the dam that created a watertight floor. Bellier stated that as the arch stress increased on the left abutment there was a reduction in permeability with a subsequent increase in uplift pressure that eventually caused rupture of the dam. Figure 5-5 shows the relationship between the geologic structure and the arch. This case history strengthens the requirement for using piezometers in rock abutments and foundations of arch dams as a positive monitoring device.

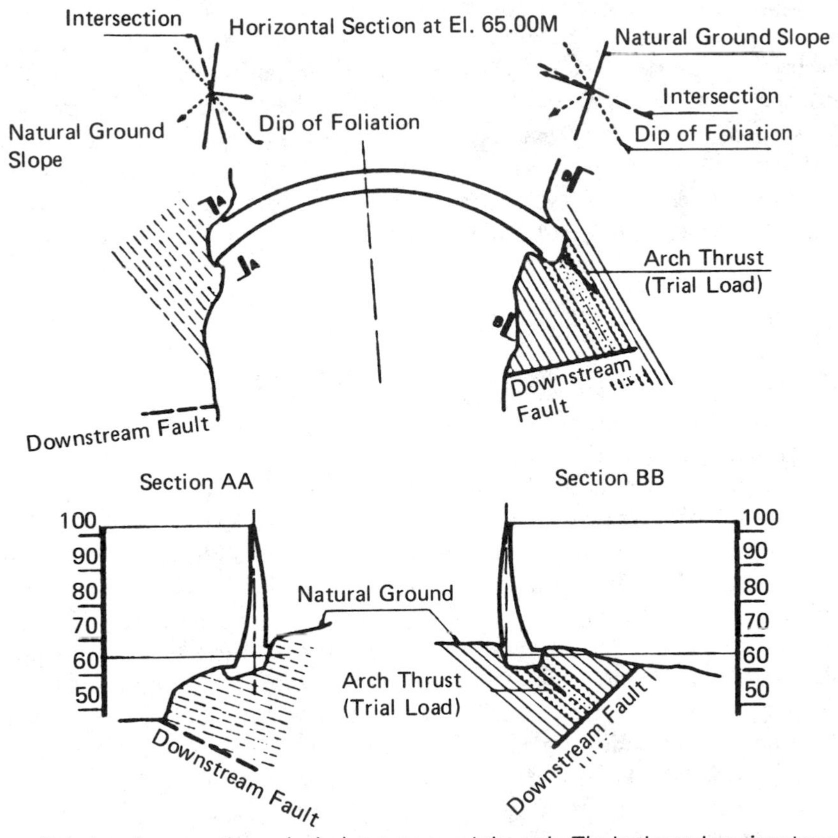


Relations between the geological structure and the arch. The horizontal section shows that the arch is in line with the foliation on the left abutment. Note how the foliation is relative to the ground surface. Conditions on the right abutment were very different.

The corresponding vertical sections clearly show the difference between the two abutments, in particular the zone compressed by the arch thrust in the left bank (Section BB). This zone could extend to great depth on account of the foliation.

FIGURE 5-5 Malpasset Dam. SOURCE: Bellier (1976).

*Uniontown Cofferdam*

It also is important to understand the structural geology of foundations for temporary structures associated with dams, such as cofferdams, which are used during initial construction or in such subsequent modifications as adding a powerhouse to an existing dam. Cofferdam failures can cause loss of life to downstream inhabitants and construction personnel and/or appreciable property damage. A cofferdam failure during construction of the Uniontown locks and dam on the Ohio River in 1971 may in part have been the result of faults in the foundation (Thomas et al. 1975). The cofferdam was completed 10 days before the failure, and the dam foundation area had been dewatered. The failure was a translational slide along a coal and underclay stratum approximately 16 feet below the top of rock. The probable cause of failure was excessive pore pressure in the underclay and coal. A fault beneath one of the cofferdam's cells appears to have provided a likely avenue of communication to water outside of the cofferdam. Intense faulting in the area contributed to the reduction in strength of the sedimentary rock.

## SOILS

The vast majority of dams are embankment dams composed of soils. Some concrete gravity dams may be in contact with soils at some portions of their foundations, particularly the abutments.

A thorough understanding of the condition of an existing dam requires detailed knowledge of the types of soils in the embankment and foundation, their spatial distribution, and their physical characteristics (moisture content, strength, permeability, and presence of discontinuities affecting permeability and strength). For most existing dams, soils data from preconstruction site exploration and testing, design, or construction are not available; therefore, this extremely important information must be cautiously inferred by visual site inspection and limited sampling and testing. Such inferred conclusions must be made in conjunction with an understanding of the geologic setting along with experience with similar soils and structures.

This section presents some very generalized information on soils. It also offers suggestions on how to obtain, from published information, more site-specific understanding of the nature and characteristics of the soils in, beneath, and around a particular dam. This approach can reduce the cost of drilling and testing and is essential for reliable interpretation of drilling and test data. It is emphasized that study of generalized information is no substitute for exploring and testing the soils in and under the specific dam.

Discussion of basic soil mechanics is not included in this section. For that information the reader is referred to texts, geotechnical engineering consultants, and sections of this book dealing with stability and seepage analyses.

### Soil Classification

Classification of soils (dividing them into systems or groups having similar characteristics) may be done in many different ways depending on the particular characteristics of interest. For example, geologic classifications tend to be focused on origin and mappability. Pedologic classifications have been developed by soil scientists with primary emphasis on agricultural qualities of the soils, with mappability being an important factor here, too. Engineering classification systems place emphasis on physical characteristics most pertinent to the particular engineering utilization at hand. Engineers, pedologists, and geologists have become increasingly aware that soil data and maps developed by each of these disciplines can be very useful to others when properly interpreted. A common denominator is that most classification systems include textural descriptions based, at least in part, on the relative abundance of different grain sizes composing the soil.

The following discussions outline the most commonly used soil classifications.

#### *Textural Classification*

A textural classification defines quantitatively the percentages of particles of various grain sizes in a soil sample. Although there are some differences among engineers, soil scientists, and geologists on the specific grain diameters defining boulders, cobbles, gravels, sand, silts, and (especially) clays, overall the differences are minor. By making appropriate sieve separation tests, a soil may be defined as, for example, 60% sand, 30% silt, and 10% clay; a more complete analysis of grain sizes can be usefully expressed on a graph (grain-size chart) showing the percentage of soils in each diameter. Engineering grain-size definitions along with a grain-size chart are shown in Figure 5-6; the figure includes some typical grain-size curves illustrating different degrees of uniformity of soil grain sizes.

#### *Descriptive Classification*

Descriptive classifications simply identify the main constituent, with less abundant or less important constituents' names being used as adjectives. For example, a soil containing mostly sand but including some silt and a little clay might be defined descriptively as a slightly clayey, silty sand.

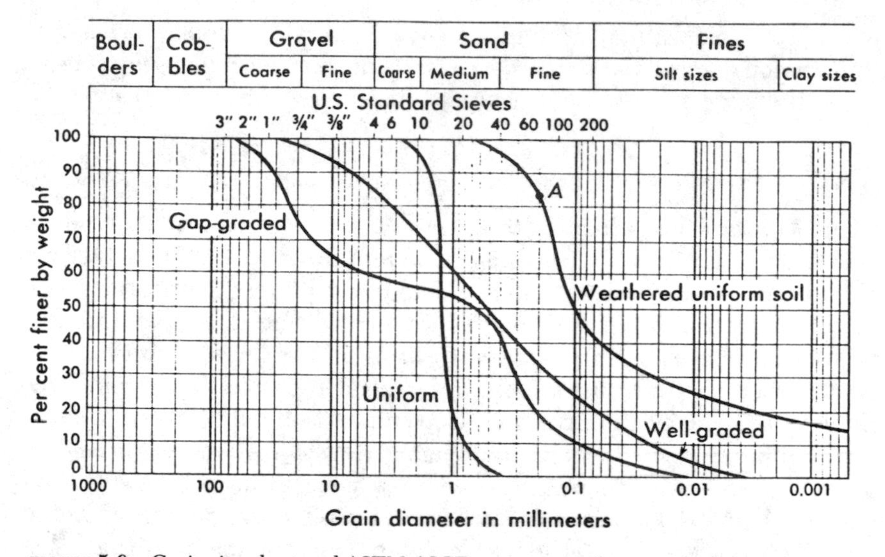


FIGURE 5-6 Grain size chart and ASTM-ASCE grain size scale. SOURCE: Sowers (1979).

Generalized terms relating to density, consistency, moisture content, color, and origin are usually included in the description as the first modifying terms. Mixtures of sand, silt, and clay are often termed *loam*, particularly by soil scientists. Definitions of various loams are shown in Figure 5-7.

### *Engineering Classification*

Several engineering soil classification systems are in use. The most common and appropriate one for dam engineering is the Unified Soil Classification System (USCS 1975). It is based on grain-size distribution and, for finer grained soils, on plasticity. Figure 5-8 shows the USCS soil classes and their definitions, and Figure 5-9 shows generalized engineering characteristics of each USCS soil class. This generalized information can be helpful in inferring information about an existing dam when used in conjunction with pedological and geologic maps and information, as discussed in the following sections.

### *Pedological Classification*

The most detailed pedological classifications focus on the agricultural characteristics of surface and near surface soils. The basic unit of classification is termed a *soil series*, and series are classified into progressively larger and inclusive families, subgroups, great groups, suborders, and orders. Any

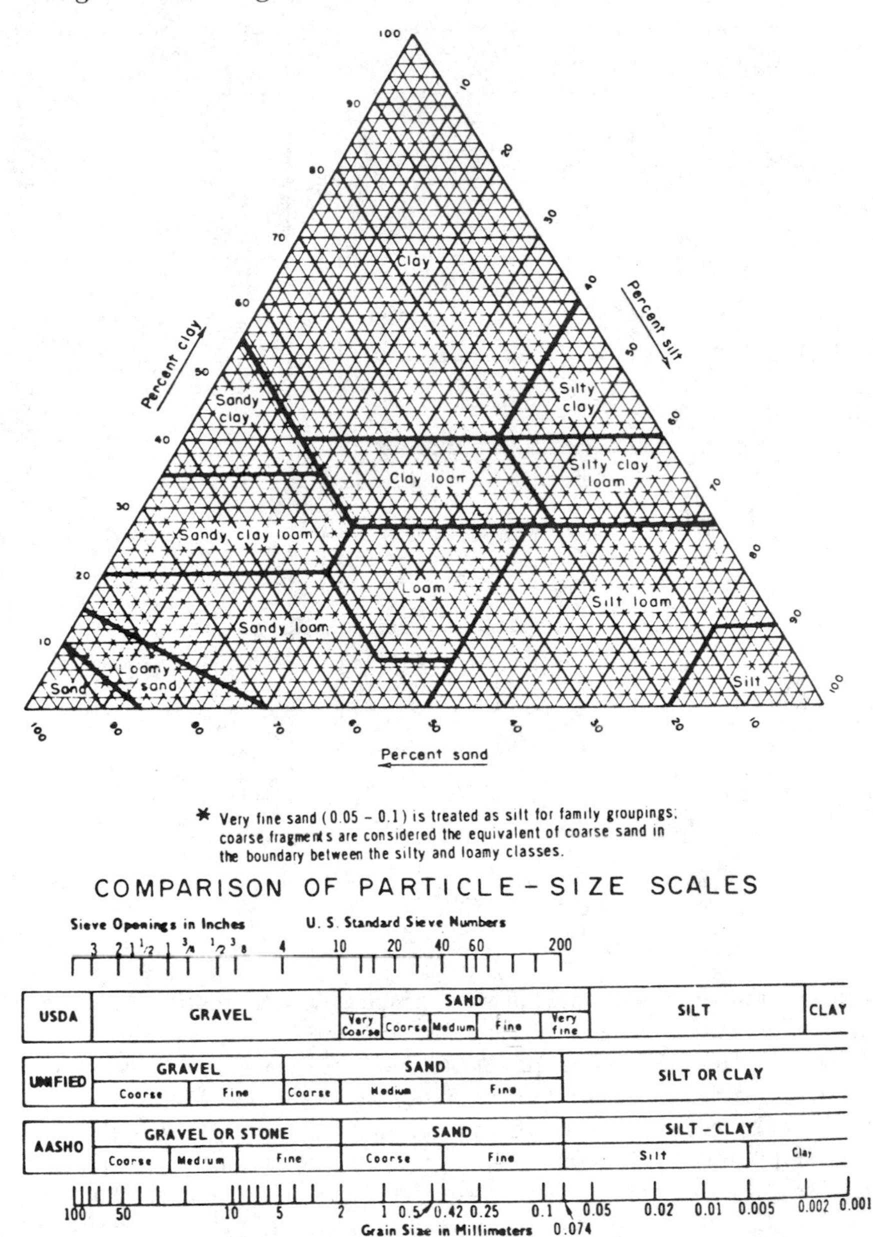


FIGURE 5-7 Soil triangle of the basic soil textural classes. SOURCE: U.S. Bureau of Reclamation (1974).

| FIELD IDENTIFICATION PROCEDURES (Excluding particles larger than 3 inches and basing fractions on estimated weights) | | | | | | GROUP SYMBOLS † | TYPICAL NAMES |
|---|---|---|---|---|---|---|---|
| COARSE GRAINED SOILS More than half of material is larger than No 200 sieve size ‡ (The No 200 sieve size is about the smallest particle visible to the naked eye) | GRAVELS More than half of coarse fraction is larger than No 4 sieve size (For visual classifications, the ¼" size may be used as equivalent to the No 4 sieve size) | CLEAN GRAVELS (Little or no fines) | Wide range in grain size and substantial amounts of all intermediate particle sizes | | | GW | Well graded gravels, gravel-sand mixtures, little or no fines |
| | | | Predominantly one size or a range of sizes with some intermediate sizes missing | | | GP | Poorly graded gravels, gravel-sand mixtures, little or no fines |
| | | GRAVELS WITH FINES (Appreciable amount of fines) | Non-plastic fines (for identification procedures see ML below) | | | GM | Silty gravels, poorly graded gravel-sand-silt mixtures |
| | | | Plastic fines (for identification procedures see CL below) | | | GC | Clayey gravels, poorly graded gravel-sand-clay mixtures |
| | SANDS More than half of coarse fraction is smaller than No 4 sieve size | CLEAN SANDS (Little or no fines) | Wide range in grain sizes and substantial amounts of all intermediate particle sizes | | | SW | Well graded sands, gravelly sands, little or no fines |
| | | | Predominantly one size or a range of sizes with some intermediate sizes missing | | | SP | Poorly graded sands, gravelly sands, little or no fines |
| | | SANDS WITH FINES (Appreciable amount of fines) | Non-plastic fines (for identification procedures see ML below) | | | SM | Silty sands, poorly graded sand-silt mixtures |
| | | | Plastic fines (for identification procedures see CL below) | | | SC | Clayey sands, poorly graded sand-clay mixtures |
| FINE GRAINED SOILS More than half of material is smaller than No 200 sieve size | IDENTIFICATION PROCEDURES ON FRACTION SMALLER THAN No. 40 SIEVE SIZE | | | | | | |
| | | | DRY STRENGTH (CRUSHING CHARACTERISTICS) | DILATANCY (REACTION TO SHAKING) | TOUGHNESS (CONSISTENCY NEAR PLASTIC LIMIT) | | |
| | SILTS AND CLAYS Liquid limit less than 50 | | None to slight | Quick to slow | None | ML | Inorganic silts and very fine sands, rock flour, silty or clayey fine sands with slight plasticity. |
| | | | Medium to high | None to very slow | Medium | CL | Inorganic clays of low to medium plasticity, gravelly clays, sandy clays, silty clays, lean clays |
| | | | Slight to medium | Slow | Slight | OL | Organic silts and organic silt-clays of low plasticity |
| | SILTS AND CLAYS Liquid limit greater than 50 | | Slight to medium | Slow to none | Slight to medium | MH | Inorganic silts, micaceous or diatomaceous fine sandy or silty soils, elastic silts |
| | | | High to very high | None | High | CH | Inorganic clays of high plasticity, fat clays |
| | | | Medium to high | None to very slow | Slight to medium | OH | Organic clays of medium to high plasticity |
| HIGHLY ORGANIC SOILS | | | Readily identified by color, odor, spongy feel and frequently by fibrous texture | | | Pt | Peat and other highly organic soils |

† Boundary classifications - Soils possessing characteristics of two groups are designated by combinations of group symbols For example GW-GC, well graded
‡ All sieve sizes on this chart are US standard

FIGURE 5-8 Unified soil classification, including identification and description. SOURCE: U.S.

particular soil series has specific visual, physical, and agricultural characteristics that can be recognized from place to place, usually within a localized geographic region. Over 10,000 soil series have been identified in the United States. Soil series are the basic mapping unit of agricultural soil scientists. All of the United States has been mapped by soil scientists at some scale, and within the last few decades many areas have been mapped in detail at scales of 1 inch = 2,000 feet or larger. In many areas the Unified Soil Classification of each soil series (and horizons within each series) has been determined, and many soil series have been subjected to other engineering tests to determine general engineering characteristics.

It is emphasized that agricultural soils maps, particularly the more modern ones where correlations have been made between soil series and engi-

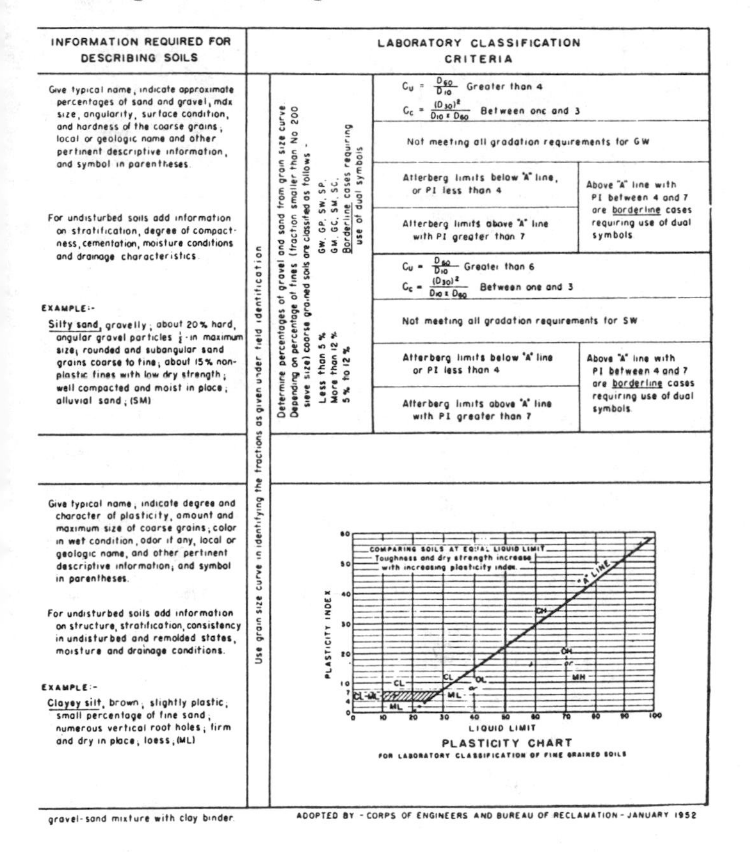

| INFORMATION REQUIRED FOR DESCRIBING SOILS | | | | LABORATORY CLASSIFICATION CRITERIA | |
|---|---|---|---|---|---|
| Give typical name, indicate approximate percentages of sand and gravel, max size, angularity, surface condition, and hardness of the coarse grains; local or geologic name and other pertinent descriptive information, and symbol in parentheses.<br><br>For undisturbed soils add information on stratification, degree of compactness, cementation, moisture conditions and drainage characteristics.<br><br>EXAMPLE:-<br>Silty sand, gravelly; about 20% hard, angular gravel particles ½-in maximum size; rounded and subangular sand grains coarse to fine; about 15% non-plastic fines with low dry strength; well compacted and moist in place; alluvial sand; (SM) | Use grain size curve in identifying the fractions as given under field identification | Determine percentages of gravel and sand from grain size curve. Depending on percentage of fines (fraction smaller than No 200 sieve size) coarse grained soils are classified as follows -<br>Less than 5% GW, GP, SW, SP.<br>More than 12% GM, GC, SM, SC.<br>5% to 12% Borderline cases requiring use of dual symbols | | $C_U = \frac{D_{60}}{D_{10}}$ Greater than 4<br>$C_C = \frac{(D_{30})^2}{D_{10} \times D_{60}}$ Between one and 3 | |
| | | | | Not meeting all gradation requirements for GW | |
| | | | | Atterberg limits below "A" line, or PI less than 4 | Above "A" line with PI between 4 and 7 are borderline cases requiring use of dual symbols |
| | | | | Atterberg limits above "A" line with PI greater than 7 | |
| | | | | $C_U = \frac{D_{60}}{D_{10}}$ Greater than 6<br>$C_C = \frac{(D_{30})^2}{D_{10} \times D_{60}}$ Between one and 3 | |
| | | | | Not meeting all gradation requirements for SW | |
| | | | | Atterberg limits below "A" line or PI less than 4 | Above "A" line with PI between 4 and 7 are borderline cases requiring use of dual symbols. |
| | | | | Atterberg limits above "A" line with PI greater than 7 | |
| | | | | | |
| Give typical name; indicate degree and character of plasticity, amount and maximum size of coarse grains; color in wet condition, odor if any, local or geologic name, and other pertinent descriptive information; and symbol in parentheses.<br><br>For undisturbed soils add information on structure, stratification, consistency in undisturbed and remolded states, moisture and drainage conditions.<br><br>EXAMPLE:-<br>Clayey silt, brown; slightly plastic; small percentage of fine sand; numerous vertical root holes; firm and dry in place; loess; (ML) | | PLASTICITY CHART FOR LABORATORY CLASSIFICATION OF FINE GRAINED SOILS | | | |

gravel-sand mixture with clay binder.

ADOPTED BY - CORPS OF ENGINEERS AND BUREAU OF RECLAMATION - JANUARY 1952



Bureau of Reclamation (1974).

neering tests, can be very helpful in inferring generalized engineering characteristics at an existing dam. Furthermore, the text accompanying the soil map may include references to actual engineering performance of the soils in the vicinity of the dam. These may help in understanding a suspected or known deficiency in the dam, its foundation, or the reservoir rim.

## *Geologic Classifications*

Soils may be classified as to their origin or mode of deposition. The broadest divisions are (a) *residual soils*, derived by in-place chemical and physical decomposition (weathering) of parent rock of soil materials and (b) *transported soils*, redistributed from their original or other site of deposition by

| Typical Names of Soil Groups | Group Symbols | Important Engineering Properties | | |
|---|---|---|---|---|
| | | Permeability When Compacted | Shear Strength When Compacted and Saturated | Compressibility When Compacted and Saturated |
| Well-graded gravels, gravel-sand mixtures, little or no fines | GW | Pervious | Excellent | Negligible |
| Poorly graded gravels, gravel-sand mixtures, little or no fines | GP | Very pervious | Good | Negligible |
| Silty gravels, poorly graded gravel-sand-silt mixtures | GM | Semipervious to impervious | Good | Negligible |
| Clayey gravels, poorly graded gravel-sand-clay mixtures | GC | Impervious | Good to fair | Very low |
| Well-graded sands, gravelly sands, little or no fines | SW | Pervious | Excellent | Negligible |
| Poorly graded sands, gravelly sands, little or no fines | SP | Pervious | Good | Very low |
| Silty sands, poorly graded sand-silt mixtures | SM | Semipervious to impervious | Good | Low |
| Clayey sands, poorly graded sand-clay mixtures | SC | Impervious | Good to fair | Low |
| Inorganic silts and very fine sands, rock flour, silty or clayey fine sands with slight plasticity | ML | Semipervious to Impervious | Fair | Medium |
| Inorganic clays of low to medium plasticity, gravelly clays, sandy clays silty clays, lean clays | CL | Impervious | Fair | Medium |
| Organic silts and organic silt-clays of low plasticity | OL | Semipervious to impervious | Poor | Medium |
| Inorganic silts, micaceous or diatomaceous fine sandy or silty soils, elastic silts | MH | Semipervious to impervious | Fair to poor | High |
| Inorganic clays of high plasticity, fat clays | CH | Impervious | Poor | High |
| Organic clays of medium to high plasticity | OH | Impervious | Poor | High |
| Peat and other highly organic soils | PT | – | – | – |

FIGURE 5-9 Soil performance in or under dams. SOURCE: U.S.

| Workability as a Construction Material | Relative Desirability for Various Uses (No. 1 is Considered the Best) | | | | | | | |
|---|---|---|---|---|---|---|---|---|
| | Rolled Earthfill Dams | | | Resistance to Piping | Foundations | | | |
| | Homogeneous Embankment | Core | Shell | | Seepage Important | Seepage Not Important | Requirements for Seepage Control: Permanent Reservoir | Requirements for Seepage Control: Floodwater Retarding |
| Excellent | – | – | 1 | Good | – | 1 | Positive cutoff or blanket | Control only within volume acceptable plus pressure relief if required |
| Good | – | – | 2 | Good | – | 3 | Positive cutoff or blanket | Control only within volume acceptable plus pressure relief if required |
| Good | 2 | 4 | – | Poor | 1 | 4 | Core trench to none | None |
| Good | 1 | 1 | – | Good | 2 | 6 | None | None |
| Excellent | – | – | 3 If gravelly | Fair | – | 2 | Positive cutoff or upstream blanket and toe drains or wells | Control only within volume acceptable plus pressure relief if required |
| Fair | – | – | 4 If gravelly | Fair to poor | – | 5 | Positive cutoff or upstream blanket and toe drains or wells | Control only within volume acceptable plus pressure relief if required |
| Fair | 4 | 5 | – | Poor to very poor | 3 | 7 | Upstream blanket and toe drains or wells | Sufficient control to prevent dangerous seepage piping |
| Good | 3 | 2 | – | Good | 4 | 8 | None | None |
| Fair | 6 | 6 | – | Poor to very poor | 6 | 9 | Positive cutoff or upstream blanket and toe drains or wells | Sufficient control to prevent dangerous seepage piping |
| Good to fair | 5 | 3 | – | Good to fair | 5 | 10 | None | None |
| Fair | 8 | 8 | – | Good to poor | 7 | 11 | None | None |
| Poor | 9 | 9 | – | Good to poor | 8 | 12 | None | None |
| Poor | 7 | 7 | – | Excellent | 9 | 13 | None | None |
| Poor | 10 | 10 | – | Good to poor | 10 | 14 | None | None |
| – | – | – | – | – | – | – | – | – |

Bureau of Reclamation (1974) and U.S. Department of Agriculture (1975).

water, gravity, wind, or ice and deposited in water or on land. The regional distribution of these soils in the United States is shown in Figure 5-10.

Discussed below are brief discussions of the major subdivisions of these two broad groups, along with examples of their implications for assessing and improving the safety of existing dams.

## Implications for Existing Dams

### *Residual Soils*

The nature of residual soils is determined by a wide array of complex variables, including the parent materials' mineralogy, texture, and structure; climate; rate of surface erosion; topography; location of groundwater table; and types of vegetation. All of these factors change over time, and since most residual soils require tens of thousands to hundreds of thousands of years for their development in any significant thickness, they have been subjected to a wide range of each of these variables. Also, because of these many variables, residual soils may be erratic laterally and vertically on a specific site.

A mature residual soil profile has the following generalized stratigraphy:

*A-Horizon.* A layer at the surface containing organic litter. It is typically relatively sandy, the clays having been removed by rainwater seeping down through the layer. It is normally only a few inches thick and is gradational down into the:

*B-Horizon.* A relatively clayey layer, the clays having evolved from chemical weathering of feldspars, micas, and other silicate minerals. Weathering has destroyed evidence of the parent materials' structure (such as layering and joints) and the soil has its own "new" texture and structure. This horizon may vary in thickness from a few inches to several feet and is gradational downward into the:

*C-Horizon.* An intermediate zone between relatively unweathered parent materials and the highly weathered B-horizon. The structure of the parent material is present to some degree (increasingly so with depth), but the mineral grains of the parent material are partially weathered, breaking or loosening the intergranular cohesion of the parent materials. This horizon is gradational downward into unweathered materials and may be inches to tens of feet in thickness.

The parent materials of residual soils may be rocks of igneous, metamorphic, or sedimentary origin; residual soil profiles may be developed on unconsolidated sediments (transported soils). Since most transported soils are relatively young, residual soils developed on them usually are thin; exceptions do exist.

Large regions of the United States are underlain by residual soils developed on metamorphic, igneous, and sedimentary rocks, and a vast number of existing dams are composed of and founded on residual soils. Some broad generalizations can be made about implications of residual soils for existing dams, recognizing that many exceptions exist. These generalizations are as follows:

- The upper soils (B-horizon) are usually more clayey and less permeable than the deeper (C-horizon) soils. Even within the C-horizon, wide ranges in permeability may exist, depending on the variability of the parent materials. Borrow pits for embankment materials may be developed to use selectively the less permeable soils for a foundation cutoff trench and for dam core materials and the more permeable soils for building the outer zones of the embankment. However, in many existing dams the proper zonation was reversed during construction, placing less permeable soils in the outer part of the slope. This creates a downstream seepage barrier that raises the phreatic surface in the dam and increases uplift pressures. This may create an unstable embankment subject to structural or seepage failure. Another common problem is horizontal interlayering of less permeable and more permeable soils during construction. This problem is found most often where large or multiple borrow pits were developed for embankment materials and earth-moving equipment was extracting soils from different depths somewhat simultaneously. Such horizontal layering of more and less permeable soils within the embankment can produce unsafe seepage pressures in the embankment, leading to structural or seepage failure.
- Relict joints, foliation or bedding planes, and faults in the C-horizon soils in a dam foundation usually control the quantities and preferential directions of seepage in the residual-soil foundation mass. Concentrated seepage along these relict structures, particularly near the outer toe of the dam, can develop into foundation piping. Furthermore, relict structures in the foundation residual soils may create zones of weakness subject to structural failure. When this is a problem, it usually develops during or soon after construction. Thus, this should be a consideration for analysis of foundation stability under earthquake loading of existing dams.
- Residual soils developed from crystalline (igneous and metamorphic) rocks may contain platy minerals, such as micas, that have important effects on the engineering characteristics of the soil, both in-place and remolded (in the embankment). One of the more important and dramatic effects of platy minerals in the soil is their tendency to rotate during compaction into horizontal positions, giving the embankment soils a higher

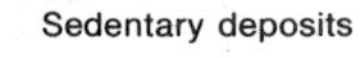
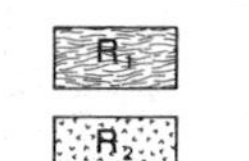

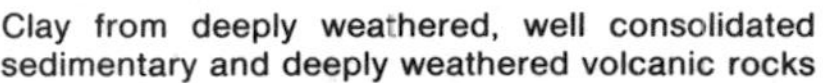
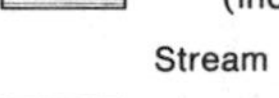

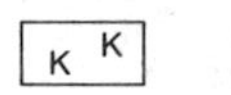
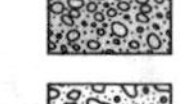

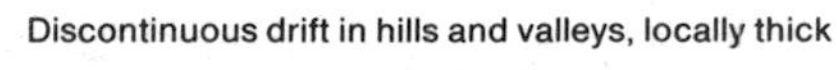

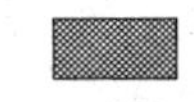


FIGURE 5-10 Surface deposits in the United States, except Alaska and Hawaii. These are the parent materials of the agriculturalist's soils; most are late Pleistocene or Holocene, and ages overlap. SOURCE: Hunt (1974).

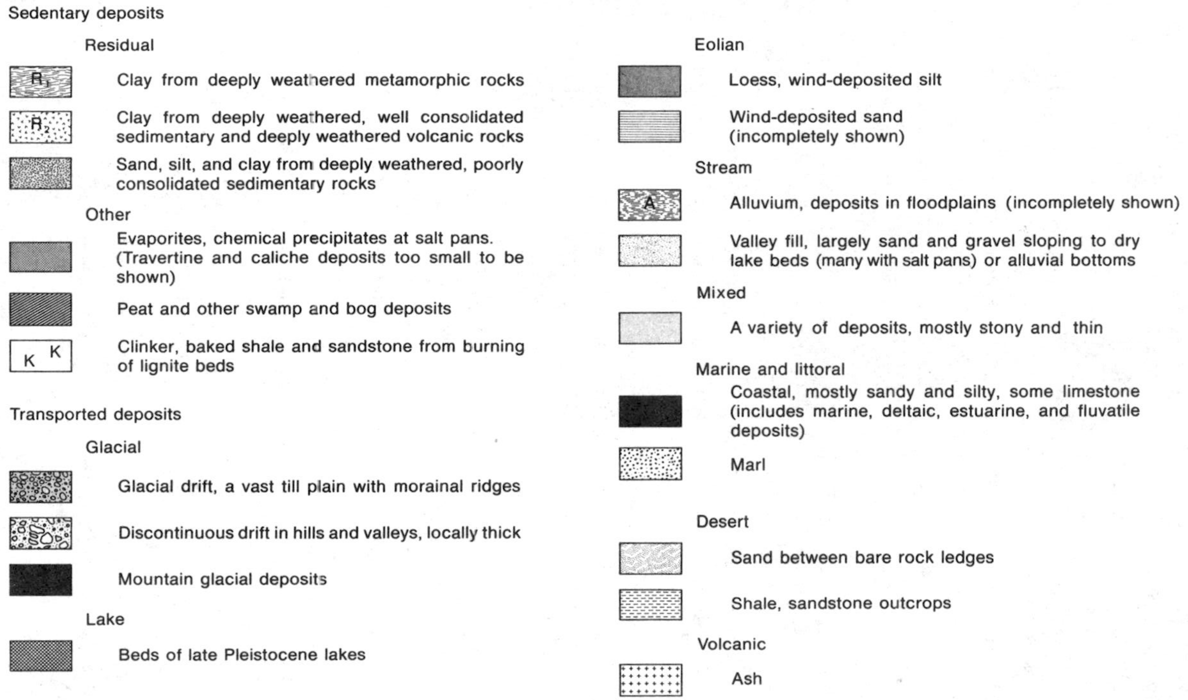

horizontal permeability compared with their vertical permeability. In most applications this is an undesirable characteristic and can lead to serious defects in the dam from both structural and seepage standpoints. When residual soil is formed from granitic rocks, the result can be a very granular material resembling sand. Such deposits, sometimes referred to as "DG" (decayed granite) have attained thicknesses of 1,500 feet. They can be very unstable when excavations are made into them; also, they are usually very pervious.

- Shales normally produce a clayey residuum, the nature of which varies widely depending on the mineralogy of the particular shale formation and many other variables. Though in the United States some shales have weathered to relatively stable soils suitable for dam construction and foundations when properly placed, some of the most treacherous soils in the nation are shale residuum and partially weathered shales, particularly those shales and soils containing significant amounts of sodium-montmorillonite clays. Among the most notable shales yielding problem soils are the widespread Pierre formation, Bearpaw formation, some other Cretaceous shales of the mid-continent and western regions, and some of the carboniferous shales in the mid-continent and Appalachian regions. Where these and similar problem shale soils exist, they are often suitable only for dam core construction due to their high-plasticity, swelling, and cracking characteristics and their sensitivity. However, they may have been incorporated in construction of existing homogeneous embankments and this can become a problem.

Another problem with many C-horizon soils developed from shale is relict planes of weakness in the soils (bedding and joints). A related problem is the tendency of these soils to break into cobble-to-gravel-sized blocks during excavation; these materials can degrade (by swelling and slaking) in the embankment and under or around spillways and other appurtenant structures, causing serious defects.

- Residual soils developed on relatively soluble rocks, such as carbonates, may contain cavities or natural "pipes" resulting from raveling of soil into cavities in the underlying rocks. These soil cavities may be open or filled with a wide variety of secondary residual and transported soils. This is obviously a serious foundation defect. A related characteristic is that the rock/residuum contact is often extremely irregular, thus causing large variations in soil thickness under the dam. Some of these soils are relatively compressible, and the resulting differential settlement of the dam can create cracks that may lead to structural or seepage instability. Some weathered limestone forms laterite, usually a reddish and very clayey soil. This soil often is erratic in its properties but is usually impervious and can be unstable in excavated slopes.

- Residuum derived from sandstones and conglomerates often has characteristics somewhat similar to those derived from some crystalline rocks. Permeability tends to increase with depth and may create foundation seepage deficiencies, especially where the original rocks contained few fine-grained minerals. Some sandstones contain significant amounts of carbonates, and their residuum and solubility may create problems similar to those outlined above for the carbonate soils.

### *Transported Soils*

Any soil that has been moved from its site of origin and redeposited is in this broad group of soils. If the soil is deposited in a stable environment it eventually becomes indurated, forming a sedimentary rock. Some of the deeper sediments along the continental margins (Coastal Plains) and in some other areas are at least partially lithified or indurated, and there is no consensus as to whether these materials should be called soils or rocks. However, for the purposes of assessing and improving the safety of existing dams, most of the transported soils of interest are relatively unconsolidated or loose surficial deposits of somewhat recent origin. The following discussions outline the major categories and some of their possible implications for existing dams.

*Alluvial (Fluvial) Soils.* Soils transported by streams and deposited in the streambed or adjacent floodplain are called alluvial soils or alluvium. They are almost universally present at dam sites. Alluvial soils may vary in thickness from a few inches to hundreds of feet and may vary in composition from large boulders to clays. Most commonly they are composed of gravels, sands, silts, and clays that have been sorted (in varying degrees) into discontinuous layers and lenses with wide ranges in horizontal permeability. They commonly are poorly consolidated (soft or loose), weak and compressible, often wet, and may contain layered or dispersed organic material. Removal of alluvial soils under at least the core of the embankment and often the entire embankment is usually required before dam construction to ensure structural and seepage stability.

Many existing dams have defects resulting from improper use of alluvial soils in the embankment or inadequate treatment of an alluvial foundation. Problems have developed where contractors have minimized excavation of a cutoff trench or an embankment foundation or have used alluvium in the future reservoir area as borrow for embankment construction. As discussed in another section of this report, the Bearwallow Dam failure was at least partly caused by placing organic alluvium in the embankment. Also, this stripping of what may have been an impervious layer over the reservoir can result in high seepage losses, such as occurred at Helena Valley Dam in Montana (Figure 5-11).

Excessive seepage through permeable alluvial layers in the foundation is a common defect of existing dams. Although this can be precluded by proper cutoff trench construction, inadequate dewatering during construction can lead to a "messy" cutoff trench that does not adequately penetrate highly permeable layers. For example, a public water supply dam in North Carolina nearly failed from foundation piping due to an inadequate seepage cutoff in its alluvial foundation. An attempt had been made to construct a slurry wall cutoff before construction; the slurry trench was excavated by a dragline, and it failed to penetrate completely a gravel layer at the base of the alluvium. After the dam was completed and the reservoir was filled, seepage through the alluvial gravels caused severe boils and piping at the dam's toe. This required immediate draining of the reservoir to prevent failure of the dam.

Problems with alluvial foundation soils are not limited to those dams that had no preconstruction subsurface investigations. Some investigators fail to fully appreciate the erratic geometry of the layers, pods, and lenses composing most alluvial deposits. Furthermore, even with the most diligent site investigation, highly permeable or weak and compressible layers may be undiscovered.

FIGURE 5-11 Erosion caused by seepage through floor of Helena Valley Reservoir.

Some existing embankment dams are composed, at least in part, of alluvial soils that would be acceptable if they had been properly placed instead of the more and less permeable soils being segregated into horizontal layers in the embankment. Drilling and sampling normally would be required to discover this defect.

*Gravity-Transported Soils.* There are two common types of gravity-transported soils: (1) colluvium and (2) talus, plus a special class (3) landslide materials.

1. *Colluvium.* Soils (and rock fragments in a soil mass) that have been transported downslope on hillsides by mass wasting and gravitational creep are colluvial soils. In addition to gravity, their movement is aided by ice heave, tree roots, animal burrows (including worms), and water. They are evidenced by bent tree trunks, hummocky or irregular slopes, landslide scars, or heterogeneous soil mixtures (lack of a mature weathering profile). In some cases the transition from residual soils to colluvium is exhibited in natural or man-made excavations by relict structural planes bent downhill and grading upward into a jumbled mass. Natural rates of movement of these materials are probably on the order of a few inches per year maximum, more typically a few inches over several hundred years. Thickness varies from a few inches to over 100 feet, and the composition varies widely from region to region. Despite local and regional variability of colluvium, all colluvial soils have one thing in common: they are inherently unstable and are likely to develop into landslides; relatively rapid movement of large masses can be triggered by minor man-made changes in topography, loading, or drainage as well as by natural events, such as heavy rainfall, snowmelt, or earthquakes. Colluvium is particularly hazardous to existing dams where landsliding may impact the reservoir, spillways, or other appurtenant structures. Furthermore, it must be ensured that portions of the dam or associated structures are not founded on colluvium in a state of creep.
2. *Talus.* Also sometimes called scree or rubble, talus is an accumulation of rock debris at the base of a steep slope. It is classically developed at the foot of mountains in relatively arid or cold regions where mechanical weathering of exposed rock slopes outstrips chemical weathering and soil formation. Talus has many of the same characteristics as colluvium, and talus deposits often creep, forming a "rock glacier" in valleys. Particularly treacherous for unwitting dam builders are creeping talus deposits that are masked by a surficial deposit of more recent soils. Defects at sites of some existing dams may include talus foundations and damaging talus slides.
3. *Landslide materials.* Relatively large masses of otherwise intact soils beneath or around a dam and its reservoir may in fact have been displaced from their site of origin by landsliding. Old landslide blocks normally come

to.rest at a state of marginal equilibrium and are subject to renewed movement when subjected to very minor changes in the environment. This may have very serious implications for existing dams, similar to the problems outlined above for colluvial and talus deposits. For example, special precautions had to be taken to protect the Oahe Dam on the Missouri River in South Dakota from landslides; the reservoir and associated highways and bridges are still being seriously affected by old landslides that were reactivated by the impoundment and by construction activities (Gardner and Allan 1979; see also Chapter 9 on reservoir problems). This instability results from high percentages of expansive clay (montmorillonite) both within the shale bedrock and between the shale layers.

*Aeolian Soils.* The most common aeolian (wind deposited) soils are loess and dunes.

Large areas in the central and western United States are covered with loess, which is a wind-deposited silt. Among the most notable areas are the Mississippi and Missouri river valleys and tributary areas where some loess deposits are tens of feet thick. Other areas covered by significant loess deposits include the High Plains, some of the Basin and Range valleys, the Snake River Plain, and the Columbia Plateau (palouse soil).

From an engineering point of view, loess deposits are characterized by their erodibility and by their tendency to collapse or subside drastically when wetted and under a structural loading. The sensitivity of loess is due to its high vertical permeability, angular grains, and weak cementation between grains. Subsidence of loess in their foundation can produce serious differential settlement in existing dams and their appurtenant structures. Where loess has been used in constructing embankments, the sensitivity may have been eliminated by remolding and compaction, but the embankment soils still may be relatively weak.

Though not nearly as widespread as loess in the United States, dunes are common in some of the western regions. Some older (Pleistocene) dunes may be covered by residual or other soils, and their presence may not be obvious. Loosely compacted, relatively clean sands in dunes are subject to liquefaction and seepage problems where they are present in the foundations of existing dams.

*Glacial Deposits.* Glacial soils produced and deposited by Pleistocene continental ice sheets are prevalent in the northern United States. Much less common are soils associated with alpine (mountain) glaciers, but they are very important at many dam sites in the Rocky Mountains, Sierra Nevada, and some of the other higher western mountains. "Glacial soils" is an extremely broad term and is used to include all forms of glacial drift, including soils moved and mixed by the large ice masses and then deposited by meltwaters (e.g., outwash plains, kames, eskers) as well as unstratified till (including moraines) deposited more directly by glaciers. Very important glacially related soils from an engineering standpoint are glaciolacustrine (lake) and glaciomarine deposits. In addition to wide variations in the origin and nature of glacial soils, the thickness of glacial deposits varies from a few inches to several hundreds of feet.

Because of such wide variations, few generalizations on the implications of glacial soils for existing dams can be made. Particularly likely to create dam defects are loose, permeable unconsolidated drift and sensitive glaciomarine soils in the dam foundation or in the reservoir margin. With this simple word of caution the reader is referred to detailed published geologic and soils maps that may be available for the locality of a particular dam. For more general background on glacial soils and their engineering implications, excellent information can be found in Legget (1961) and Krynine and Judd (1957).

### *Dispersive Clays*

Of particular importance for some existing dams are clay deposits that disperse (deflocculate, disaggregate) rapidly in water; they may be either residual or transported in origin and are discussed here as a special class. A few dam failures and many accidents or near-failures have been attributed to these soils.

Problems with dispersive clays include surface erosion, and particularly unusual gully erosion in natural slopes, cut slopes and embankments. Erosion along shrinkage cracks in embankments sometimes produces tunneling and jugging, common terms applied to piping and sinkhole development. In 1973, 20 dams in Mississippi were discovered to have unusually severe erosion problems from this cause, after a period of heavy rainfall following a dry period.

Probably even more important is the susceptibility of dispersive clays to piping at the seepage exits of the embankment, particularly where cracks (even minute cracks) provide avenues for concentrated seepage and dispersed clay removal. Piping failure of the dam can develop rapidly under these conditions.

An excellent treatise on dispersive clays is that by Sherard and Decker (1977). It is a collection of 32 research papers. Some of the conclusions of the editors are paraphrased as follows:

- Limited data indicate that dispersive soils are most commonly found as alluvium, lacustrine, slope wash, or weathered loessial deposits. How-

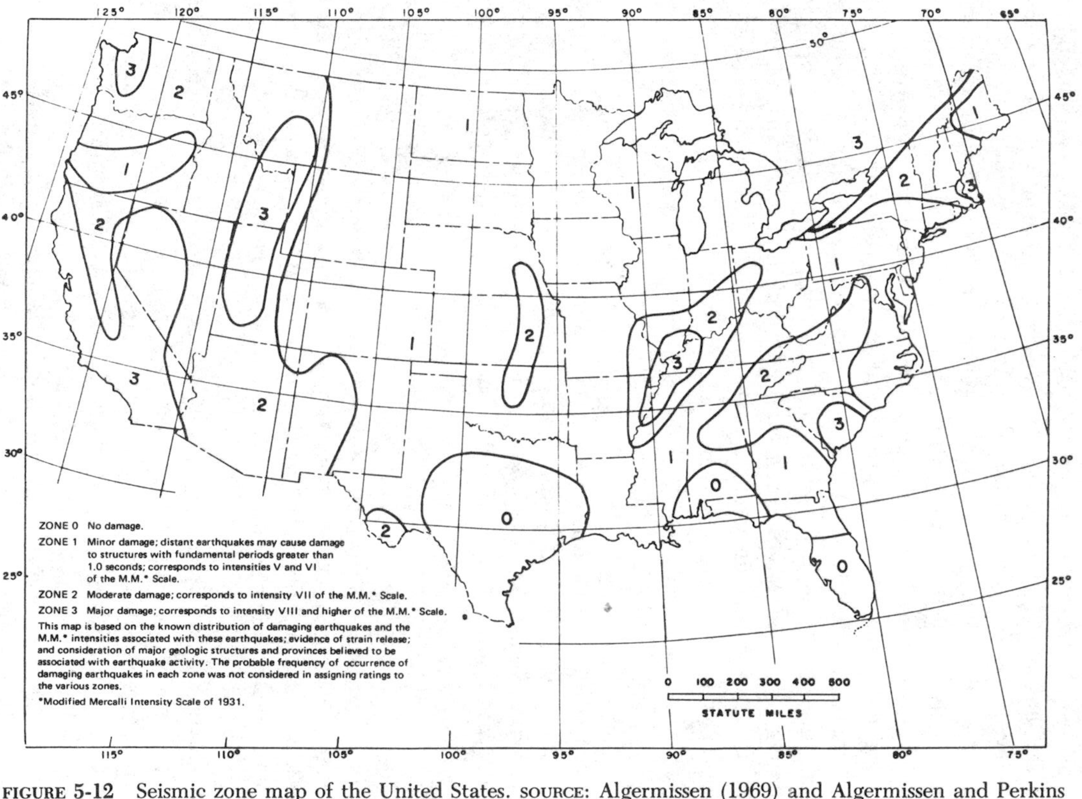
ZONE 0 No damage.
ZONE 1 Minor damage; distant earthquakes may cause damage to structures with fundamental periods greater than 1.0 seconds; corresponds to intensities V and VI of the M.M.* Scale.
ZONE 2 Moderate damage; corresponds to intensity VII of the M.M.* Scale.
ZONE 3 Major damage; corresponds to intensity VIII and higher of the M.M.* Scale.
This map is based on the known distribution of damaging earthquakes and the M.M.* intensities associated with these earthquakes; evidence of strain release; and consideration of major geologic structures and provinces believed to be associated with earthquake activity. The probable frequency of occurrence of damaging earthquakes in each zone was not considered in assigning ratings to the various zones.
*Modified Mercalli Intensity Scale of 1931.
0 100 200 300 400 500
STATUTE MILES

FIGURE 5-12 Seismic zone map of the United States. SOURCE: Algermissen (1969) and Algermissen and Perkins (1976).

ever, they have been found in residual soils developed from igneous, metamorphic, and sedimentary deposits. According to Sherard and Decker: "It must be anticipated that they could be found anywhere."

- A simple field test is the Emerson Crumb test, conducted by dropping a small ball of soil into a container of water and observing the relative speed of dispersion. However, although field tests may serve as indicators, none is absolutely conclusive in identifying highly dispersive soils.
- To determine the presence of dispersive clays, four tests should be performed: (1) the Pinhole test, (2) tests of dissolved salts in pore water (Na/total dissolved salts ratio), (3) SCS dispersion test, and (4) the Emerson Crumb test. No single one of these tests is always conclusive.
- There must be some exit of concentrated leakage (such as cracks) for piping to develop in dispersive clays.
- Piping can be prevented by installing a filter containing significant amounts of fine sand. Another method is to treat the upper first foot of the embankment with calcium (such as lime or gypsum).
- Erosion of the upstream face can be prevented by lime treatment or a protective blanket of nondispersive soil.

### Sources of Geologic and Pedological Soils Maps

Geologic maps covering the dam and surrounding areas may be obtained through the State Geological Surveys. These agencies have maps developed by their own staff, who are knowledgeable of other geologic maps that may be available from the U.S. Geological Survey, universities, and elsewhere.

Agricultural soils survey maps may be obtained from the local Soil and Water Conservation District, the U.S. Department of Agriculture Soil Conservation Service, or the local Agricultural Extension Agent.

## EARTHQUAKE CONSIDERATIONS

In any rock formation, high strains within the crust may induce stresses higher than the strength of the rock. This can result in a sudden release of stored energy that causes a fracture (fault) that may extend over a considerable area. For example, the 1906 San Francisco earthquake produced a break over 300 miles long. The energy from such a sudden rupture spreads out in all directions with decreasing ground motion at increasing distance from the break. Major earthquakes seem to occur where they have happened before. However, it is possible to have an earthquake almost anywhere. Hence, from studies of past earthquakes a given region can be identified as having a high or low probability of earthquake, such as is shown in Figures 5-12 and 5-13.

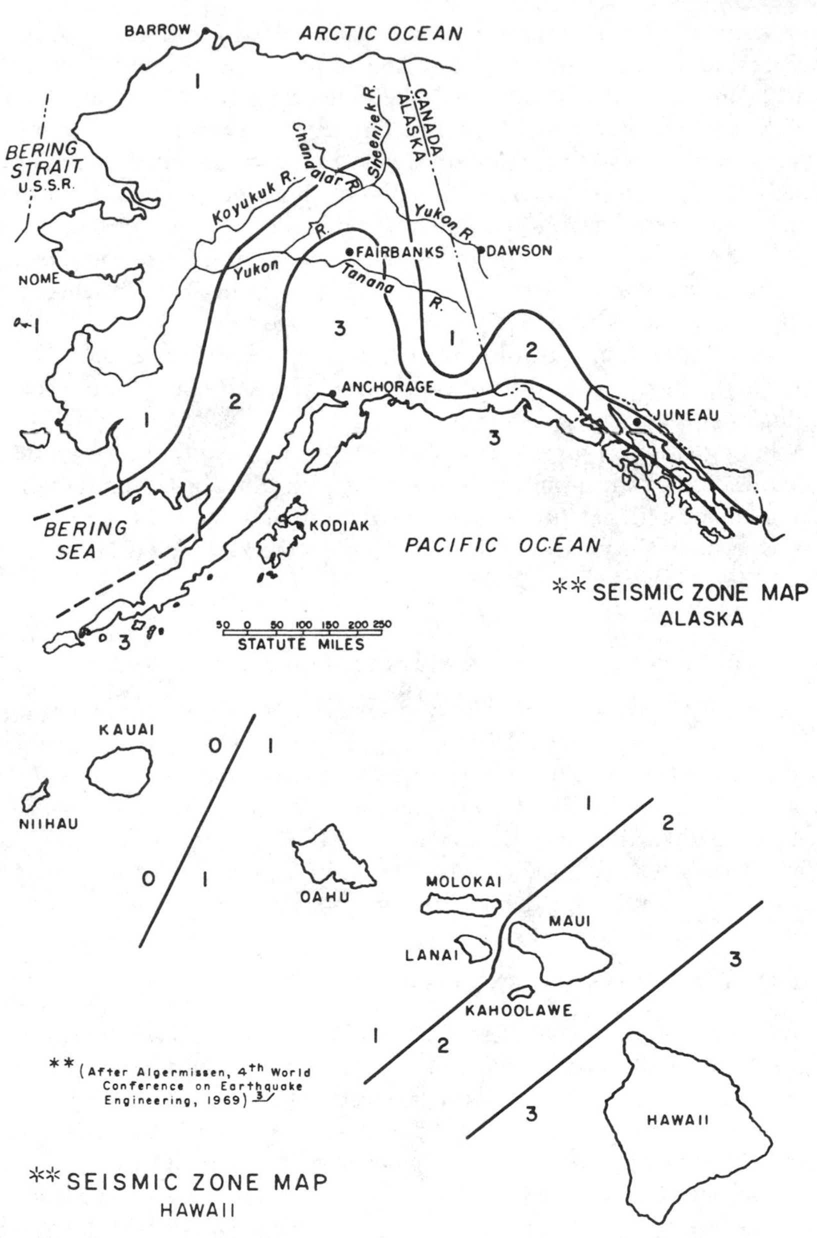


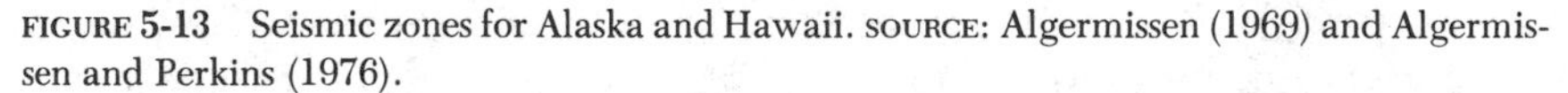
FIGURE 5-13 Seismic zones for Alaska and Hawaii. SOURCE: Algermissen (1969) and Algermissen and Perkins (1976).

At any location on the earth's surface remote from the actual source or epicenter of an earthquake, energy is felt as a series of shocks in all directions. From seismometers installed at various locations, a record of these shocks can be obtained in terms of the acceleration, usually expressed as a fractional equivalent of the acceleration of gravity, e.g., 0.20 acceleration due to gravity. The shocks can be expected to vary in intensity with time in an irregular manner. The greatest accelerations usually are in a horizontal direction, and, because of the inertia effect, a horizontal shear force is set up at the base of the dam and then successively upward throughout the dam as it responds to the shear force at the base. However, it is acknowledged that in some major events the vertical acceleration has exceeded the horizontal (Kerr 1980). If the dam is moving toward the reservoir the force exerted on the surface of the dam by the water is momentarily increased because of the inertia of the water. Conversely, when the dam moves away from the reservoir this hydrodynamic force tends to decrease the water pressure. Thus the effect of the earthquake in the dam is twofold: the body force due to the inertia of the dam and the hydrodynamic force caused by the interaction of the dam and reservoir.

## Ground Motion Analyses

To obtain data essential for estimating the performance of a dam during an earthquake, it is necessary to adopt a methodical approach, as is shown in Figure 5-14.

Determination of ground motions at a dam site requires estimates to be made of the following: (1) magnitude and epicenter of earthquakes that are expected to affect the site and (2) the ground motion that can be produced by the estimated earthquakes. Obtaining such information will require a detailed study of seismological records for the region in combination with a study of the regional geology and the immediate site geology. The regional studies should be on the area within a 200-mile radius of the dam. A first step is to determine the seismic zone within which the dam is located (see Figures 5-12 and 5-13).

A commonly used method of measuring and expressing ground motion involves the following four steps:

1. modified mercalli intensity (see Table 5-4),
2. individual ground motion parameters (maximum acceleration and velocity, predominant period of the wave, and duration of the strong shaking),
3. response spectra, and
4. accelerograms.

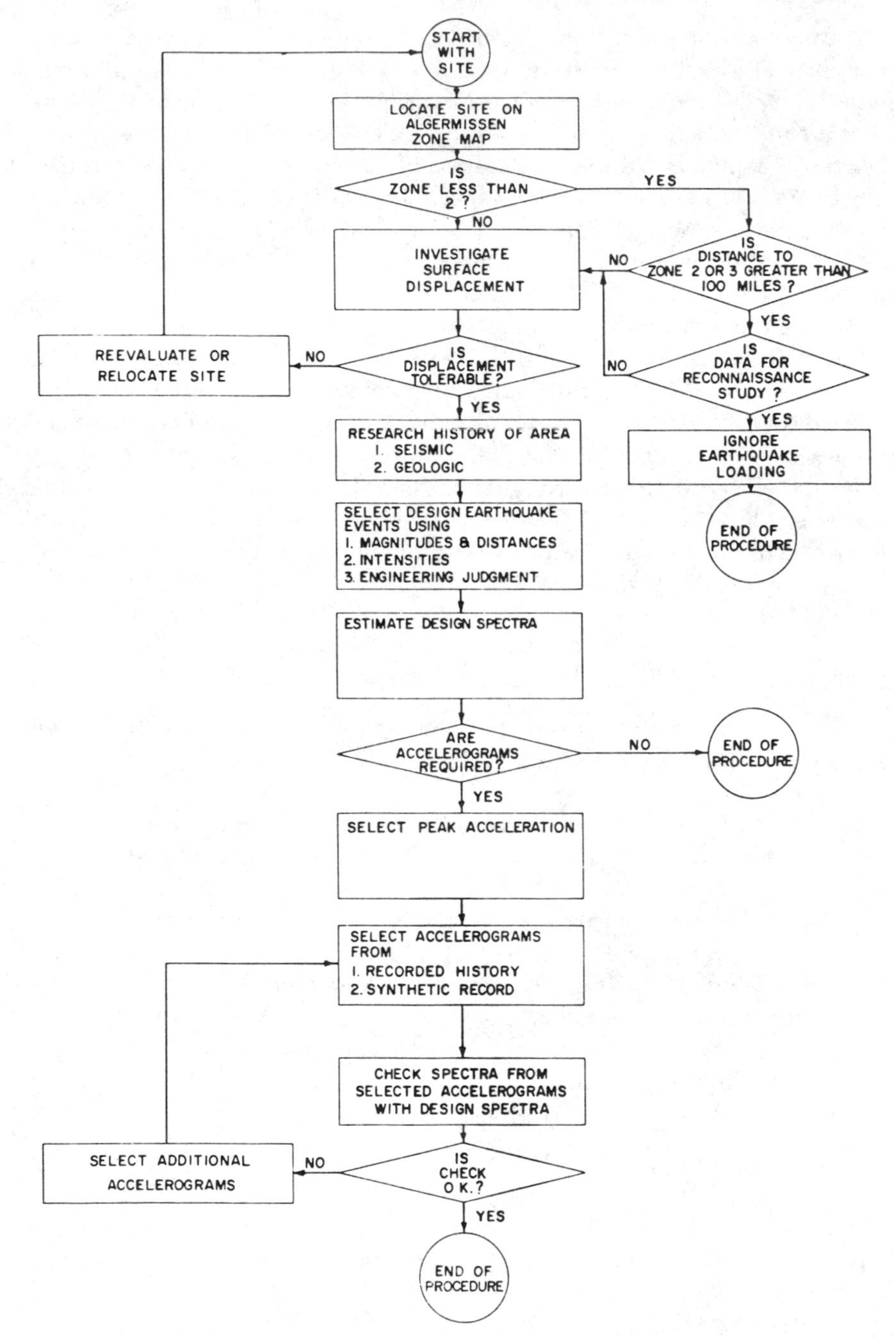


FIGURE 5-14 Procedure to determine ground motion at a site. SOURCE: Boggs et al. (1972).

Intensity data can be developed by interviewing people that live within the 200-mile radius. From them it is necessary to obtain descriptions of any damage to structures, any earthquake movements they felt while living in the region, and displacement or damage that these movements may have caused to objects within residences or other buildings.

Acceleration is one of the ground motion parameters and has an approximate relationship to magnitude and intensity (Table 5-4). Generally acceleration does not indicate the frequency or the duration of the shaking. Therefore, it is necessary to consider acceleration period and duration in order to describe acceleration. Numerous methods have been proposed for this purpose. The objective is, in terms of acceleration, to identify the maximum credible earthquake (MCE) and the operating basis earthquake (OBE). Both the MCE and the OBE are considered in the design and evaluation of dams. The latter is some fraction of the MCE, perhaps one-half. The OBE may be selected on a probabilistic basis from regional and local geology and seismology studies as being likely to occur during the life of the project. Generally, it is at least as large as earthquakes that have occurred in the seismotectonic province in which the site is located. Under the MCE the dam, if correctly designed, would suffer a small amount of damage, but there would be no release of reservoir water. Under the OBE there should be no permanent damage, and the dam should be able to resume operation with a minimum of delay after the earthquake. A first estimate of maximum acceleration can be obtained from the USGS Open File Report 76-416. A more precise determination of the maximum acceleration to be expected at a given locality requires the combined consideration of the following:

1. the earthquake record for the region,
2. the length and the depth of all major faults,
3. whether the foundation material is rock or soil, and
4. the distance of the dam site from the faults.

After this information has been obtained, it is necessary to make three independent determinations to estimate the effect of an earthquake on a specific dam:

1. the amount of earthquake force,
2. the maximum stresses in the dam, and
3. the material strength required to resist these stresses.

These determinations can be accomplished in two steps: (1) a Phase I analysis that uses a series of empirical parameters in a pseudo-static linear analysis (this gives a comparatively quick approximation of behavior expressed as maximum stresses or safety factors) and (2) a Phase II analysis

**TABLE 5-4** Approximate Relationships: Earthquake Intensity, Acceleration, and Magnitude

Modified Mercalli Intensity Scale

I Not felt. Marginal and long-period effects of large earthquakes.
II Felt by persons at rest, on upper floors, or favorably placed.
III Felt indoors. Hanging objects swing. Vibration like passing of light trucks. Duration estimated. May not be recognized as an earthquake.
IV Hanging objects swing. Vibration like passing of heavy trucks. Standing motor cars rock. Windows, dishes, doors rattle. Glasses clink. In upper range of IV, wooden walls and frame crack.
V Felt outdoors; direction estimated. Sleepers wakened. Liquids disturbed. Doors swing. Shutters, pictures move. Pendulum clocks stop, start, change rate.
VI Felt by all. Many frightened and run outdoors. Persons walk unsteadily. Windows, dishes, glassware broken. Books off shelves. Pictures off walls. Furniture moved or overturned. Weak plaster and adobe crack. Small bells ring.
VII Difficult to stand. Noticed by drivers of motor cars. Fall of plaster, loose bricks, tiles, etc. Some cracks in masonry. Waves on ponds; water turbid. Small slides, caving of sand or gravel banks. Large bells ring. Concrete irrigation ditches damaged.
VIII Steering of motor cars affected. Damage to masonry; some partial collapse. Twisting, fall of chimneys, monuments, elevated tanks. Frame houses moved if not bolted down. Branches broken. Changes in springs and wells. Cracks in wet ground and on steep slopes.
IX General panic. Weak masonry destroyed, good masonry seriously damaged. Frame structures, if not bolted, shift off foundations. Frames racked. Serious damage to reservoirs. Underground pipes broken. Ground cracked, sand and mud ejected, earthquake fountains, sand craters.
X Most masonry and frame structures destroyed with their foundations. Serious damage to embankments. Large landslides. Water thrown on banks of canals, rivers, lakes, etc. Rails bent slightly.
XI Rails bent greatly. Underground pipelines completely out of service.
XII Damage nearly total. Large rock masses displaced. Lines of sight and level distorted. Objects thrown into the air.

SOURCE: Boggs et al. (1972).

that will focus on questionable details and determine maximum stresses more accurately. For most dams the Phase I analysis should be sufficient. In borderline cases the Phase II analysis may be required.

As previously noted, there are numerous methods for estimating ground motion values for magnitude and distance from a fault. A system that has been used frequently by the U.S. Bureau of Reclamation is based on work by Schnabel and Seed (1972). The relationship between acceleration and fault location is demonstrated by the curves in Figure 5-15.

Because of the complex nature of earthquake-induced ground motion and its interaction with structures, the concept of response spectrum has been developed. This is a plot of the maximum values of acceleration, velocity, and displacement that will be experienced by a family of single-degree-of-freedom systems subjected to a time-history of ground motion. Maximum values of the parameters are expressed as a function of the natural period in damping of the system. The response spectrum for a given earthquake can be estimated directly if the magnitude and distance or the maximum ground acceleration, velocity, and displacement are known. It is

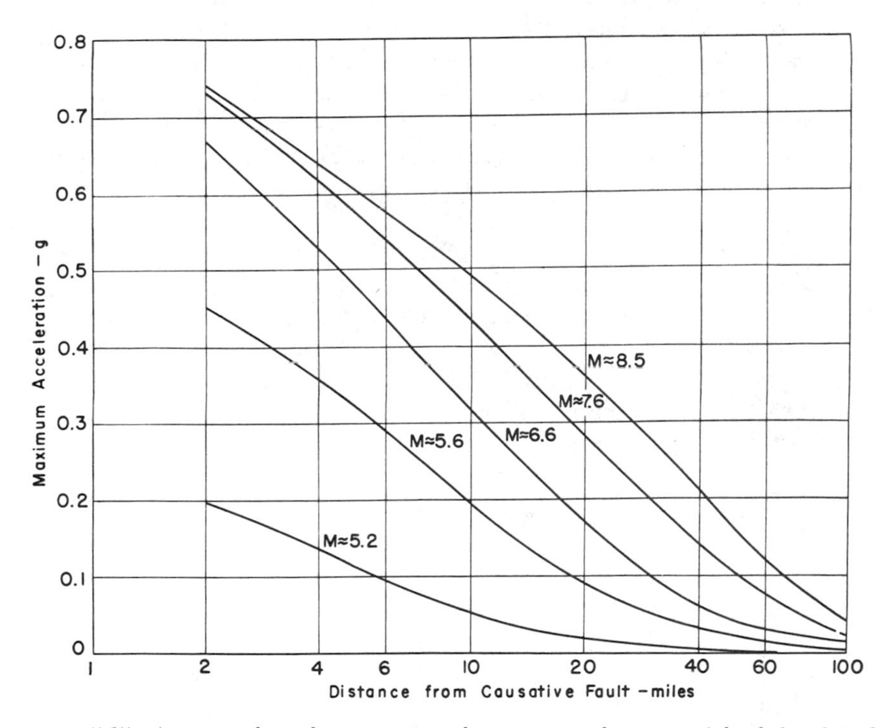


FIGURE 5-15 Average values of maximum accelerations in rock. SOURCE: Schnabel and Seed (1972) and Boggs et al. (1972).

also possible to estimate the response spectrum indirectly from scaled accelerograms. If only intensity data are available, an estimate must be made of the magnitude and distance of an earthquake that would produce the predicted site intensity. In these determinations the selected accelerogram records should be compatible with ground motions expected from the design earthquake. The accelerograms can be obtained from historical records, from artificially generated accelerograms, or from specialists in this type of analysis. In all cases the records should be evaluated by specialists in the field of earthquake engineering. The first step in determining ground motion is to locate the most recently available map. Estimates of the MCE and OBE from these maps are indicated in Table 5-5.

Magnitude estimates can be made directly from instrument and intensity data and surface faulting. Indirect estimates can be developed from the site intensity and acceleration relationships (plus an attenuation factor) that determine the epicentral intensity. The design accelerograms are obtained by selecting real or synthetic accelerograms that can be adjusted to approximate the peak acceleration. Response spectra from design accelerograms must be similar to the design response spectra. Thus, several design accelerograms should be used for the analysis of any structure. Behavior of earth embankments and mechanical equipment depends not only on the magnitude of the seismic event but also on its duration. This means that the design and condition of the structures must be considered along with the response spectrum. A typical response spectrum is shown in Figure 5-16, which compares the natural period of the earthquake with the maximum acceleration.

The earthquake magnitude can be correlated in a general way with the length of rupture along a fault. Generally, the greater the rupture length the larger will be the earthquake magnitude. Figure 5-17 depicts such relationships for a number of specific earthquakes and faults. (Each number on the figure refers to a specific fault and associated earthquake.) As can be seen on this figure, there is considerable scatter to the data; this indicates that expert knowledge is required to extrapolate this type of information

**TABLE 5-5** Earthquake Acceleration

| Region | MCE | OBE |
|---|---|---|
| 0 | 0 g | 0 g |
| 1 | 0.1 g | 0.05 g |
| 2 | 0.2 g | 0.1 g |
| 3 | 0.4 g | 0.2 g |

NOTE: g = acceleration due to gravity.

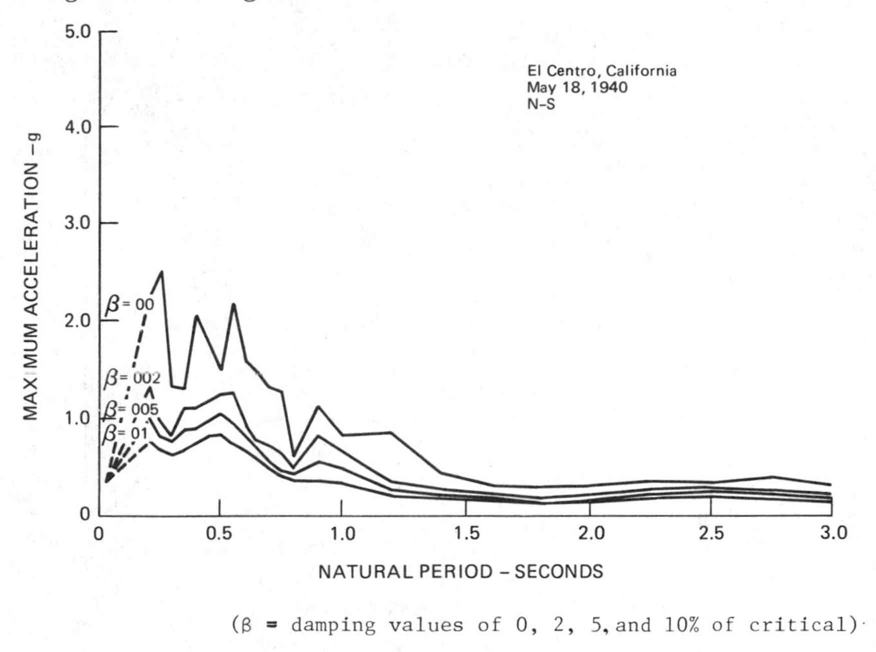


FIGURE 5-16 Linear plot of response spectra. SOURCE: Boggs et al. (1972).

for any specific dam site. A detailed evaluation of the relationship between faults and earthquake magnitude is given in Slemmons (1977).

*Analysis.* The analysis of the behavior of a dam under the selected ground motions requires considerable experience and specialization. Generally speaking there are two methods used: (1) the quasi-static method wherein the behavior of the dam is studied under the action of the reservoir-water forces that are induced by an earthquake and (2) the dynamic analysis that considers both maximum horizontal and vertical accelerations and the frequency components. The second method requires a more complex and costly analysis that would be used only where the cost and type of structure would justify it. The objective of such analyses is to determine the dam response to different types of earthquake loadings. For example, in an arch dam, major consideration is given to the various movements of the rings and cantilevers that constitute an arch dam; for a gravity dam, attention also must be given to the shearing forces within the dam or along its base. For a concrete dam the forces are applied in upstream, downstream, and vertical directions in a mode that will produce stress everywhere in the structure and the maxima of these stresses are collected for study. For earth

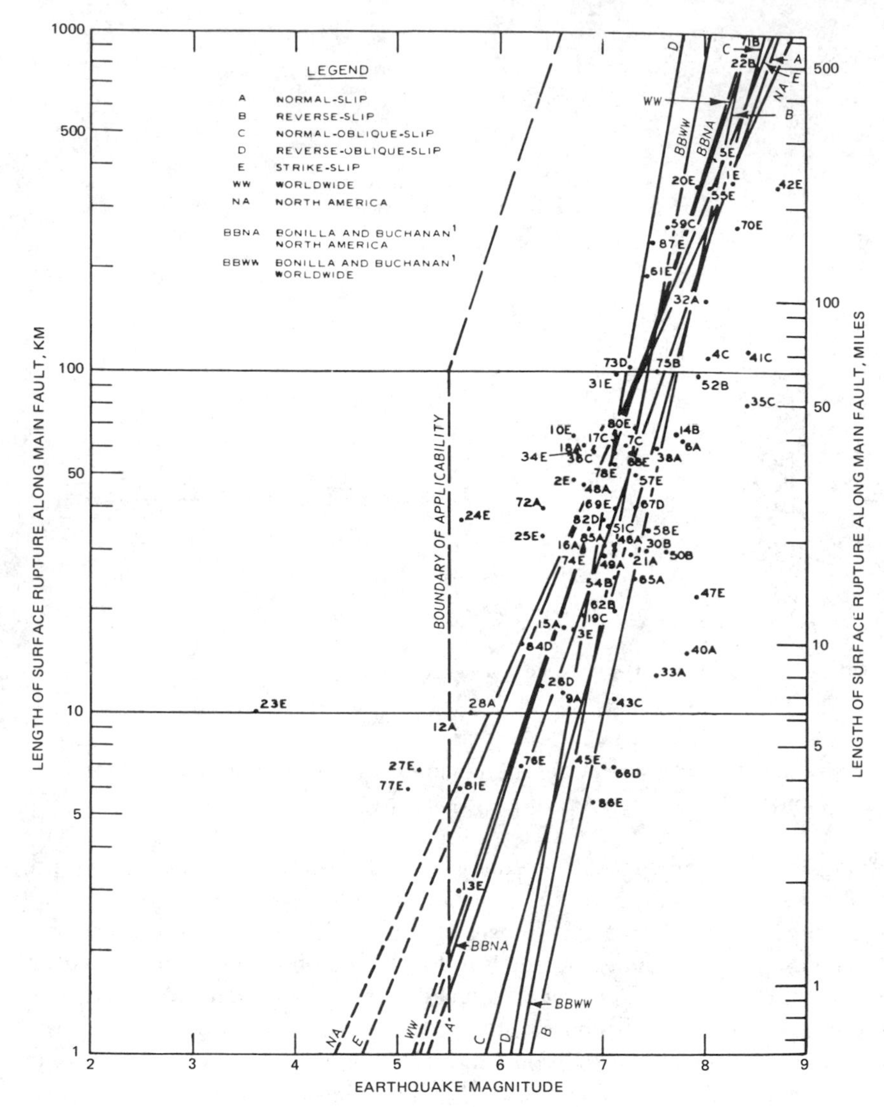


FIGURE 5-17 Relationship of earthquake magnitude to length of surface rupture along the main fault zone. SOURCE: Slemmons (1977).

dams the choice of method is more involved, and the reader is referred to the detailed discussions presented in Seed (1979) and Seed (1983) for a presentation of the latest state-of-the-art approaches.

The above information and procedures are relevant to the Phase I analysis. In the Phase II analysis a much more detailed study of dam behavior must be made. For example, the following information is needed:

1. Location of all active or capable faults in the vicinity.
2. The length and typical depth of each fault.
3. The product of the values in step 2 to find the energy released from the fault.
4. The attenuated energy received at the dam site (the perpendicular distance from the dam site to the fault is used).
5. The MCE, which will be the maximum of any of the energies determined in step 4.
6. The OBE, which can be estimated at one-half MCE.

The duration of strong shaking can be estimated from empirical data, such as shown in Figure 5-18. The shaking duration is generally assumed to continue so long as faulting is occurring.

Attenuation is the relationship between the amount of energy produced at the source (hypocenter) of the earthquake and the amount of energy available at some specific distance from this source. It is usually expressed as an attenuation factor, which is the acceleration at a site divided by the acceleration at the epicenter. Several approaches have been used, and the relatively wide range of these results is indicated in Figure 5-19. Generally, the empirical studies indicate that the range can be from a rapid attenuation in the first 20 miles to a more gradual attenuation at greater distances from the epicenter. At large epicentral distances the attenuation factors

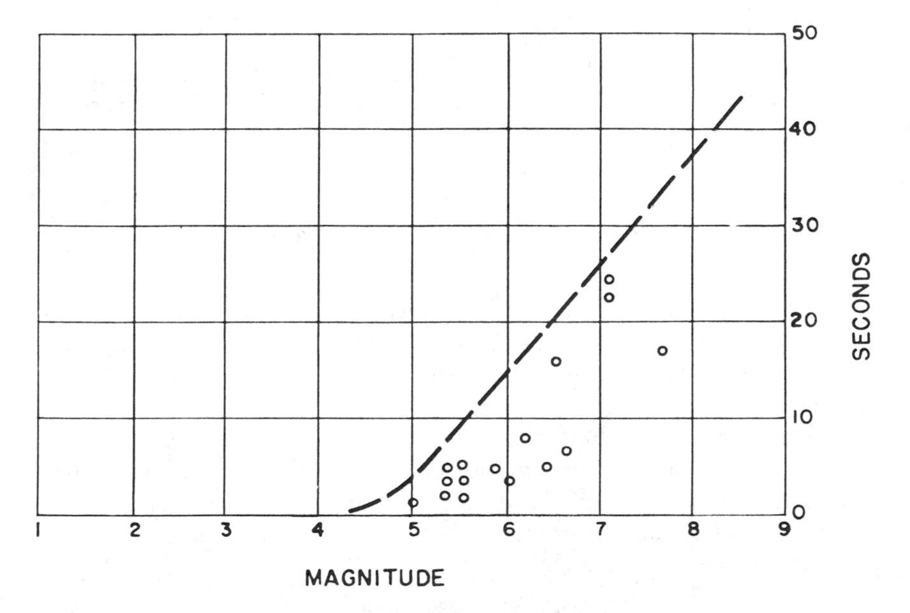


FIGURE 5-18 Duration of strong shaking. SOURCE: Boggs et al. (1972).

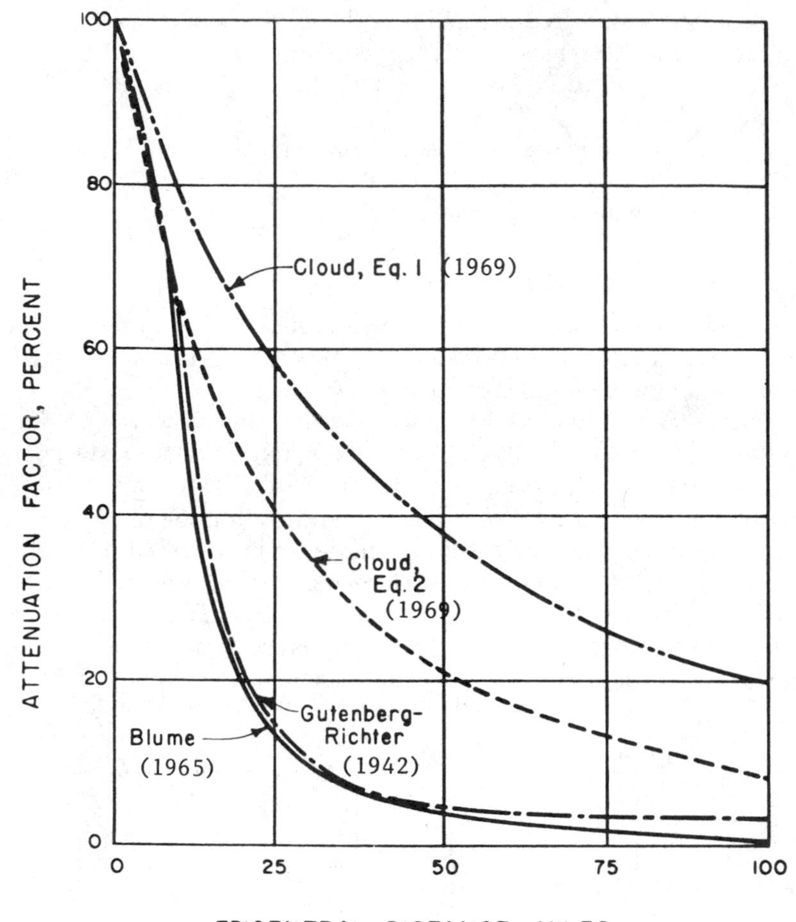

FIGURE 5-19 Attenuation factor versus distance. SOURCE: Boggs et al. (1972).

may vary by several magnitudes. This means that engineering judgment must be used to consider the effects of focal depth, unusual geologic structure, unequal distribution of energy radiation with direction from the epicenter, etc.

To determine the possible frequency of earthquake occurrence at a site it is necessary to use probability methods; the input data are obtained from existing earthquake records. There is some agreement (Boggs et al. 1972) that there is a definable relationship between earthquake magnitude and frequency of occurrence, such as:

$$\log N = a - bM$$

where $N$ equals the number of earthquakes, $M$ equals magnitude, and $a$ and $b$ are constants established by observations.

## REFERENCES


Algermissen, S. T. (1969) *Seismic Risk Studies in the United States*, 4th World Conference on Earthquake Engineering, Santiago, Chile, Vol. I.

Algermissen, S. T., and Perkins, D. M. (1976) *A Probabilistic Estimate of Maximum Acceleration in Rock in the Contiguous United States*, U.S. Geological Survey Open File Report 76-416.

Beene, R. R. W. (1967) "Waco Dam Slide," *Journal of the Soil Mechanics and Foundation Division*, Proceedings of the ASCE, Vol. 92, No. SM4, July.

Bellier, J. (1976) *The Malpasset Dam*, The Evaluation of Dam Safety, Engineering Foundation Conference Proceedings, November 28–December 3, ASCE.

Blume, J. (1965) "Earthquake Ground Motion and Engineering Procedures for Important Installations Near Active Faults", Proceedings 3rd World Conference on Earthquake Engineering, New Zealand, Vol. III.

Boggs, H. L., et al. (1972) *Method for Estimating Design Earthquake Rock Motions*, U.S. Bureau of Reclamation Engineering and Research Center, Denver, Colo. (revised).

Cloud, W. K., and Perez, V. (1969) "Strong Motion Records and Acceleration," *Proceedings, 4th World Conference on Earthquake Engineering*, Chile, Vol. I.

Gardner, Charles H., and Tice, T. Allan (1979) *The Forrest City Creep Landslide on US-212, Oahe Reservoir, Missouri River, South Dakota*, Proceedings, National Highway Geology Symposium.

Gutenberg, B. and Richter, C. F. (1942) "Earthquake Magnitude, Intensity, Energy and Acceleration," *Bulletin Seismological Society of America*, Vol. 32, No. 3, pp. 163–191.

Hunt, C. B. (1974) *Natural Regions of the United States and Canada*, W. H. Freeman, New York.

Independent Panel to Review the Cause of Teton Dam Failure (1976) *Failure of Teton Dam*, Report to U.S. Department of Interior and State of Idaho.

Jansen, R. B., Dukleth, G. W., Gordon, B. B., James, L. B., and Shields, C. E. (1967) "Earth Movement at Baldwin Hills Reservoir," *Journal of the Soils Mechanics and Foundation Division*, Proceedings of the ASCE, Vol. 93, No. SM4, July.

Judd, W. R. (1969) *Statistical Methods to Compile and Correlate Rock Properties and Preliminary Results*, Purdue University Technical Report No. 2, Office of Chief of Engineers, Department of the Army, NTIS AD701086.

Judd, W. R. (1971) *Statistical Relationships for Certain Rock Properties*, Purdue University Technical Report No. 6, Office of Chief of Engineers, Department of the Army, Washington, D.C.

Kerr, R. A. (1980) "How Much Is Too Much When the Earth Quakes?" *Science*, Vol. 209, August.

Krynine, D. P., and Judd, W. R. (1957) *Principles of Engineering Geology and Geotechnics*, McGraw-Hill, New York.

Legget, R. F., ed. (1961) *Soils in Canada*, The Royal Society of Canada Special Publications, No. 3, University of Toronto Press.

Leps, T. M. (1972) *Analysis of Failure of Baldwin Hills Reservoir*, Proceedings, Specialty Conference on Performance of Earth and Earth-Supported Structures, Purdue University and ASCE.

Morrison, P., Maley, R., Brady, G., and Porcella, R. (1977) *Earthquake Recordings On or Near Dams*, U.S. Committee on Large Dams Committee on Earthquakes, California Institute of Technology.

Schnabel, P. B., and Seed, H. B. (1972) *Accelerations in Rock for Earthquakes in the Western United States*, Earthquake Engineering Research Center Report 72-2, University of California, Berkeley.

Seed, H. B. (1979) "Considerations in the Earthquake-Resistant Design of Earth and Rockfill Dams," *Geotechnique*, Vol. 29, No. 3, pp. 215–263.

Seed, H. B. (1983) "Earthquake-Resistant Design of Earth Dams" in Seismic Design of Embankments and Caverns, Proceedings of a Symposium Sponsored by the ASCE Geotechnical Engineering Division in conjunction with the ASCE National Convention, Philadelphia, Pa., May 16–20, 1983 (Terry R. Howard, ed.).

Sherard, J. L., and Decker, R. S., eds. (1977) *Dispersive Clays, Related Piping, and Erosion in Geotechnical Projects*, ASTM Special Technical Publication 623, American Society for Testing and Materials.

Slemmons, D. B. (1977) *Faults and Earthquake Magnitude*, U.S. Army Engineer Waterways Experiment Station Miscellaneous Paper S-73-1, Vicksburg, Miss.

Sowers, G. F. (1971) *Introductory Soil Mechanics and Foundations*, Macmillan Publishing Co., New York.

Thomas, H. E., Miller, E. J., and Speaker, J. J. (1975) Difficult Dam Problems—Cofferdam Failure," *Civil Engineering*, August.

Touloukian, Y. S., Judd, W. R., and Roy, R. F. (1981) *Physical Properties of Rocks and Minerals*, McGraw-Hill/Cindas Data Series on Material Properties, Vol. II-2, McGraw-Hill, New York.

U.S. Bureau of Reclamation (1974) *Earth Manual*, 2d ed., Government Printing Office, Washington, D.C.

U.S. Department of Agriculture, Soil Conservation Service (1975) *Engineering Field Manual for Conservation Practices*, NTIS Publ. PB 244668.

## RECOMMENDED READING

Buol, S. W., Hole, F. D., and McCracken, R. J. (1973) *Soil Genesis and Classification*, Iowa State University Press, Ames.

Gary, M., McAffee, R., Jr., and Wolf, C. L., eds. (1972) *Glossary of Geology*, American Geological Institute, Washington, D.C.

Hunt, C. B. (1977) *Geology of Soils*, W. H. Freeman, New York.

Jaeger, C. (1972) *Rock Mechanics and Engineering*, Cambridge University Press.

Legget, R. F. (1962) *Geology and Engineering*, McGraw-Hill, New York.

National Academy of Sciences, Committee on Seismology (1976) *Predicting Earthquakes*, NAS, Washington, D.C.

Sopher, C. D., and Baird, J. V. (1982) *Soils and Soil Management*, Reston Publishing Company, Reston, Va.

Terzaghi, K., and Peck, R. B. (1967) *Soil Mechanics in Engineering Practice*, John Wiley & Sons, New York.

Woods, K. B., Miles, R. D., and Lovell, C. W., Jr. (1962) "Origin, Formation, and Distribution of Soils in North America" in *Foundation Engineering*, G. A. Leonards, ed., McGraw-Hill, New York.

# 6
# Concrete and Masonry Dams

## GRAVITY DAMS

Gravity dams (see Figure 6-1) are the most common of the concrete and masonry types and the simplest type to design and build. A gravity dam depends on its weight to withstand the forces imposed on it. It generally is constructed of unreinforced blocks of concrete with flexible seals in the joints between the blocks. The most common types of failure are overturning or sliding on the foundation.

The foundation for a gravity dam must be capable of resisting the applied forces without overstressing of the dam or its foundation. The horizontal forces on the dam tend to make it slide in a downstream direction, which results in horizontal stresses at the base of the dam. These in turn may try to induce shear failure in the concrete at the base or along the concrete-rock contact or within the rock foundation. Uplift forces, in combination with other loads, tend to overturn the dam, which in turn may cause crushing of the rock along the toe of the dam.

There are a number of older dams in existence constructed of rock and cement or concrete masonry. These generally have been relatively small and are usually of some form of gravity-type configuration. Their greatest weakness generally lies in the tendency for the masonry or cement between blocks to deteriorate with resultant leakage, deformation, and general disintegration.

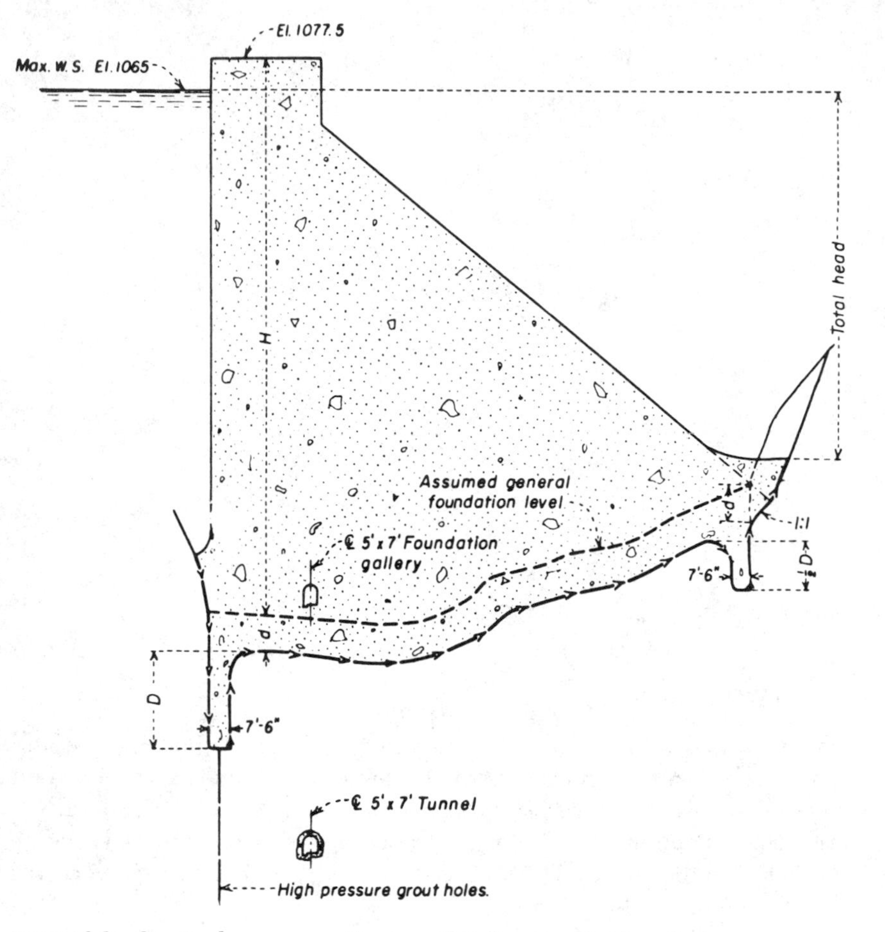

FIGURE 6-1 Gravity dam. SOURCE: Courtesy, U.S. Bureau of Reclamation.

## Buttress Dams

This is a form of gravity (see Figure 6-2) dam so far as the force distribution is concerned. It consists of a sloping slab of concrete that rests on vertical buttresses. Because of its shape there are high unit loads underneath the buttresses; thus, the foundation must not undergo unacceptable settlement or shearing.

In addition to the factors mentioned for gravity dams, particular attention must be paid to the quality and performance of the concrete in the face slab. Because of its relative thinness it cannot withstand excessive deterioration, pitting, or spalling that will decrease the strength of the slab and increase its potential for seepage through the concrete. The buttresses also must be designed to withstand overturning forces. If their footings are too small, the resulting high unit loads can induce crushing in the rock.

Because of their shape, buttress dams usually do not require extensive, if any, drainage systems, and drainage galleries within the dam would not be feasible.

## Arch Dams

Arch dams (see Figure 6-3) are relatively thin compared with gravity dams. The forces imposed on such a dam are, for the most part, carried into the abutments, and the foundation is required only to carry the weight of the structure. The shape of the dam may resemble a portion of a circle, an

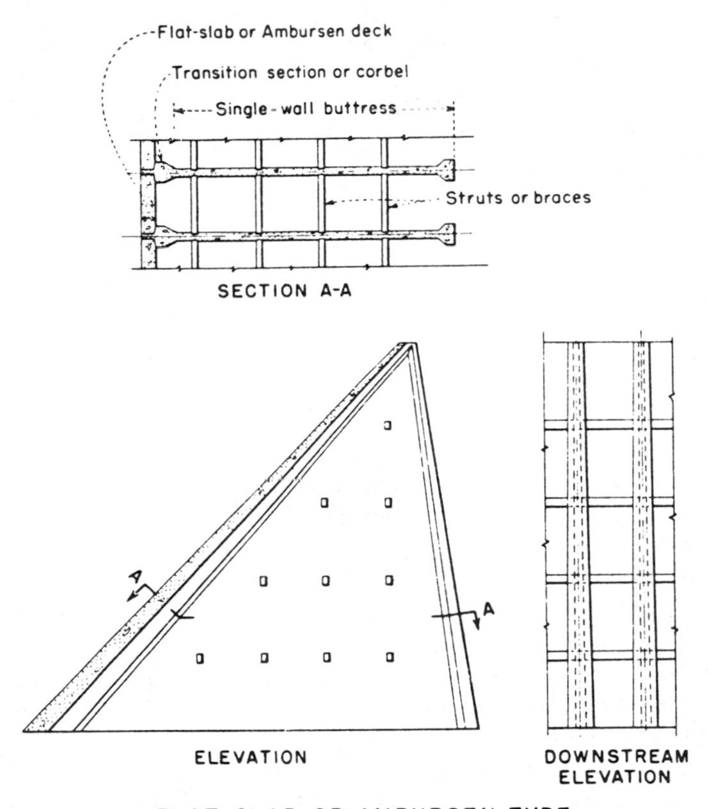


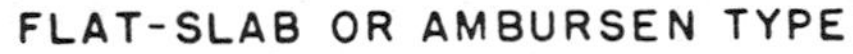


FIGURE 6-2 Simple buttress dam. SOURCE: Courtesy, U.S. Bureau of Reclamation.

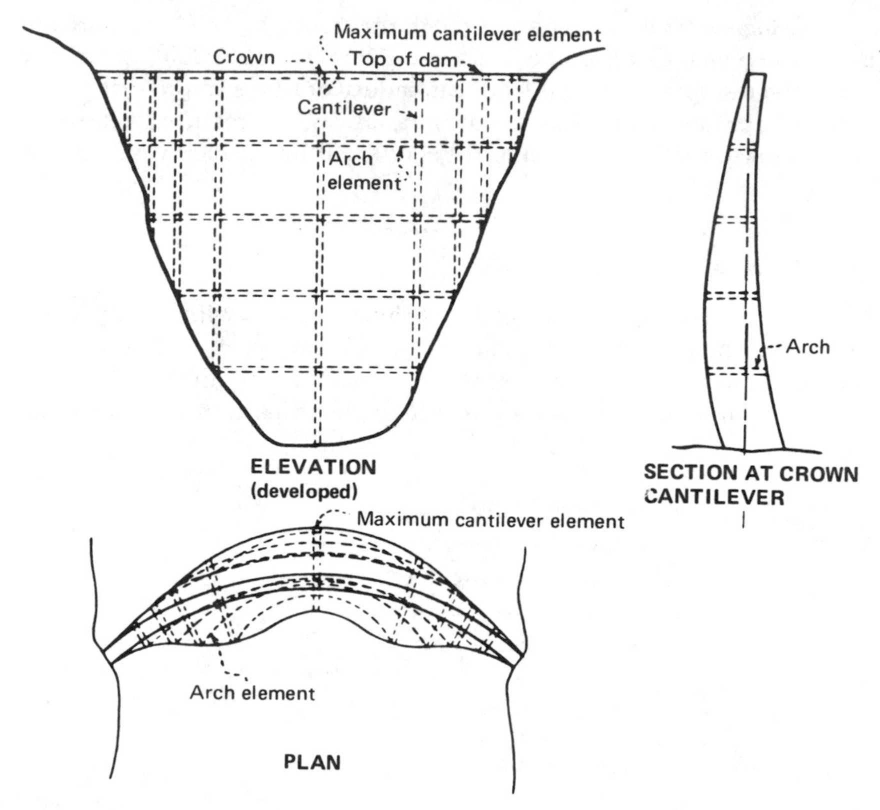


FIGURE 6-3 Plan, profile, and section of a symmetrical arch dam. SOURCE: Courtesy, U.S. Bureau of Reclamation.

ellipse, or some combination thereof. The dam usually is constructed of a series of relatively thin blocks that are keyed together (see Figure 6-4). The construction joints that result may be grouted during or after construction or left open. In the latter case it is expected they will close under the reservoir load. Occasionally, flexible seals may be installed in the vertical joints between the blocks.

Because of the translation of imposed forces into the abutments, the design must consider the amount of deformation (modulus of deformation) that will occur in the abutments when the various loads are imposed on the dam. If the deformation exceeds design criteria, tension cracking can occur in the concrete. (See the section Abutment or Foundation Deformation.) Because the design is predicated on the flexibility of an arch, it is generally desirable that the modulus of elasticity of the rock abutments be less than that of the dam concrete.

Although controversial, some designs do consider the possibility of uplift. Thus, there may be drainage galleries and their appurtenant drain holes within the dam; drainage galleries and drain holes are generally installed in the abutments.

Possible failure modes in an arch dam are overturning, excessive abutment movement causing tension cracks in the concrete and subsequent rupture of the dam, mass movement of the abutments causing dam failure or disruptive stresses in the dam, and excessive uplift in the foundation that causes movement of rock blocks in the foundation and/or overturning of the dam.

### Arch-Gravity Dams

In arch-gravity dams imposed loads are carried partially by the foundation and partially by the abutments. These dams are of block construction and have a cross section that has a mass somewhere between that of an arch and

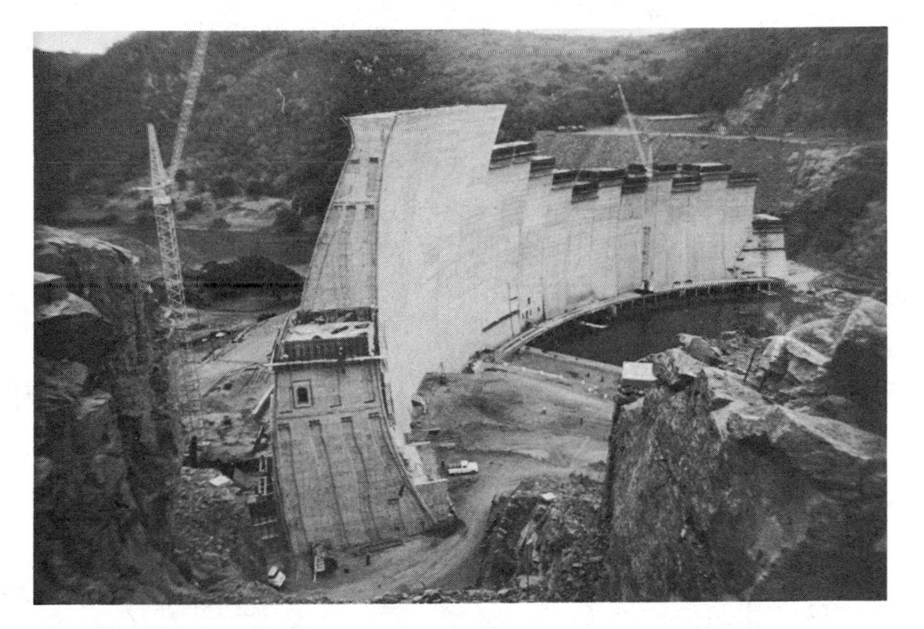

FIGURE 6-4 Concrete arch dam under construction; shows keys between blocks.

that of a gravity dam. The comments made earlier for arch and gravity dams are applicable to this type of structure, too.

### Miscellaneous Types

Various combinations of the types of dams described above may be designed for unique site situations. These include multiple arch (see Figure 6-5), multiple dome, compound arch, and gravity-buttress. The type of dam indicates the mode of distribution of the forces imposed on it.

## COMMON DEFECTS AND REMEDIES

The following discussions are intended, first, to emphasize the defects and remedies that generally could be relevant to any type of concrete dam and,

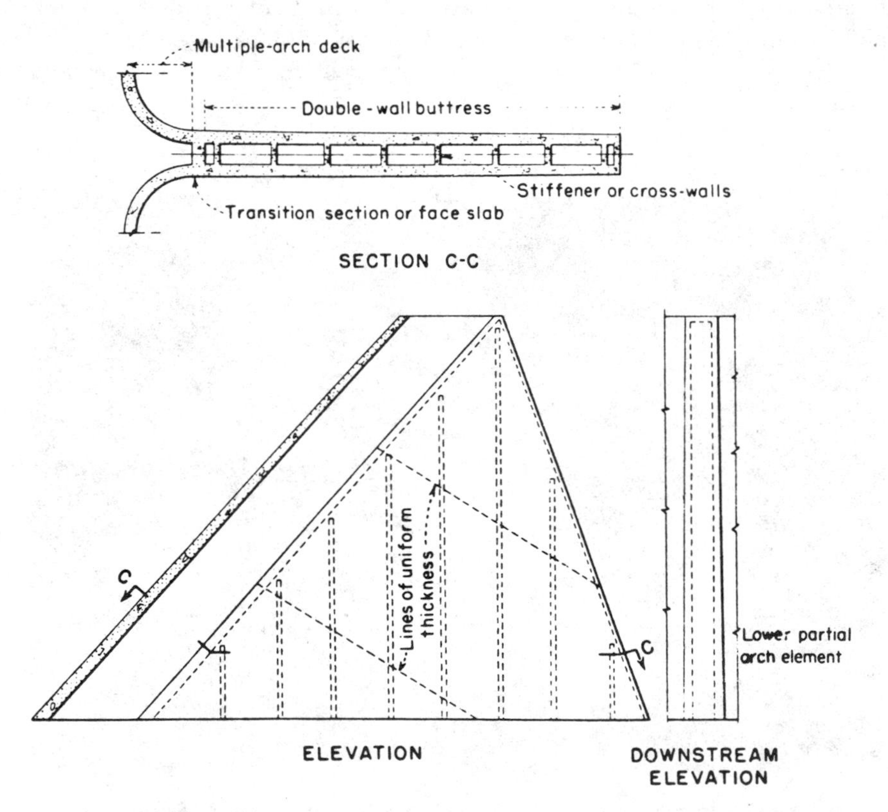


FIGURE 6-5 Double-wall buttress multiple-arch type. SOURCE: Courtesy, U.S. Bureau of Reclamation.

second, to indicate those remedies that are applicable primarily to a specific type of dam. A summary of the discussions is presented in a matrix format in Table 6-1.

## Abutments

### *Joints, Fractures, Faults, and Shear Zones*

The orientation of major discontinuities in abutments is critical in relation to the distribution of stresses from an arch dam but not as critical for a gravity structure. For an arch dam the main consideration is whether the direction of such discontinuities is parallel to or closely parallel to the directions of thrust from an arch (see Figure 6-6). If so, movements can occur that would result in weakening or possible loss of large blocks in the abutment. For a gravity dam the potential for sliding may be greatest when the foundation rock has horizontal bedding, particularly where combined with slick bedding planes. Consideration also must be given to a zone within the foundation rock that is peculiarly susceptible to the development of unacceptable uplift forces.

The presence and behavior of large faults or shear zones in those abutment areas within the zone of stress influence of the structure is of potential concern. Mass abutment movement may occur because percolation of water through these zones or water-softening of the rock material may reduce the shearing strength or cause consolidation of the rock. If at the upstream side of the dam the zone is more pervious than at the downstream side, uplift or pressure buildup can occur.

### *Seepage or Leakage*

Seepage developing in the abutments for any type of concrete dam can produce a critical condition. It usually is associated with fractures or shear zones. Of particular note is whether such seepage at the outlet is clear or contains silt or rock fragments. If the water is cloudy, silty, or muddy the water flow may be eroding the rock material itself or washing out clay or other impervious material that has been in the rock cracks. Continuation of this process (piping) can weaken the overall strength of the abutment or can produce increasingly large channels for water flow. If left untreated, the openings can enlarge sufficiently to cause abutment collapse or major movement of the abutment with the creation of unacceptable stresses in the body of the arch. Clear water leakage may be of concern if the quantity represents an unacceptable loss of reservoir storage, or the water may lubricate rock surfaces or reduce the strength of the rock element or discontinui-

**TABLE 6-1** Evaluation Matrix of Masonry Dams

| Indicator | Possible Causes | Possible Effects | Potential Remedial Measures (listed roughly in order of recommended action) |
|---|---|---|---|
| (A) Concrete (general)<br>Cracking (shallow)<br>Crazing<br>Spalling | Freeze-thaw cycling<br>Reactivity<br>Sulfate attach<br>Leaching<br>Aging | Accelerated deterioration<br>Reduction of allowable stresses<br>Reduction of effective section<br>Increased stresses<br>Loss of weight<br>Increased leakage | Determine concrete qualities by testing.<br>Coring<br>Petrographic<br>Density<br>Sonic (geophysical)<br>Porosity and permeability<br>Impact<br>Modulus of elasticity<br>Determine loss of section and weight.<br>Perform stress/stability analysis.<br>Protect (seal) surfaces from exposure and water.<br>Coatings<br>Gunite<br>Concrete<br>Steel<br>Remove and replace affected sections if cost-effective and if moisture can be kept out.<br>Remove (in extreme cases only). |
| (B) Concrete (local)<br>Spalling and cracking | Stress concentrations<br>Freeze-thaw action<br>Differential movement | Progressive deterioration<br>Increase leakage<br>Loss of section<br>Stress concentrations | Conduct survey and establish movement monitoring system.<br>Install pins, monuments, or other devices to accurately measure opening and closing of joints.<br>Determine quality of deteriorated concrete similarly to (A).<br>Remove and repair deteriorated sections.<br>Protect other surfaces with coatings or cover. |

**TABLE 6-1** Evaluation Matrix of Masonry Dams (*continued*)

| Indicator | Possible Causes | Possible Effects | Potential Remedial Measures (listed roughly in order of recommended action) |
|---|---|---|---|
| (C) Concrete | | | |
| Deep cracking | Excessive loading<br>Overstress<br>Uplift<br>Shrinkage (usually occurs early in life)<br>Expansion<br>Foundation movement<br>Seismic activity<br>Loss of strength<br>Concrete creep | Increased leakage<br>Accelerated deterioration<br>Progressive cracking<br>Stress redistribution<br>Increased stresses<br>Reduced stability<br>Differential movement | Determine depth/extent of cracking.<br>Sonic testing<br>Coring<br>Interior inspection, from galleries if present<br>Seal or grout cracks.<br>Evaluate short- and long-term effects.<br>Assess effects on stresses and stress redistribution.<br>Assess potential for leakage and consequent results.<br>Determine cause.<br>Check for movement.<br>Perform loading analysis.<br>Perform stress analysis.<br>Perform stability analysis.<br>Eliminate cause if feasible.<br>Increase drainage.<br>Seal upstream face. |
| (D) Leakage | | | |
| Moist or wet surfaces on concrete | Cracks<br>Deteriorated concrete<br>Porous concrete | Increased rate of deterioration<br>Leaching<br>Loss of weight<br>Loss of strength<br>Increased leakage | Review to determine if causes relating to (A) apply and pursue same remedial measures.<br>Determine depth and extent of cracks and see (C) for possible remedial measures. |
| (E) Leakage | | | |
| Concentrated through concrete | Cracks<br>Differential movement<br>Open joints<br>High uplift<br>Leaking pipes and conduits<br>Plugged drains | Loss of concrete matrix<br>Loss of structural integrity<br>Increased uplift | Map location of all leaks.<br>Monitor quantities and relate to reservoir elevation and other potential influencing conditions.<br>Determine path of water if possible. |

**TABLE 6-1** Evaluation Matrix of Masonry Dams (*continued*)

| Indicator | Possible Causes | Possible Effects | Potential Remedial Measures (listed roughly in order of recommended action) |
|---|---|---|---|
| | Erosion or cavitation of concrete<br>Leaching | | Detail inspection<br>Dye tests<br>Check condition of pipes, conduits, drains, etc. and repair if necessary.<br>Assess short- and long-term consequences.<br>After determining source, try to plug or seal the crack or opening at upstream side.<br>Determine basic cause, e.g., movement, stress conditions, and correct. |
| (F) Leakage<br>Through concrete (noticeable change) | Self-sealing of cracks<br>Plugged drains<br>Broken drains<br>Differential movement<br>Concrete failure | Increased uplift<br>Loss of concrete<br>Stress redistribution | Pursue essentially same measures as for (E).<br>Improve drainage. |
| (G) Leakage<br>Foundation and abutments | Foundation deterioration<br>Inadequate drains<br>Opening of joints, seams, shears, etc.<br>Movement | Foundation weakening with potential failure<br>Piping through foundation<br>Increased uplift<br>Loss of stability<br>Differential movement of dam<br>Loss of revenue/water<br>Loss of storage | Map location of all peaks.<br>Observe vegetation or other signs of moisture. Infrared film a possibility<br>Pursue measures similar to (E)<br>Specifically assess hazards associated with slides, piping, or sloughing.<br>Seal source of leakage with impervious membrane.<br>Seal with sand-cement, chemical grout, or other cutoff.<br>Provide controlled drainage system.<br>Add free-draining stability material on downstream side. |

**TABLE 6-1** Evaluation Matrix of Masonry Dams (*continued*)

| Indicator | Possible Causes | Possible Effects | Potential Remedial Measures (listed roughly in order of recommended action) |
|---|---|---|---|
| (H) Movement | Foundation settlement or heave<br>Abutment movement<br>Seismic activity<br>Overtopping<br>Excessive loading or uplift<br>Concrete expansion due to chemical action | Increased leakage<br>Inoperable appurtenances<br>Severe cracking<br>Stress redistribution<br>Reduction in stability<br>Anomalous changes in section or plan | Establish survey control system.<br>Monuments for horizontal control—some must be sufficiently far from dam to be out of influence zone.<br>Monuments for vertical control.<br>Pins, monuments, plates, gages, etc., across joints.<br>Inspect after each seismic event.<br>Establish photographic record.<br>Check for changes in leakage.<br>Isolate whether cause is in foundation/abutment or dam.<br>Review loadings.<br>Analyze foundation or abutment similarly to embankment dam.<br>Remedial measures are highly dependent on results of above. |
| (I) Development of offsets | Foundation movement<br>Differential movement<br>Seismic activity<br>Unforeseen loads | Increased cracking and spalling<br>Increased leaks<br>Binding of gates and operators | Same measures as for (H). |
| (J) Erosion and loss of foundation at toe or at outlets and spillway | Inadequate channel capacity<br>Channelization of water (spills or stream flow)<br>Lack of protection<br>Overtopping | Undermining<br>Loss of stability<br>Complete failure of appurtenances | Channel uncontrolled flows.<br>Improve drainage with pipes, lined ditches, etc.<br>Protect eroded area with concrete, gunite, rock or gabions as appropriate. |

**TABLE 6-1** Evaluation Matrix of Masonry Dams (*continued*)

| Indicator | Possible Causes | Possible Effects | Potential Remedial Measures (listed roughly in order of recommended action) |
|---|---|---|---|
| | Poor energy dissipation<br>Poor foundation<br>Piping or leakage<br>Poor drainage<br>Normal weathering | | Rock bolt blocky or slabby rock.<br>Increase spillway capacity to prevent overtopping.<br>Control spills and provide proper energy dissipation. |
| (K) Inoperability of gates and valves | Failed parts<br>Corrosion<br>Build-up of mineral deposits<br>Blockages<br>Debris<br>Silt deposits<br>Ice<br>Differential movements | Inability to operate<br>Reduced capacity of spillways/outlets<br>Increased probability of overtopping | Inspects operating parts and repair or replace.<br>If capacity to release water is inhibited, consider temporary change in reservoir operations.<br>Methodically and systematically determine cause.<br>Provide corrosion protection.<br>If ice is problem, provide barriers or aeration, see (P).<br>If due to silt, debris, or other blockage, remove that cause.<br>Provide log booms, debris barriers, trash racks or other facility to alleviate blockage. |
| (L) Reservoir slides | Unstable geology<br>Saturation<br>High runoff<br>Sloughing | Sudden high waves with resultant overtopping<br>Siltation<br>Blockage of outlets and spillways<br>Increased loading<br>Reduction of reservoir capacity | Determine potential for waves and damage to dam.<br>Stabilize slides (see Chapter 7).<br>Modify reservoir operation. |
| (M) Siltation | Geology<br>Normal or abnormal inflow<br>Cultivation upstream<br>Vegetation removal | Increased loads<br>Reduced stability<br>Plugging of outlets<br>Reduction of reservoir capacity | Dredge reservoir (usually economic only for small reservoirs).<br>Provide upstream siltation ponds.<br>Increase upstream vegetation. |

**TABLE 6-1** Evaluation Matrix of Masonry Dams (*continued*)

| Indicator | Possible Causes | Possible Effects | Potential Remedial Measures (listed roughly in order of recommended action) |
|---|---|---|---|
| | | | Review loading and dam stability and correct. Prestressed hold-downs. Remove only silt close to dam (temporary measure). Increase sluicing (will usually affect only area close to outlet). |
| (N) Debris | Floods<br>Logging<br>Vegetation | Plugging of spillways<br>Plugging of outlets<br>Damage to trash racks and equipment | Regularly remove debris from reservoir.<br>Provide log/debris boom.<br>Provide sluicing gate or chute to dispose of material over the dam.<br>Simplify spillway arrangement so debris can be passed without plugging. |
| (O) High waves | Wind<br>Reservoir slides | Overtopping<br>Damage to equipment<br>Undermining of banks | Increase freeboard to prevent overtopping.<br>Protect equipment against high water.<br>Design parapet wall to deflect waves back to reservoir.<br>Provide emergency spill to skim off high water.<br>Treat potential slides (see Chapter 7). |
| (P) Ice | Cold weather | Accelerated deterioration<br>Blockage of spillways and outlets<br>Damage to piping and equipment<br>Misoperation of gates<br>Damage to trash racks<br>Parapet damage<br>Increased loading | Operate reservoir to keep ice at level where damage will be minimal.<br>Keep spillway gates open.<br>Simplify spillway so ice can pass without restriction.<br>Provide aeration near operating equipment.<br>Review loadings on dam resulting from ice and assure dam can tolerate. |

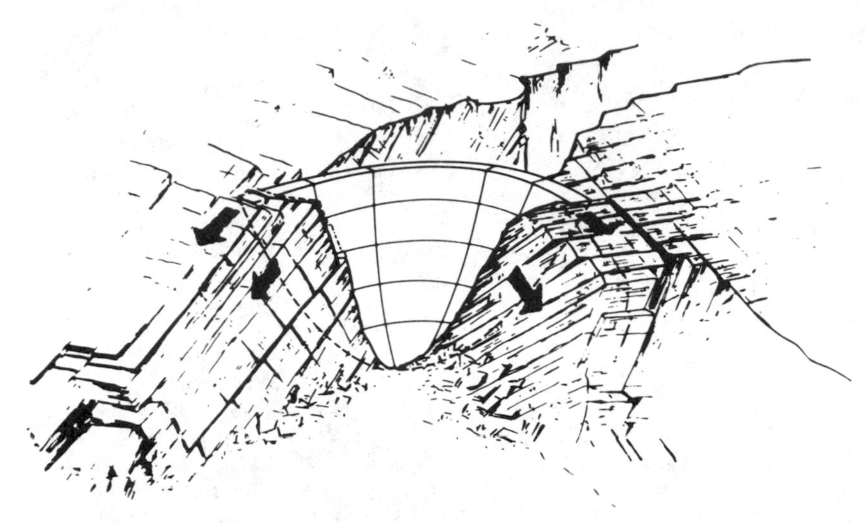

FIGURE 6-6 Relation of geologic structure to arch thrust.

ties in the rock system. Clear or muddy water may indicate the development of uplift forces in the abutment that were not contemplated in the design of the dam. (See the section Uplift.) The Malpasset Dam failure in France (International Committee on Large Dams 1973) is an example of failure due to foundation water pressures and lubrication (see Figure 5-5).

Various methods of leakage control have been used.

- Channelizing the seepage flow so increased head does not develop as a result of erosion.
- Installation of sand filters in the flow channels at the point of egress to prevent piping.
- Grouting with cement or other sealing materials to provide a barrier to the flow. Such barriers should be created near the upstream face of the dam, not downstream from the line of intersection between the dam axis and the abutment because it could result in unacceptable uplift in the dam or the abutment.
- Sealing the entrances to such cracks. Sealing materials would include bitumens, epoxies or other resins, cement grout, bentonite, concrete, and impervious soil blankets.

*Abutment or Foundation Deformation*

This is particularly critical for an arch dam because excessive deformation can produce unacceptable tensile stresses. Its effect may appear as tension cracks in the concrete or as high tensile stress measurements if instruments are installed in the extrados of the arch. In addition, the closing and possible crushing of faces of discontinuities, such as joints and fractures, might be observable. Also, there may be anomalous decreases in abutment seepage rates as the result of the closure of openings in the rock mass; such decreases may cause undesirable uplift pressures.

Unusual movements in the rock mass of the abutment may result in loosening of large blocks or possible slides of both upstream and downstream abutments of the dam. In a gravity structure, incipient sliding motion may develop at the contact between the concrete and the rock. The evidence for this might be a zone of freshly exposed rock observable on the upstream face of the contact. Particularly critical are the slopes adjacent to spillways; slide blockage of an intake, chute, or spillway basin could be disastrous if it occurs during the operation of a spillway.

In the case of an arch dam care must be taken to ensure that stresses developing from the dam have a sufficient mass of stable rock available to accept such stresses without undesirable displacements occurring in the rock mass. That is, there must be no topographic reentries immediately downstream from and within the influence of the abutment thrust area (see Figure 6-7). If such reentries occur, the possibility exists that the entire abutment mass may move in a downstream direction. The major area for the acceptance of thrust from an arch is often within an acute triangle that has its apex at the concrete-rock contact; the internal angle is about 15°, and the river side of the triangle is parallel to the thalweg of the river valley.

In examining abutment areas consideration must be given to the fact that minimum safety may exist in the upper part of double curvature arches because the upper parts of valley walls are generally looser than the lower walls and earthquakes would induce stronger reactions in the upper area of both the dam and its abutments.

Drainage galleries in the abutments should be examined carefully to determine if all drains are open and operating or whether some have been plugged by mineral deposits and/or silt. Pressure gages on drains should indicate if excessive uplift forces are developing.

Abutment deformation can be recognized by fresh cracks in the rock surface, blocks falling from abutments, or displacement of vegetation. Recording instruments or surface survey markers may be installed in the abutment whereby both vertical and horizontal movement may be detected. Cracks in the dam concrete where it joins the rock also may be evidence of deformation. (It should be noted that such cracking or crushing may also be caused by excessive deformation of the arch caused by something other than abutment movement.)

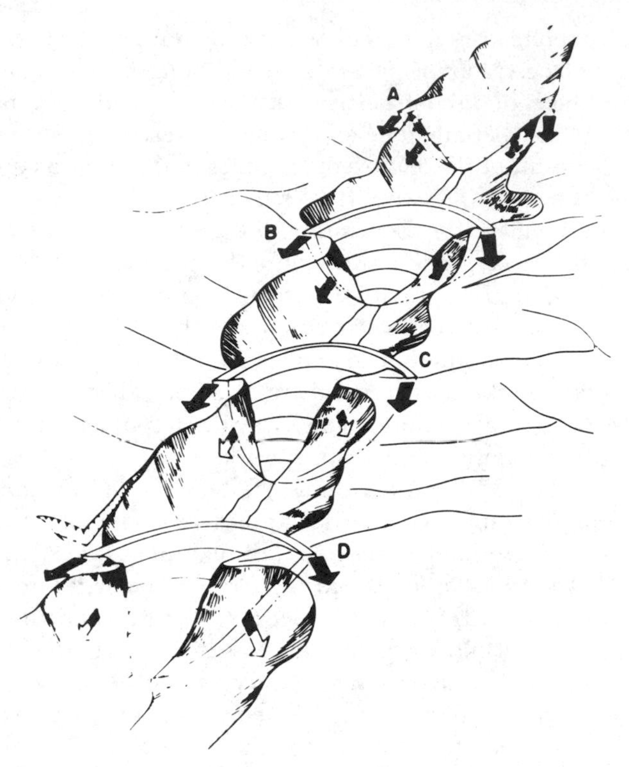


FIGURE 6-7 Favorable topography for arch dams (A and B); unfavorable topography-unshaded parts of arrows indicate part of thrust that is daylighting (C and D).

Deformation may be caused by reservoir load, earthquake forces, arch dam thrust, unfavorable orientation of the cracks in the rock with respect to the load directions imposed by the dam or reservoir, large ranges of temperature, freezing and thawing of water within the cracks in the rock, softening of the contact surfaces between rock blocks caused by water or other weathering forces, an abutment mass insufficient to withstand the overall thrust forces from the arch dam, presence of shear or fault zones that were not contemplated in the design, or excessive uplift forces.

Any unforeseen movement in the abutment will induce stresses in an arch dam or buttress that may not have been considered in its design. Loosened blocks can endanger any structures in their potential fall path. If sufficient rock falls into the channel below the dam, it can cause blockage of the outlets to waterways. Rock falls into the reservoir immediately upstream from the dam may interfere with operation of outlets, penstocks, or spillways. Rock falling onto the dam can endanger personnel and damage the concrete and roadways or other structures on top of the dam.

Motion of the abutment can cause leakage with a potential for lubrication or pressure buildup and ultimately can cause complete failure of the dam. If ancillary structures are located on the abutment (or within it), the rock movement can interfere with their operation and/or cause damage to them.

Remedies are highly dependent on the cause of the deformation or rock falls. Possibilities include any one or combination of the following:

- Deep rock anchors or rock bolts to tie together and strengthen the abutment and reduce or prevent further deformation.
- Horizontal or vertical concrete beams across the rock mass and anchored by rock bolts. Such anchors can be tensioned and grouted in or tensioned by future movement of the rock mass.
- If the rock falls are caused by abutment deformation and the deformation cannot be reduced, some relief can be achieved by extensive scaling of the abutment (removal of potentially loose blocks). If continuous loosening of blocks is expected, it may be desirable to build diverter walls. These are massive concrete walls located such that any loose rock falling from above will hit the wall and be diverted away from critical structures below the wall. This system was used successfully at Kortes Dam in Wyoming.
- If water is the causative agent, elimination of the flow of water into the abutment is needed.
- Extensive grouting to improve the modulus of deformation of the rock mass.
- Placing of buttressing rock, concrete, or gabions.

Abutments are particularly susceptible to damage by earthquake forces. As previously noted, generally the upper part of valley walls are looser than the lower part of the walls; thus, earthquakes can induce much stronger reactions in the upper portion of the abutments.

One California owner of a concrete gravity dam with a suspect abutment has used a number of measures to monitor and stabilize an abutment including regrading part of the abutment (upstream) and installing horizontal drains and inclinometers, installing drains under the dam to reduce uplift, and construction of concrete crib-wall backfilled with rock just downstream of the abutment. Stabilizing rock placed against the dam in the vicinity of the abutment and an elaborate survey control system and inclinometers have been installed. The foundation also has been heavily cement grouted. Other effective stabilizing measures are erosion control, rock bolting, deep anchorage, chemical grouting, installation of galleries and piles or other reinforcing members, and resloping of abutments. Considerable experience in grouting, both cement and chemical, is recorded in the ASCE (1982).

### Uplift

Any estimate of uplift should be based on the current effectiveness of the foundation drainage system, as indicated by measurements, the quality of maintenance, and other influencing features, such as the possible presence of a silt layer on the reservoir floor that is more impervious than the rock foundation.

The existence of a reliable foundation drainage system might justify relaxation of the design criterion published by several federal dam building agencies to the effect that full reservoir pressure should be assumed under the portion of the dam base not in compression. That criterion was originally adopted without the support of substantiating field measurements and without regard to specific characteristics of dam and foundation. However, it was considered appropriate for incorporation into conservative design criteria for new dams. Agencies promulgating such design criteria intend them to be guides to uniformly safe design rather than restrictions on the designer and intend to permit variations wherever warranted. Thus, it is only reasonable that the evaluation of existing dams consider the influence of actual site features and characteristics.

In a recent stability evaluation for an overtopping flood condition (Claytor Dam, New River, Virginia), an effective foundation drainage system was considered as warranting deviation from the assumption of full reservoir pressure under any portion of the dam base not in compression. Papers describing the Claytor Dam studies were presented by representatives of the American Electric Power Service Corporation on October 28, 1982, at the annual ASCE Convention in New Orleans. A supporting paper (Goodman et al. 1982) was presented. Essentially, the studies found that the foundation drains would continue to function under loading conditions causing a lack of compression on a portion of the base and would preclude a buildup of uplift to full reservoir pressure under the dam.

Experience at the U.S. Army Corps of Engineers' Dworshak Dam on the North Fork of the Clearwater River, Idaho, attests to the effectiveness of drains in reducing hydrostatic pressures in narrow cracks. During and shortly after reservoir filling, vertical cracks striking upstream-downstream occurred in the center of 9 monoliths of the concrete gravity structure. They extended from the base of dam to heights ranging to almost 400 feet and propagated downstream past the drainage gallery. Drain holes (1.5 inches in diameter) were drilled into the cracks from the galleries and angled to intersect the cracks at 5-foot vertical intervals along a line about

30 feet from the upstream face. These drain holes successfully reduced hydrostatic pressures in the cracks, resulting in the reduction of crack widths and arresting of the downstream propagation of the cracks.

Costly remedial measures might be avoided where foundation drainage, either existing or added, can be relied on to preclude the buildup of uplift pressures to full reservoir magnitudes in narrow cracks of the size comparable to those created by structure rotation under the increased flood loading. In such cases uplift can be represented by a linear distribution from tailwater pressure at the toe to tailwater pressure plus a percentage of the difference between the headwater and tailwater pressures at the line of drains and thence to headwater pressure at the heel of structure. The percentage of the difference factor that determines the uplift at the drains should be based on pressure measurements that can be obtained during normal pool levels.

Generally, the most effective and economical solution to reduction of uplift forces is the installation of drains. Where they already exist, their monitoring and maintenance are essential. Regular drain flow observations must be part of any surveillance program. Accumulation of deposits in the drains is monitored by periodic probing to determine location and characteristics of the obstructing material. When uplift is steadily increasing or when seepage flows have decreased substantially, the need for cleaning drains or drilling new ones is indicated. When drains become so obstructed as to impair their function, and the deposits are relatively soft, they can be cleaned by washing. However, this is often only a temporary remedy. A better solution is to redrill the old drain or to drill new drains (Abraham and Lundin 1976). Where drains do not exist or are inadequate, new ones can often be drilled into the foundation from existing galleries or from the downstream face. At the California dam referred to earlier, some old drains leading to a gallery were cleaned, new drains were drilled from the gallery into the foundation, the drains under the spillway bucket were cleaned, and new drains were drilled at a flat angle underneath the spillway bucket.

## Concrete Quality

For existing dams a great deal can be learned about concrete quality by visual observation. Surfaces subject to rapidly flowing water, such as spillways or outlet chutes, must be examined carefully. Silty or sandy water can erode concrete. Water moving rapidly past abrupt surface changes creates regions of negative pressure which may cause cavitation erosion as evidenced by increasingly deep holes in the concrete. Vibration also can result from pressure fluctuations and high-velocity impingement. Where small amounts of material have been removed, simple repairs have been made at some dams with a very smooth epoxy coating. Large repairs have needed extra strong concrete, such as fiber-reinforced concrete. Steel plates have in some cases been installed on cavitated surfaces. Also, cavitation erosion often can be prevented by the introduction, under the water flow, of air through slots or other openings.

Where strength is a question, nondestructive tests, such as the rebound hammer or sonic velocity measurements, are only qualitative. The most accurate evaluation of strength can be made by extracting cores of a diameter two to three times the size of the largest particles in the concrete. These can be tested for compressive and tensile strength, for modulus of elasticity and Poisson's ratio, and for density. All of these properties are needed for any analysis of the behavior of a dam.

Careful attention should be paid to the appearance of weathered concrete surfaces. Pattern cracking may denote either drying shrinkage or, in extreme cases, alkali-aggregate reaction. Heavy surface scaling may indicate freeze-thaw effects or insufficient cement and, consequently, low strength.

### *Experience with Deterioration at Drum Afterbay Dam*

The story of Drum Afterbay Dam is a good example of the detection and investigation of a dam with deteriorating concrete. Built in 1924, this dam was a thin arch structure, 95 feet high, situated at elevation 3,200 feet on the western slope of the Sierra Nevada Mountains in California. Aggregate for the concrete was crushed from the rock (schist) at the dam site, which turned out to be an unfortunate decision. Twenty years after construction the downstream face showed visible signs of deterioration due to frost action, with particularly noticeable deterioration in the horizontal joints between lifts. At that time some repairs were made by chipping out poor concrete and filling with gunite. After another 20 years it was apparent that a more thorough investigation should be made to pinpoint the causes of the worsening deterioration. This later study found, in addition to freeze-thaw action, visible signs of a possible alkali-aggregate reaction. At this time a more elaborate study was made, utilizing 6-inch and NX cores and sonic velocity measurements. From the cores, measurements were made of strength, modulus of elasticity, Poisson's ratio, density and thermal diffusivity; also, a careful petrographic examination was made. Correlations between pulse velocity measurements and strength were used to target the areas of generally deteriorated concrete, which by this time had reached strengths as low as 1,400 psi. The petrographic examinations showed that the principal culprit was pyrites in the aggregate, which in combination with the lime from the hydrating cement set up new compounds of low strength. After this the prognosis for the concrete was more of the same or worse. The dam was deteriorating at an accelerating rate, and the decision was made to replace the dam entirely (Pirtz et al. 1970).

### *Experience with Synthetic Materials for Concrete Repairs*

The strength and exceptional adhesive ability of certain synthetic materials have led to their application in repairing concrete both for surface treatment and for injection to seal cracks. Resins with low sensitivity to water have been used as bonding agents between old and new concrete. Epoxy-based and polyester-based resins have been widely used for facing on dams and other hydraulic structures. Epoxy-based resins of appropriate mix have been found to be more effective on damp concrete than polyester-based resins. The viscosity of resins used for injection can be varied from pumpable mortar to very thin grout. Careful workmanship is required to ensure lasting protection by resins.

The Southern California Edison Company has made effective use of synthetics in sealing concrete surfaces. For example, the upstream face of Rush Meadows Dam, a concrete arch at high elevation in the Sierra Nevada, was coated in 1977 with a layer of gunite covered by two coats of polysulfide. The first layer of polysulfide was thin, placed over a primer, and was followed by a thicker final layer. The treated face effected a substantial reduction in seepage and has shown no signs of distress, neither peeling nor general deterioration. Edison has made such applications on other dams with comparable success. Pacific Gas and Electric Company also has used similar techniques successfully.

Repair of concrete by injection of synthetics has a less extensive record but holds promise in special cases. At the Corbara Dam in Italy an experimental attempt was made to seal cracks in buttresses (due to thermal shrinkage) by application of epoxy resins. Remedial work was done in the winter to ensure the widest opening of the cracks. The work entailed drilling, chemical washing of cracks, blowing with air, placing small copper pipes to drain and control grouting, superficial mortaring, and grouting at about 60 to 70 psi. Some of the work was done by flowing warm air into the crack prior to the injection. Several difficulties were incurred at some cracks, such as only partial penetration due to excessive viscosity or inadequate adhesion because of moisture or unfavorable temperature. However, there was an appreciable improvement in shear strength along the cracks sufficiently treated with the resin.

For internal remedy of general fine cracking in concrete structures, the potential for success can be enhanced by injection of resins into boreholes, with careful temperature control, drying with hot air, and proper venting (Vallino and Forgano 1982).

### *Experience with Steel-Fiber Concrete*

Where concrete is subjected to high impact or erosion or cavitation, improvement can be obtained by removing damaged material and replacing it with a mix containing randomly distributed steel fibers. This was successfully accomplished by the U.S. Army Corps of Engineers at Dworshak Dam in Idaho in the stilling basin and at a sluice. The fibrous concrete had a low water/cement ratio and a high cement factor and was placed in the more deeply damaged areas. Some surfaces were polymerized to improve durability. Fibrous concrete was used similarly for remedial work on the stilling basin at Libby Dam in Montana. Additionally, certain areas of floor slabs in the stilling basin were polymerized. Shallower repairs at Dworshak were done with epoxy mortar but did not prove satisfactory; most of it failed after a rather short period of service. Nonetheless, in other projects with less demanding service conditions, epxoy mortar has provided effective repair.

## STABILITY ANALYSES

Concrete and masonry dams must interact with the rock foundation to withstand loads from the weight of the structure, forces from volume change due to temperature, internal water pressures (uplift), external water pressures, backfill, silt, ice, earthquake forces, and equipment (see Figure 6-8). Uplift pressures used in stability analyses should be compatible with drainage provisions and uplift measurements if available. Dams should be capable of resisting all appropriate load combinations and have adequate strength and stability with acceptable factors of safety. The factors of safety recommended for various loading combinations are given in U.S. Bureau of Reclamation (1976, 1977).

The foundation has a significant influence in the stability evaluation of masonry structures. It must have adequate strength to support the heavy loads of the structure without excessive displacement. In addition, it must function as the water barrier with adequate provisions for drainage and relief of uplift. It should also be as free as practicable of such weaknesses as extensive weathering, faults, jointing, and clay seams. The existence of such defects at existing dams should be evaluated carefully to determine if they require treatment.

Gravity dams can be analyzed by the gravity method, trial-load twist analysis, or the beam and cantilever method, depending on the configuration of the dam, the continuity between the blocks, and the degree of re-

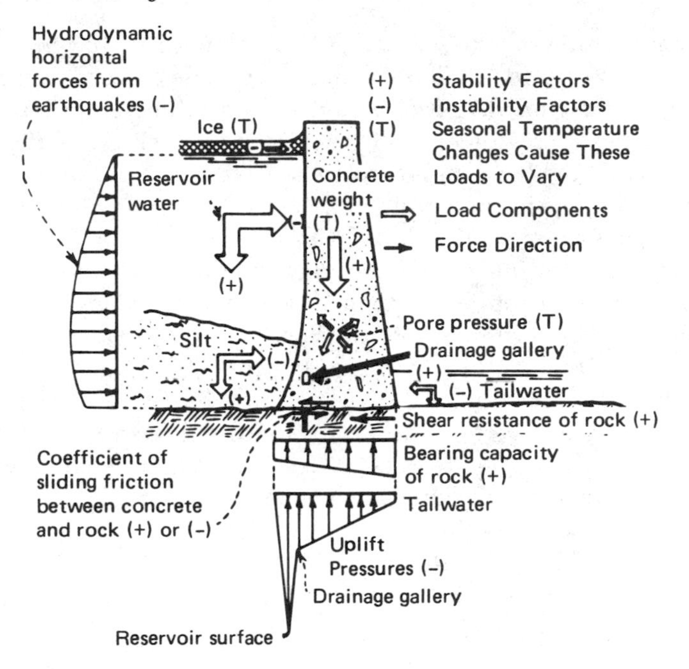


FIGURE 6-8 Expected loads on a concrete dam.

finement required. The gravity method is the most common and is applicable when the vertical joints between individual monoliths are not keyed or grouted. Trial-load twist analysis and the beam and cantilever method are appropriate when the monolith joints are keyed and grouted; however, the gravity method can be used in this situation for an approximate or preliminary analysis. Descriptions of these methods, together with safety factors and allowable stresses, can be found in U.S. Army Corps of Engineers (1958–1960) and U.S. Bureau of Reclamation (1976).

Arch dams are usually analyzed by the independent arch theory (limited to relatively small structures or analyses preliminary to more refined methods) or by trial-load methods. Both two- and three-dimensional finite element methods of analysis are available and can be used to perform trial-load analysis or other stress-determination methods. Details of some methods, with appropriate safety factors and allowable stresses, can be found in U.S. Bureau of Reclamation (1977).

## Flood Loading

The evaluation of stability of gravity dams during a spillway design flood is necessary in deciding whether modifications, such as added spilling capacity or strengthening measures, should be accomplished. In most existing dams a probable maximum flood would overtop the dam. However, concrete gravity dams on firm rock foundations are inherently resistant to overtopping flows provided stability against overturning and sliding are ensured and that the groin and foundation downstream of the dam are capable of resisting erosion and disintegration resulting from impingement of the overtopping water.

An analysis to determine stability during great floods should be based on conservative estimates of headwater and tailwater elevations. The analysis must consider site-specific conditions, such as quality of materials in the dam and auxiliary structures, foundation permeability and competence, and overturning. However, extensive damage to the structural components may be acceptable in certain cases for this extreme event. In addition to estimates of headwater and tailwater elevations, it is necessary to estimate the possible increase in the uplift loading on the structure.

### Seismic Loading

#### *Ground Motions*

The ground motions to be used in an analysis of the seismic load conditions are discussed in Chapter 5.

#### *Concrete Dam Response*

The way in which a dam responds to an earthquake is complex and varies with the type of dam and its foundation. For example, at a concrete dam on a rock foundation the earthquake motion is first felt at the foundation as rapidly changing motions in all directions, and many motions per second. Usually the horizontal accelerations are stronger than the vertical components of the motion, but all are present. The vertical acceleration adds to or subtracts from the weight of the dam. The dam responds by deforming elastically and developing stress. For a given seismic record methods now exist for determining these stresses and deformations.

The computed stresses developed by the earthquake are compared with the strength of concrete cores obtained from the dam. In the latter circumstance an allowance must be made for the rapidity of loading and the linearity of the analysis. Fresh cores must be used in these strength tests.

Some concrete dams have been damaged by earthquakes; others have been left untouched. For example, Koyna Dam, a concrete gravity dam in India, suffered a number of major cracks near the top after the Koyna earthquake in 1967 (Chopra and Chakrabarti 1973). However, these cracks are confined mainly to horizontal construction joints. For safety the dam was later buttressed with additional concrete. On the other hand, Pacoima Dam, a concrete arch dam in California, sitting practically on the epicenter of the San Fernando earthquake in 1971, was undamaged from a shock measured at over 1.2 acceleration due to gravity on one abutment. [The recording at the abutment may be of questionable validity. However, peak horizontal acceleration at the dam base may have been on the order of 0.75 acceleration due to gravity (Seed et al. 1973)].

### *Methodology*

Most existing concrete dams in potentially seismic zones were designed for seismic loads by using equivalent static forces. These forces were obtained by multiplying dam weight by a seismic coefficient. It is generally agreed that this method is adequate for structures located in seismic zones below 3 (see Figures 5-12 and 5-13). In zones 3 and 4, or in other locations where the proximity to active faults warrants, a dynamic analysis should be made using, at a minimum, a response spectrum analysis. A time history analysis should be made where stress variations with time are critical. Descriptions of these methods can be found in Chopra and Chakrabarti (1973), Chopra and Corns (1979), U.S. Army Corps of Engineers (1958–1960), and U.S. Bureau of Reclamation (1976).

## IMPROVING STABILITY

### General Measures

An existing gravity dam that has questionable resistance against sliding or overturning may be improved in various ways, depending on the suspected cause of instability. If excessive uplift is a problem, foundation drainage can be improved by cleaning drains and/or adding more drains. Increased positive resistance has been accomplished by stressed tendons anchored in the foundation rock, addition of concrete mass, construction of concrete buttresses, or placing a buttressing embankment against the downstream face.

### Buttress and Multiple-Arch Dams

Slab and buttress and multiple-arch dams built 50 to 70 years ago were designed on principles that may not meet modern standards. Many of these structures have been modified to overcome questionable stability, especially in resistance to lateral loading, such as earthquake acceleration. In some cases the strength of the concrete also has been found to be low. Cracks in various elements have indicated serious overstressing, even under normal loading. The arches forming the faces of some old multiple-arch dams have small central angles, so that the arches impose considerable thrust on the buttresses normal to their center lines. Such forces must be resisted partly by the adjoining arches if the buttresses are insufficiently braced. Some dams of this type originally had no steel reinforcement in the buttresses. In such cases cracking has typically been observed to extend diagonally downward from the upstream to the downstream face, being open at the juncture with the arches but terminating in hairline cracks at the downstream extremity, suggesting a slight rotational movement of the buttress about its toe. Micrometer gage readings have generally not shown appreciable movement at such cracks after their initial formation.

Stability analyses of slender buttressed concrete dams with minimal reinforcement and bracing have disclosed typically that, even with relatively low seismic accelerations, the buttresses could be unstable. These weaknesses can be overcome by reinforcing the buttresses in various ways. Successful methods of strengthening include posttensioning the buttresses and constructing bracing members between them. The addition of shear walls in alternate panels or bays has in some cases provided effective lateral resistance. Shear keys have typically been provided at the joints, and bolts have been extended through the buttresses, with large bearing plates on the back side to distribute the bolt load on the old concrete. Horizontal beams bolted between buttresses also have served effectively. A basic requisite is that the connections between bracing elements and buttresses be detailed in such a way that lateral loads are transferred safely. Otherwise, the struts might be of less benefit than intended.

In an investigation of buttressed concrete dams in California, concrete strengths in some of these structures were found to average less than 2,000 psi. For example, the compressive strength of concrete cylinders taken from one old multiple-arch dam averaged 1,889 psi and varied from 1,225 psi to 3,185 psi. This wide variation was attributed primarily to deficiencies in quality control during construction. No evidence of alkali-aggregate reaction was found. Where such chemical activity has been involved, even broader ranges of strength have been observed, with the minimum being less than one-fourth the maximum in some cases. In such cases the principal emphasis must be on the low-strength zones of the dam. A complete determination of structural adequacy necessitates data on the whole strength envelope, including both the range and distribution of values. Lack of uniformity of strength may induce excessive stress concentrations in low-strength areas, particularly if the weak zones are large. A concrete dam may have the capability to bridge across defects of limited extent.

## Rollcrete

The threshold of a new technology was recently crossed by design and construction of Willow Creek Dam in Oregon, by the U.S. Army Corps of Engineers. This dam is made entirely of roller-compacted concrete, which is essentially a well-graded gravel fill containing cement. Other dams of this type are in the design phase. The costs of concrete placement and construction time are reduced substantially by using this method. Rollcrete differs from soil cement in several important respects, primarily related to the mix, although both are placed in layers and are compacted by rollers. The cement content of soil cement may range as high as 18% by weight. In contrast, the cement content of rollcrete may be between 2.5 and 7% by weight. Compared with regular concrete, rollcrete requires less cement to attain equal strength, and its mix demands less strict processing and gradation. Compaction ensures denser concrete. The promise of this new technology may be greatest in construction of large new structures, where the potential economies of scale are obvious. However, it would appear to offer advantages also in remedial work on existing dams. For example, it would have useful applications where mass concrete sections have to be enlarged or where spillways and other channels need to be extended.

## Experience at Condit Dam

The 125-foot-high Condit Dam in the State of Washington was rehabilitated in 1972 by improved drainage facilities and by installation of steel anchors (deSousa 1973). This concrete gravity structure, 60 years old at that time, had been determined to have a marginal factor of safety under normal loading conditions and to have inadequate resistance to extreme loadings by flood or earthquake. The concrete was in satisfactory condition, but the drains were only partially effective. Nearly full uplift pressure occurred at the midpoint of the dam base. In one phase of the remedial program a series of new drain holes was drilled radially from two sluice pipes that pass through the dam at low level. This reduced the uplift pressure from a maximum of 33 psi to less than 8 psi. Concrete cores recovered from the drilling had test strengths varying from 2,760 psi to 6,690 psi, with an average of 4,470 psi.

Under extreme loading the dam would have been stressed in tension at the heel up to unacceptable levels. Therefore, as an additional remedial measure, steel anchors were installed to limit tension to 20 psi. Twenty-two posttensioned anchors were installed in the dam, and 3 were placed in the spillway foundation. The anchors varied in length from 50 to 100 feet and, each was loaded to about 300 kips. The typical depth of embedment in the basaltic foundation was 25 feet. Other corrective work at the Condit Dam included pumping of 470 cubic yards of concrete into a fissure under the spillway structure and drilling six drain holes radially upward from the diversion tunnel to the base of the dam.

### Experiences at Spaulding Dams

Three separate concrete dams form Pacific Gas and Electric Co.'s Lake Spaulding in California. The main dam is a 276-foot-high arch-gravity dam. Dam 2 (the main spillway dam) is a 42-foot-high gated gravity structure. Dam 3 is a 91-foot-high gravity buttress dam. All three dams were built between 1912 and 1919, and the concrete had deteriorated significantly. The investigations and improvements to these dams illustrate some varied economical solutions to different problems.

Investigation of the concrete in the main dam included determination of concrete strength, density, modulus of elasticity, and overall quality as determined by sonic velocity testing. Coring was done to determine depth of cracks and deterioration as well as the bond between lifts. Chemical analyses of reservoir and leakage water were made. Recording thermometers were installed in the dam to determine seasonal concrete temperatures. Stress/strain gages were applied to the surfaces to record these values for comparison to water loading and temperatures.

Stress analyses were conducted using both two- and three-dimensional finite element methods of analysis. Various input parameters for concrete and foundation properties were used. For dams 2 and 3 conventional static stability analyses were made. Dam 3 was found to be marginally stable. Dam 2 was stable for existing loads but required anchors to accommodate loads resulting from increased flood loading.

The main dam was improved by constructing a 12-inch-thick reinforced concrete membrane over most of the upstream face after the old deteriorated concrete was removed. Vertical drains were placed between new and old concrete at the vertical joints. This membrane reduced leakage and prevented further deterioration of old concrete. The dam crest was raised slightly to increase the spillway capacity at dam 2. At floods greater than the 1:500-year occurrence level the main dam will overtop, so protection for downstream appurtenances was provided.

Two radial gates were added at dam 2 to increase spillway capacity. Posttensioned anchors were installed to improve stability under the increased water level conditions. An epoxy coating was applied to the concrete to prevent further deterioration.

Dam 3 was partially reconstructed with a trippable flashboard type spillway at its lower end. In its higher reaches, where overtopping could not be tolerated, the crest was raised slightly. A reinforced concrete membrane was constructed on the entire upstream face, and rockfill was placed against both upstream and downstream sides for part of their height, in order to increase stability.

This is an example of a fully integrated approach to resolve deterioration, stability, and spillway capacity problems.

## REFERENCES

Abraham, T. J., and Lundin, L. W. (1976) *T.V.A.'s Design Practices and Experiences in Dam and Foundation Drainage Systems*, Transactions, ICOLD.

American Society of Civil Engineers (ASCE) (1982) Proceedings, *Grouting in Geotechnical Engineering*, W. H. Baker, ed., New Orleans.

Chakrabarti, P., and Chopra, A. K. (1973) "Earthquake Analysis of Gravity Dams Including Hydrodynamic Interaction," *International Journal of Earthquake Engineering and Structural Dynamics*, Vol. 2, No. 2 (October–December), pp. 143–160.

Chopra, A. K., and Chakrabarti, P. (1973) "The Koyna Earthquake and the Damage to Koyna Dam," *Bulletin of the Seismological Society of America*, Vol. 63, No. 2, pp. 381–397.

Chopra, A. K., and Corns, C. F. (1979) *Dynamic Method for Earthquake Resistant Design and Safety Evaluation of Concrete Gravity Dams*, Transactions of ICOLD Congress, New Delhi.

deSousa, S. A. (1973) *Rehabilitation of an Old Concrete Dam*, Proceedings of Engineering Foundation Conference on Inspection, Maintenance and Rehabilitation of Old Dams, Pacific Grove, California.

Goodman, R. E., Amadei, B., and Sitar, N. (in press) Analysis of Uplift Pressure in a Crack Below a Dam, paper given at ASCE Annual Convention, New Orleans.

Pirtz, D., Strassburger, A. G., and Mielenz, R. C. (1970) "Investigation of Deteriorated Concrete Arch Dam," *American Society of Civil Engineers, Power Division Journal*, January.

Seed, H. B., Lee, K. L., Idriss, I. M., and Makdisi, F. (1973) *Analysis of the Slides in the San Fernando Dams During the Earthquake of February 9, 1971*, Earthquake Engineering Research Center, Report No. EERC 73-2, University of California at Berkeley.

U.S. Army Corps of Engineers (1958-1960) *Gravity Dam Design*, EM 1110-2-2200, Government Printing Office, Washington, D.C.

U.S. Bureau of Reclamation (1976) *Design of Gravity Dams, Design Manual for Concrete Gravity Dams*, Government Printing Office, Washington, D.C.

U.S. Bureau of Reclamation (1977) *Design of Arch Dams, Design Manual for Concrete Arch Dams*, Government Printing Office, Washington, D.C.

U.S. Committee on Large Dams (1975) *Lessons from Dam Incidents, USA*, ASCE, New York.

Vallino, G., and Forgano, G. (1982) *Design Criteria for Improvement of the Concrete Buttresses of Corbara Dam*, Transactions of 14th Congress, Rio de Janeiro, ICOLD.

## RECOMMENDED READING

Chopra, A. K. (1970) "Earthquake Response of Concrete Gravity Dams," *Journal of the Engineering Mechanics Division*, ASCE, Vol. 96, No. EM-4 (August), pp. 443–454.

Dungar, R., and Severe, R. T. (1968) *Dynamic Analysis of Arch Dams*, Paper No. 7, Symposium on Arch Dams, Institution of Civil Engineers.

Golze, A. R., et al. (1977) *Handbook of Dam Engineering*, Van Nostrand Reinhold, New York.

Howell, C. H., and Jaquith, A. C. (1928) "Analysis of Arch Dams by Trial-Load Method," ASCE Conference Proceedings.

International Commission on Large Dams (1970) Proceedings, Xth Congress, Montreal, *Recent Developments in the Design and Construction of Concrete Dams.*

International Commission on Large Dams (1979) Proceedings, XIII Congress, New Delhi, "Deterioration or Failures of Dams."

Jansen, R. B. (1980) *Dams and Public Safety*, U.S. Department of the Interior, Denver, Colo.

Proceedings of the Engineering Foundation Conference (1973) *Inspection, Maintenance and Rehabilitation of Old Dams*, Pacific Grove, Calif.

Proceedings of the Engineering Foundation Conference (1974) *Foundations for Dams*, Pacific Grove, Calif.

Proceedings of the Engineering Foundation Conference (1976) *The Evaluation of Dam Safety*, Pacific Grove, Calif.

Severn, R. T. (1976) "The Aseismic Design of Concrete Dams," *Water Power and Dam Construction*, pp. 37–38 (January), pp. 41–46 (February).

Structural Engineers Association of California (1975) *Recommended Lateral Force Requirements and Commentary*, San Francisco.

Thomas, H. H. (1976) *The Engineering of Large Dams*, Vol. I, John Wiley & Sons, New York.

U.S. Bureau of Reclamation (1977) *Design of Small Dams*, Government Printing Office, Washington, D.C.

Westergaard, H. M. (1933) "Water Pressure on Dams During Earthquakes," *Transactions ASCE*, Vol. 98.

# 7
# Embankment Dams

## TYPES OF DAMS AND FOUNDATIONS

Embankment-type dams have been classified in a number of different ways, but various authorities have not always been in agreement on terminology. Classification generally recognizes (1) the predominant material comprising the embankment, either earth or rock; (2) the method by which the materials were placed in the embankment; and (3) the geometric configuration or internal zoning of the cross-section. A classification modifier is often included to denote the purpose or use of the dam, such as diversion dam, storage dam, coffer dam, tailings dam, afterbay dam, etc.

A formal, rigid classification is less important than an understanding of the performance characteristics and purposes of the zones and components forming the total dam.

Embankment dams are constructed of natural materials obtained from borrows and quarries and from waste materials obtained from mining and milling operations. The two primary types are the earthfill dam, an embankment dam in which more than one-half of the total volume is formed by compacted or sluiced fine-grained material, and the rockfill dam in which more than one-half of the total volume is formed by compacted or dumped pervious natural or quarried stone.

### Earthfill Dams

#### *Homogeneous Earthfill Dams*

Homogenous earthfill dams are composed of materials having essentially the same physical properties throughout the cross-section. Modern homogeneous dams usually incorporate some form of drainage zones or other elements for controlling internal saturation and seepage forces; however, many small older dams do not have these provisions. Rock toes, horizontal blanket drains, vertical and inclined chimney drains, line drains, and finger drains or a combination of these various forms have been used for these purposes. The drainage facilities are composed of pervious sand, gravel, or rock fragments separated from direct contact with the main body of the dam by properly graded filter zones to prevent migration of fine-grained soils into the drain elements and to reduce rapidly the hydraulic gradient of the seepage flow.

### *Hydraulic Fill Dams*

Hydraulic fill dams are constructed of materials that are conveyed into their final position in the dam by suspension in flowing water. Originally this sluicing was assumed to sort out and deposit the coarser materials near the faces of the dam and the finer materials near the center of the cross-section. With few exceptions, dams of this type have not been constructed in the United States since about 1940 mainly because the development of large, efficient earth-moving machines has made other types of embankment dams more economical and because the seepage and structural performance of these other types are more predictable (Jansen et al. 1976). The experience record during the period 1920-1940 demonstrates the unreliability of the theory of idealized grading and sorting into pervious shells and impervious cores and the propensity for failure during construction.

The vulnerability of hydraulic fill dams to accidents and failures from long-duration seismic ground motions was vividly demonstrated during the 1971 San Fernando, California, earthquake. Consequently, many old hydraulic fill dams in California have either been replaced or extremely modified and strengthened. Others, after site-specific investigations, have been declared safe and are in service. Hydraulic fill dams and earthquakes are not confined to California. Although they are no longer favored in the United States, a substantial number of hydraulic fill dams are in service in the United States and require surveillance and safety evaluation (Leps et al. 1978).

### *Zoned Earthfill Dams*

Zoned earthfill dams are composed of an impervious zone or core of fine-grained soils located within the interior of the cross-section and supported by outer zones or shells of more pervious sand, gravel, cobbles, or rock fragments. Transition zones of intermediate permeability are frequently included between the core and the shells for economic utilization of all materials that must be excavated for the project and to prevent intermingling or transport of materials at the zone interfaces.

Various configurations and positions of the core zone have been used. The zone may be centered on the dam axis with positive slopes or it may have a vertical or overhanging downstream slope. The selection of the various shapes is controlled by the properties and quantities of the available construction materials and the stability and seepage control objectives of the design.

### *Diaphragm Earth Dams*

Diaphragm earth dams consist of a pervious or semipervious embankment together with an impermeable barrier formed by a thin membrane or wall. The diaphragm may be positioned in the embankment along the axis or on the upstream face of the embankment. The stability of the dam is supplied by the mass of the embankment, and the water retention capability is supplied by the diaphragm. Cement concrete, asphaltic concrete, and steel plate have been used for diaphragms.

### *Stonewall-Earth Dams*

Stonewall-earth dams are composed of rubble-masonry walls and an earth filling. This type of dam is generally quite old—100 or more years—and of modest height. Some have only a downstream wall, in which case the upstream face of the earth filling is sloped. Others have both an upstream and a downstream wall that retains the interposed filling. The exposed surfaces of the walls are usually vertical or near vertical. The filled surfaces are sometimes battered or sloped. The walls are usually dry rubble but may occasionally be mortared.

## Rockfill Dams

### *Faced Rockfill Dams*

Faced rockfill dams consist of a pervious rock embankment with an impermeable membrane on the upstream face. The rock mass provides stability and the membrane, or facing, retains the water. Older faced rockfill dams were constructed by dumping the rock in relatively high lifts or tips. Sometimes the rock was sluiced in an attempt to reduce settlement by washing the rock fines and spalls into the interstices of the mass and creating direct contact between the larger blocks of rock. An upstream narrow zone of derrick-placed stone was commonly used to create a uniform surface to support the facing and to reduce the amount of movement and distress in the facing.

Since about 1960 the construction procedures for faced rockfill dams have been improved considerably, through the efforts of J. B. Cooke and others, resulting in less embankment settlement and less damage to the facing. A large percentage of the rock is placed and compacted in horizontal lifts. A special zone of selected small-size rock is used to support the face. The main body of the embankment is zoned with the rock sizes in the zones increasing toward the downstream face. All but the zones of larger-sized rock are compacted by vibratory rollers or rubber-tired compactors. The largest-size rock is usually dumped in lifts of moderate height.

The facings consist of reinforced portland cement concrete, asphaltic concrete, reinforced or unreinforced gunite or shotcrete, and timber. Different thicknesses, joint details, and spacing for concrete facings have been developed over the years.

Newer dams of this type have been constructed with thinner slabs, reduced amount of reinforcement, minimum joint spacing, and closed vertical construction joints. Horizontal joints have been limited to those required for construction purposes. A zone of compacted fine-grained soil has been placed over the lower elevations of the facing when the dam site is V-shaped or where there is an inner gorge (Davis and Sorenson 1969).

### *Impervious Core Rockfill Dams*

Impervious core rockfill dams consist of an interior impervious zone or element supported by zones of dumped or compacted rock. The interior element controls the retention of the water and is usually a compacted impervious soil protected by filter or thin transition zones. A few old dams have thin vertical concrete core walls located on the central dam axis. Depending on the position and configuration of the core, these dams are usually classified as central core, inclined core, or sloping core rockfills, and each has its own stability, seepage control, construction advantages, and site compatibility characteristics.

The composition and construction of the filter and transition zones are especially critical in this type of dam because of the relative thinness of the core and the magnitude of the hydraulic gradient.

### *Rockfilled Crib Dams*

Rockfilled crib dams consist of a framework of interlocked timbers or concrete prismatic bars that confine rock blocks and fragments. The water facing and overpour surfaces are usually timber fastened to the crib members.

Construction of this type of dam was common around the turn of the century, and some were later modified by the construction of concrete superstructures. The stability characteristics of a crib dam resemble those of a concrete gravity section; however, it is listed here because of its rock composition. Timber crib dams are sometimes constructed for diversion purposes.

## Foundations

Embankment dams can be constructed on foundations that would be unsuitable for concrete dams. The foundation requirements for earthfill dams are less stringent than those for rockfill dams (Engineering Foundation 1974). Foundations for embankment dams must provide stable support under all conditions of saturation and loading without undergoing excessive deformation or settlement. The foundation must also provide sufficient resistance to leakage where excessive loss of water would be uneconomic.

Foundations are extremely variable in their geologic, topographical, strength, and water retention characteristics. Each is unique and is an integral part of a dam. During design and construction the foundation characteristics can be modified and improved by such treatments as excavation, shaping, curtain and consolidation grouting, blanketing, densification, installation of sheet piling, prewetting, etc. These various forms of treatment are primarily for the purposes of (1) strengthening, (2) safely controlling seepage and leakage, and (3) limiting the influence of the foundation on embankment deformations. However, for an existing dam one can only evaluate the effectiveness of the treatment from the construction record and observable performance.

Based on strength and resistance to seepage and leakage, foundations can be typified as (1) rock, (2) sand and gravel, and (3) silt and clay or a combination thereof. Earthfill dams have been adapted to all three of these types of foundations. Types 2 and 3 have generally been determined unsuitable for faced rockfill dams. Type 3 has been determined unsuitable for impervious core rockfill dams without extensive foundation excavation and treatment.

The foundation types have been treated in a variety of ways depending on the designers' versatility and objectives and the type and configuration of the dam. The foundations of many existing dams will not have received any special treatment and present safety concerns. Where treatment was afforded, it varies under the different zones of the dam, depending on the intended functions of the zones and the foundation type.

Foundations have been treated for seepage control by (1) earth backfilled cutoffs, (2) concrete or sheet piling cutoff walls, (3) slurry walls, (4) grout curtains, (5) vertical drains, (6) relief wells, and (7) impervious earth blankets or a combination of these methods (Wilson and Marsal 1979). Foundations have been treated for strengthening by (1) excavating weak materials and formations; (2) consolidation grouting; (3) prewetting collapsible soils; (4) installing vertical drains to accelerate consolidation and accompanying strength gain during embankment placement; and (5) to a limited extent, vibratory densification. Foundations containing saturated, fine, cohesionless sand of low density are suspect, especially in regions of higher seismicity, because of the tendency of the sand to collapse and liquefy during long-duration ground shaking from earthquakes. Many foundations of this type have probably received no treatment for such a condition.

## DEFECTS AND REMEDIES

It has been emphasized that dam failures are usually caused by a complex chain of events that involves one or more defects and that failure can be averted by properly identifying and remedying the defects. For embankment dams the major nonhydraulic defects causing failure ultimately involve slope or foundation *structural* instability and/or slope or foundation *seepage* instability. Closely associated defects are excessive settlement, slope erosion, malfunctioning drains, problems at the abutment or foundation/embankment interface, and/or excessive vegetation and rodent activity.

Equally important threats to the overall structural or seepage stability of the dam are defects in appurtenant structures, such as spillways and conduits, and associated outlet works, such as gates, hoists, and valves.

The following sections include discussions of common defects that can cause partial or total failure of the dam, indicators of these defects, possible causes of each defect, effects on the dam, methods of investigating the defects, and potential remedial measures, with brief examples of actual applications on existing dams. Table 7-1 is an evaluation matrix for embankment dams that briefly summarizes these discussions.

In applying these and other remedies the complex interrelationships between the dam and its foundation, appurtenant structures, and reservoir margin must be considered. Furthermore, extreme caution must be exercised to avoid creating a new defect in the process of remedying an existing one.

### Slope and Foundation Instability

Instability of embankment dams or their foundations may occur as a result of (1) extended periods of high reservoir level that result in high pore pressures within the embankment, (2) rapid drawdown of the reservoir from a high level, (3) earthquake shaking, or (4) deterioration of effectiveness of drains and other factors. Each of these conditions deserves careful attention when dam safety is evaluated.

Unless an embankment shows signs of instability or high pore pressures during normal operations, there is no way to determine by inspection whether it will be stable under the above described loading conditions. Evaluating the stability of an embankment for such conditions requires determination of the strengths of the embankment and foundation materials and comparison of these strengths with the stresses that result from the loading.

Failures of dams due to extended periods of seepage at high pool have occurred in at least one large dam and in a number of smaller dams. If a dam is found to be unstable for steady seepage at high pool level, the most common remedy is to install drains, relief wells, or other seepage control measures to reduce the magnitudes of the pore pressures within the embankment and/or its foundation.

Rapid drawdown has caused instability in the upstream slopes of many dams, including Pilarcitos Dam and San Luis Dam in California and many others. Rapid drawdown slides in embankments ordinarily do not have any significant potential for loss of impoundment, because they usually involve sliding at a depth of only a few feet in the upstream slope and do not extend through the top of the dam. In the case of San Luis Dam, however, the sliding was considerably deeper, extending into a layer of highly plastic clay in the foundation. Although the slide at San Luis Dam did not extend through the top, deep-seated failures of this type do have a potential for doing so, and this possibility needs to be considered when stability during drawdown is evaluated. Stability during drawdown can be improved by flattening the upstream slope or by adding a layer of free-draining material to blanket the upstream slope.

Earthquakes have caused instability and complete failures in dams built of loose, cohesionless (sandy or silty) soils and dams built on foundations containing such soils, which can "liquefy" or lose all strength under cyclic loading. Examples include Sheffield Dam and Lower San Fernando Dam in California. Sheffield Dam failed completely as a result of liquefaction of loose sands in the foundation, and the entire reservoir was released. Lower San Fernando Dam suffered a deep-seated upstream slide that extended through the top of the dam and lowered the top to within about 3 feet of reservoir level. The reservoir was lowered as quickly as possible, and complete failure was avoided by a narrow margin.

Usually when a dam is found to be unsafe during an earthquake because of a liquefiable foundation, the remedy is to build another, more stable

**TABLE 7-1** Evaluation Matrix of Embankment Dams

| Defect | Possible Indicators | Possible Causes | Effects | Potential Remedial Measures |
|---|---|---|---|---|
| (A) Embankment mass movement (slope failure) | Slumps on embankment face<br>Longitudinal cracks<br>Arcuate cracks<br>Hummocky (irregular) slope<br>Bulge in slope<br>Sag in crest<br>Bent tree trunks<br>Misaligned guard rails or similar structures | Inadequate strength<br>Slopes too steep<br>Phreatic surface too high<br>Cracking due to differential settlement<br>Earthquake<br>Rapid drawdown of reservoir or tailwater<br>Large trees on dam overturned<br>Spillway or surface drainage discharge eroding embankment<br>Temporary saturation due to rain storms, snowmelt, or high tailwater<br>Decaying organic material in embankment | Possible massive failure of dam<br>Damage to spillway or outlet works, resulting in dam failure | Determine specific cause(s) by test borings, strength tests, and piezometers. Based on test results, design appropriate remedies. Some alternatives are:<br>*Free-draining downstream buttress*<br>Flatten slopes<br>Lower the phreatic surface (upstream barrier, internal slurry wall or membrane cutoff, grouting)<br>Remove and replace weak soils<br>Control surface erosion with riprap or other means<br>Realign-relocate appurtenant structures as required<br>Permanent partial reduction in pool level<br>In some cases total draining and breaching are required for safety or are more economical |
| (B) Embankment excessive | Seepage carrying soil fines | Lack of appropriate internal drainage | Dam failure by internal erosion | Distinguishing unsafe seepage from normal seepage requires |

| | | | | |
|---|---|---|---|---|
| seepage | Sinkholes on embankment face<br>Boils<br>Concentrated seepage<br>Unusual wetness on embankment slope<br>Unusually soft or quick embankment slope<br>Marsh-type vegetation on embankment slope | Inadequate core or cutoff<br>Inappropriate embankment material<br>Layering of relatively permeable zones in embankment<br>Inadequate compaction<br>Clogging of drains or filters<br>Burrows caused by muskrats, beavers, groundhogs, foxes, moles, chipmunks<br>Surface erosion gullies intersecting seepage zone<br>Temporary saturation due to rain storms, snowmelt<br>Seepage into, out of, or along conduits and drains | Structural failure due to uplift of embankment or appurtenant structures<br>Loss of storage | considerable judgment.<br>Amount of change in the rate of seepage is an important factor. May require installation of piezometers to help determine seriousness. Highly concentrated seepage *or* evidence of internal erosion or mass movement definitely requires treatment. If it appears that seepage line is high enough to threaten mass stability, consider steps under mass movement above. *If mass movement is not indicated, a filtered drain in the area(s) of concern is usually most appropriate*. Other alternatives:<br>Upstream seepage barrier (blanket)<br>Install seepage cutoff beneath crest, such as slurry wall, thin membrane wall, grouting<br>Filtered relief wells<br>Fill gullies with filtered drain, riprap, prevent further erosion<br>Remove trees, replace soil<br>Trap and remove animals<br>In some cases total draining and breaching is the most economical safe action |

**TABLE 7-1** Evaluation Matrix of Embankment Dams (*continued*)

| Defect | Possible Indicators | Possible Causes | Effects | Potential Remedial Measures |
|---|---|---|---|---|
| (C) Foundation movement | Heave of foundation near embankment toe<br>Sinkholes<br>Transverse *or* longitudinal cracks in embankment<br>Sags in dam crest | Consolidation settlement<br>Collapse of cavities (limestone terrane)<br>Shear failure (usually occurs during construction and thus is usually not a problem with existing dams)<br>Liquefaction<br>Earthquake | Embankment failure due to loss of support, cracking, piping, mass movement<br>Misalignment of appurtenant structures<br>Cracking of appurtenant structures<br>Loss of freeboard (storage) due to sags in crest | Increase embankment mass with free-draining massive downstream addition (subsurface data needed for optimal safe design)<br>Regrade crest<br>Realign appurtenant structures<br>Repair appurtenant structures |
| (D) Foundation excessive seepage | Seepage carrying soil fines<br>Sinkholes<br>Boils at toe and downstream<br>Concentrated seepage<br>Unusually soft or quick ground | Inadequate cutoff<br>(Re)opening of cavities (limestone terrane)<br>Cracks due to differential settlement<br>Fractures in foundation rock or soils | Embankment failure due to internal erosion in foundation, loss of support, collapse<br>Loss of storage | See measures for embankment seepage (above)<br>Downstream filtered drain trench or relief wells<br>Upstream blanket<br>Grouting<br>Slurry wall or membrane<br>Permanent reduction in reservoir pool level |
| (E) Unprotected slopes | Obvious visual indicators | Undersized material<br>Disintegrating riprap<br>Surface not properly graded<br>Obstructed or improperly located surface drain outfalls | Deep gullying<br>Beached upstream slope<br>Reduced cross section can cause structural or seepage failure | Place or augment riprap<br>Backfill and regrade surface<br>Place granular downstream slope protection<br>Realign and extend discharge of spillway and surface drains as required |

| | | | | |
|---|---|---|---|---|
| (F) Spillways | | | | |
| F1. Erosion of spillway | Obvious visual indicators | Inappropriate materials (soils)<br>Too frequently used spillway<br>Improper design section in spillway or stilling basin<br>Lack of maintenance | Failure of spillway, then embankment | Reduce frequency of use by providing other spillway/storage capacity<br>Riprap spillway<br>Vegetate spillway<br>Pave spillway (with underdrainage)<br>Provide appropriate energy dissipator |
| F2. Uplift | Vertical separation along joints<br>Cracking of slabs<br>Cracking of walls<br>Lateral movement or tilting of walls | Inadequate or improperly designed underdrainage | Failure of spillway, then embankment | Provide relief drains<br>Provide underdrains |
| F3. Undermining | May be difficult to detect<br>Seepage/soil losses at toe of spillway<br>Core holes and pressure tests<br>Sonic tests<br>Sometimes reflected by slab misalignment, cracking (due to loss of support), cavities | Piping under spillway slab due to inadequate or improperly designed underdrainage | Failure of spillway, then embankment | Grout existing voids<br>Seal cracks<br>Seal joints<br>Provide filtered underdrains<br>Replace spillway |
| F4. Concrete components; cracking, displacement, overstressing instability | Obvious visual indications<br>Also, see section on masonry dams | See Chapter 6 | Failure of spillway, then embankment<br>Failure of gate piers<br>Reduced spillway capacity | See Chapter 6 |

**TABLE 7-1** Evaluation Matrix of Embankment Dams (*continued*)

| Defect | Possible Indicators | Possible Causes | Effects | Potential Remedial Measures |
|---|---|---|---|---|
| F5. Obstruction | Obvious visual indicators | Lack of log booms or trash racks<br>Improper sizing<br>Unanticipated trash burden | Overtopping of dam or spillway walls<br>Damage to spillway structures | Establish maintenance/inspection program<br>Trash racks and log booms<br>Modify sizing<br>Remove source of obstructing material |
| (G) Spillways and outlets | | | | |
| G1. Faulty gates and hoists | Normally, gates and hoists are operated to test working order<br>Visual inspection for corrosion and to ensure that all components are present and in good order | Settlement<br>Corrosion<br>Initial misalignment<br>Vandalism | Loss of control of spillway release<br>Could result in inability to drain reservoir to prevent structural or seepage failure<br>Could result in overtopping failure | Realign and replace as necessary<br>Provide protection against tampering and vandalism |
| G2. Obstruction | Obvious visual indicators | Inadequate trash rack, log booms<br>Collapse (see above possible causes)<br>Vandalism | Loss of outlet capacity could result in overtopping failure of dam (see above effects) | Establish maintenance/inspection program<br>Trash racks, log booms |
| G3. Inaccessible gate controls | Obvious visual indicators | Poor design<br>Bridge overload<br>Pier displacement from earth loads<br>Unstable gate tower<br>Inundated or blocked roads | Loss of control of reservoir releases<br>Loss of control of spillway releases | Replace or strengthen structural component<br>Counteract earth load<br>Convert dry tower to wet tower<br>Provide alternate access |

| | | | | |
|---|---|---|---|---|
| G4. Project insecurity from tampering and vandalism | Obvious visual indicators<br>Rocks in stilling basin<br>Missing or damaged control stem wheels, etc. | Lack of barriers<br>Infrequent site visitations by owner | Structural damage<br>Interference with flow<br>Inoperable controls could preclude emergency spillway releases or opening bottom drain, resulting in dam failure | Security fences<br>Intrusion alarms<br>Periodic site visits<br>Putting locks on emergency controls usually *not* advisable |
| (H) Outlet works | | | | |
| H1. Leaking conduit | Visual inspection or water leaks into pipe<br>Visual inspection for soil deposits in pipe<br>TV inspection for above if pipe not accessible to direct inspection<br>Sinkholes in embankment surface over or near alignment of conduit<br>Discharge temperature/chemistry measurements (rarely conclusive)<br>Pipe thickness measurements (ultrasonic) | Misaligned pipe due to settlement or poor placement<br>Inappropriate connections<br>Improper bedding<br>Improper backfill compaction<br>Corrosion of pipe<br>Electrolytic loss of pipe metal<br>Erosion of conduit<br>Uplift of drop inlet<br>Vibration (water impact or water hammer)<br>Downstream control only | Piping of embankment soil, may result in sudden and complete failure of dam<br>Loss of storage<br>Loss of conduit | Sleeve conduit and grout annulus<br>Grout entire conduit and provide new outlet<br>Grout outside perimeter of conduit<br>Move control gate to upstream end to depressurize<br>For eroding pipe, remove source of eroding material and/or pave bottom of pipe<br>Provide additional anchorage if needed to resist uplift<br>Modify hydraulically to eliminate vibration<br>Provide cathodic protection |

**TABLE 7-1** Evaluation Matrix of Embankment Dams (*continued*)

| Defect | Possible Indicators | Possible Causes | Effects | Potential Remedial Measures |
|---|---|---|---|---|
| H2. Piping along conduit | May be difficult to detect<br>Seepage/soil deposits at downstream end of pipe<br>Sinkholes in embankment surface over or near alignment of conduit<br>Deformation of conduit due to loss of soil support<br>Voids between pipe and soil at outlet | Inadequate compaction of backfill<br>Improper bedding<br>Leaks in conduit (see above possible causes) | Possible sudden and complete failure of dam<br>Loss of storage<br>Loss of conduit | Install filtered drain around outlet of conduit<br>Grout outside perimeter of conduit if significant loss of embankment soil is thought to have occurred<br>Provide cathodic protection |
| (I) Reservoir margin | | | | |
| I1. Mass movement (landslides) | Indications of cracking, creep, distortion of reservoir margin | Complex array of possible geologic factors<br>Earthquake triggering<br>Rainfall triggering<br>Reservoir rise | Obstruction of emergency spillway<br>Damage to outlet works<br>Wave damage to dam<br>In severe cases, damage (movement) of embankment | Removal of unstable material<br>Rock bolting<br>Buttresses<br>Other classical landslide treatment techniques |
| I2. Seepage from margin of reservoir | Loss of storage<br>Emergence of seepage downstream of reservoir | Complex array of possible geologic factors | Loss of storage<br>In rare cases, failure of reservoir margin by piping | Blanket on reservoir bottom<br>In rare cases, grouting<br>In rare cases, provide filtered drain to prevent piping |

NOTE: This table considers only the major nonhydraulic considerations for existing embankment dams. It is not meant to be comprehensive. Also, the complex interrelationships between the embankment, foundation, appurtenant structures, and reservoir margin must be kept in mind.

dam downstream, abandoning the original embankment. If an embankment itself is potentially unstable during an earthquake, it may be replaced, as was the case with Upper San Leandro Dam (Gordon et al. 1973), or it may be possible to strengthen the dam with a buttress fill, as was done for San Pablo and Henshaw dams. In the case of Henshaw Dam in southern California, the reservoir level was lowered and a reinforced rockfill embankment was constructed to buttress the downstream slope of the dam and to retard outflow from the reservoir in case the old dam failed. A foundation that contains loose liquefiable soils may be densified by deep compaction techniques, such as vibroflotation, compaction piles, or heavy tamping, or it may be excavated and replaced. Drainage to reduce liquefaction susceptibility has been considered in some cases.

Dams built of cohesive soils on stable foundations have been found to perform quite well during earthquakes, and they pose a much smaller hazard than do dams constructed of or founded on loose, liquefiable, cohesionless soils (Seed et al. 1977). Even though they may remain stable during an earthquake, cohesive soil embankments may suffer permanent deformations as a consequence of earthquake shaking, which may take the form of bulging of the slopes, bodily movement of the dam, and possibly settlement of the top.

### *Causes of Slope Instability During Operation*

Once an embankment dam has been constructed to full height and has been stable for a period of time, there are (at least) three conditions that may result in subsequent instability:

1. Rapid drawdown of the reservoir may result in instability of the upstream slope. There are no reported cases where this type of failure led to loss of impoundment.
2. High reservoir levels for extended periods may result in high pore pressures in the embankment and its foundation and in instability in the downstream slope. One such case has been reported where this type of failure led to loss of impoundment, although the failure might also have been due to seepage erosion and piping (ASCE/USCOLD 1975).
3. Earthquakes subject the embankment to transient forces and may also result in loss of strength in loose cohesionless soils.

### *Methods of Assessing Stability of Existing Dams*

The fact that a dam has been subjected to the most severe conditions (drawdown, sustained high reservoir level, or earthquake) expected during its life without suffering failure or accident is the strongest possible indicator of its ability to withstand similar conditions in the future. Even then, a careful examination of the dam, searching for signs of actual or incipient instability, is an absolutely essential part of any evaluation of stability for an existing dam.

If a dam has not been subjected to the most severe conditions expected, evaluation of its safety requires collection of data to estimate the strength properties of the embankment and the water pressures to be expected in the embankment during the event.

These data can be obtained through examination of as-built drawings, construction records, and test results for record samples, if available, combined with field explorations, field tests, and laboratory tests. Existing piezometer records can be evaluated and new piezometers installed if necessary to obtain needed information. In a case where investigations indicate that design assumptions were conservative or correct with regard to strength and pore pressures, additional stability analyses are not necessary. If the investigations indicate lower strength or higher pore pressure than used in design analyses, additional analyses will be needed to determine if the stability of the embankment will be adequate. Analyses used to determine the stability of embankment dams are discussed in the Stability Analyses section.

### *Remedies*

Some earthfill dams were constructed with unreinforced concrete face linings under the assumption that the embankment was much more pervious than the lining and would thus result in a low phreatic surface in the embankment. These linings have cracked and admitted water in sufficient quantities into the embankment so that the raised phreatic surface reduces the embankment stability unacceptably. The design concept is no longer accepted as valid.

In some cases the stability has been restored to the desired degree by installing a filter and drain blanket beneath a new portland cement or asphaltic concrete lining so that seepage through the lining no longer saturates the embankment. In another case a terminal storage reservoir was completely lined with a compacted impervious clay lining, which was protected from weathering by a 3-inch asphaltic concrete covering. The impervious earth lining was used as protection against seepage losses and instability in the geologic formations comprising the reservoir bowl. The purpose of the asphaltic covering was to preserve water quality and to aid in periodic cleaning of the reservoir. Large areas of the lining failed by sliding due to reservoir drawdown. The slide material was removed and a system of line drains and toe collector drains installed in the formation. The system was drained through an existing drain outfall serving the upstream impervious zone of the earthfill dam. The clay blanket and asphaltic covering were then replaced.

### Excessive Settlement

Settlement due to consolidation or compression of embankment and foundation soils usually continues at a slowly decreasing rate after construction of an embankment, and this action may lower the top below its design elevation. Such a condition can be detected readily by level surveys of the dam crest or by settlement observations. Usually dams are constructed with excess freeboard or "camber," in order that some amount of settlement may occur without lowering the top below design level. If excessive settlement occurs and the top is too low, it can be raised by adding fill or a concrete parapet wall to the top of the dam.

In extreme cases the foundation soils may be so weak and compressible that it is not possible to raise the top to the desired level simply by adding fill, because the embankment would be unstable. This was the case at Mohicanville Dike No. 2 (appurtenant to Mohicanville Dam in Ohio), which was constructed on soft peat and clay and had been about 12 feet below its design level for many years. Methods considered to raise the dike to design elevation were (1) stage construction to raise the embankment slowly, while the foundation soils gained strength through consolidation; (2) construction of a concrete parapet wall atop the dike, together with a concrete diaphragm wall cutoff; and (3) use of geofabric reinforcing in the embankment to permit construction of a less massive dike imposing smaller loads on the foundation, together with an upstream seepage cutoff.

Transverse vertical cracks in homogenous earthfill dams have resulted from differential settlement caused by foundation profile irregularities and by consolidation upon saturation of the embankment or of a granular foundation. Nonplastic soils placed at below-optimum water content are especially vulnerable because they are brittle and highly erodible.

One successful remedy has been the placement of a flexible, impervious asphaltic membrane on the upstream face together with a shallow trench at the upstream edge of the dam crest backfilled with a compacted mixture of soil and bentonite. The backfilled trench serves to intercept the wider, more defined cracks that are usually confined to shallow depths, while the membrane serves to seal off any narrow or incipient cracks that might extend more deeply. The embankment face is shallowly stripped, rolled, and bladed. A heavy asphaltic impregnated hemp mesh is anchored and a sprayed asphaltic emulsion is applied, followed by a thin sand coating and a second penetration coat of sprayed asphalt. A 2-foot normal thickness protective blanket of fine-grained soil is then placed over the membrane commencing at the bottom of the slope.

Older dumped rockfill dams have characteristically settled relatively large amounts, especially from the initial application of the water load during first filling. Where the facing is reinforced concrete, this settlement has often cracked the slabs, closed and spalled the central vertical joints and the horizontal joints, torn waterstops, and opened the terminal vertical joints and any perimetric joints. This disruption of the facing has usually been most severe in the lower elevations and has resulted in leakage that in some instances has been very large (hundreds of second-feet). Fortunately, this type of embankment dam is very resistant to large leakage flows as long as the foundation is nonerodible.

Where there is concern for safety or economic loss of water or where circumstances are psychologically disturbing, the leakage can be reduced and controlled by placing a compacted berm of a well-graded clay-silt-sand-gravel mixture on the facing to an intermediate height. If the reservoir cannot be taken out of service, a gravel-clay-bentonite slurry can be deposited under water on the face with a wheeled enclosed skip cart. The skip cart travels on the face and is remotely opened when at the discharge position. The conventional tremie method is suitable when the required volume of the mixture is smaller.

## Slope Erosion

Embankment dams, particularly if they are relatively old or did not have proper slope protection measures incorporated into their original construction, are subject to slope deterioration from erosion of both their upstream and downstream faces. Such erosion does not necessarily lead to catastrophic dam failure or even a major safety problem. However, if it continues to occur and corrective measures are not taken, serious consequences could develop because the embankment cross section would continue to be reduced, often at the most critical elevations. As noted in Chapter 5, the potential for rapid erosion of dispersive clays calls for special care in investigation where the presence of such clays is a possibility.

The erosional problems most commonly encountered are those created by wave action on the upstream dam slope and by improperly controlled runoff from precipitation on the dam top and/or downstream slope. Areas of contact between the embankment and abutments and the embankment and appurtenant structures are especially vulnerable. However, if the public has access to a particular site for recreation purposes, erosion of both upstream and downstream slopes and the dam crest can be aggravated by repeated foot or vehicular traffic, causing rutting and loosening of soils, which become susceptible to further erosion during succeeding runoff from rainfall or snowmelt.

Downstream slope erosion is readily recognizable at any given time in the form of shallow or deep gullying caused by runoff concentrations on the slope or at the embankment-abutment contact. On the other hand, upstream slope erosion may or may not be readily recognizable, depending on the level of the reservoir as related to the eroded area. Improperly protected upstream slopes can erode rapidly and severely from periods of heavy wave action, but it is possible that such erosion could temporarily go undetected, particularly if the reservoir is rising, until the water level later drops to expose the damaged area. This latter type of erosion usually results in combined "benching" and oversteepening of portions of the upstream slope and could lead to more serious slope instability if not corrected in time.

### *Remedies*

Upstream slopes severely benched by erosion can be restored by removal of loose surface materials, slope grading, and replacement with compacted fill. A cushion or bedding layer of properly graded sand and gravel or small rock is then placed on the restored slope and covered with a layer of sound, durable riprap (graded rockfill), properly sized and dimensioned to suit the wave conditions expected to be encountered.

As a substitute for large, graded rockfill, which may not be readily available in the proper size and soundness, commercially available gabions or slope protection mattresses can be placed on the embankment face and filled with smaller rock to provide equivalent protection. If erosion of the underlying embankment is a concern, a suitable geotextile filter fabric can first be placed on the slope to prevent migration of the fine soil particles from beneath the gabions.

The general characteristics and uses of geotextiles and allied products are briefly described in the section Seepage and Piping. Fabric forms have been successfully used for erosion-controlling revetments on embankment, excavated, or natural slopes instead of the more expensive or unavailable conventional rock riprap (Lamberton 1980).

The revetments are made with a two-layer fabric woven together at tie points to form a quilt-like envelope. The tie points can be spaced on grids from 5 to 10 inches on centers to create "pillows" analogous to stone of different sizes. An open weave is used at the tie points to join the layers together and to form weep holes for relieving hydrostatic pressure behind the revetment.

The revetment is installed by first anchoring the upper edge in a trench along the top of the slope with mortar injected into the end of the fabric envelope. The fabric is then rolled down the slope and pumped full of mortar, expanding the envelope into a pattern of individual nodules or pillows. The mortar may be either a sand-cement or a pea gravel-sand-cement mix.

Although the upper layer of fabric may be gradually degraded by ultraviolet radiation and abraded by erosion, the lower layer is bonded to the mortar and provides both a filter to retain the underlying soil and flexible tensile reinforcement to help hold the mortar pillows in position. For some subsoil conditions a filter layer may be required below the fabric.

Other common means of upstream slope protection in areas where rockfill is not readily available are asphalt or portland cement concrete and soil cement. Each type of protection requires its own unique design and bedding conditions. In some smaller less important storage structures, under certain conditions of reservoir size, operating conditions, dam face slope, and protection from prevailing winds, grass slope vegetation has been used as a relatively inexpensive erosion protection measure. However, this latter means must be viewed as a less reliable long-term approach and would normally require more frequent maintenance. Other, less common and usually more expensive means of upstream slope protection, such as commercially manufactured liner fabrics or sheeting, are also available and are certainly worth considering under the proper site conditions and environment. Pacific Gas and Electric Co. has had good experience with a number of installations of a gunite membrane. This is also highly effective as a rodent inhibitor.

Downstream slope erosion, in the form of gullying, can be repaired and its recurrence prevented by excavating the eroded areas to provide working room, refilling and compacting the eroded and excavated areas (placement of impervious materials over large areas on the downstream slope should be avoided), then placing a protective layer of angular rock or stream gravels and cobbles on the slope. Often, vegetative slope protection, in coordination with proper slope drainage, will provide a relatively inexpensive means of slope protection. A system of concrete- or asphalt-lined surface drains, either preformed or cast in place, can be installed on narrow berms and used in conjunction with a protective cover of planted grasses to provide the required level of protection. Maintenance, in the form of initial watering and periodic mowing, is often necessary with this type of slope cover.

The means of erosion protection discussed above for both upstream and downstream embankment slopes can be used equally as effectively on native soil slopes adjacent to the dam embankment if erosion of such slopes threatens to undermine the dam embankment or create a more serious slope stability problem related to the dam abutment or associated hydraulic structures.

Any surface erosion that seems to be resulting from water seeping through the dam embankment or abutment may be indicative of a much more serious dam stability problem and should be quickly and thoroughly checked by a qualified engineer to determine the need for further study and remedial action.

## Seepage and Piping

All dams, regardless of type, leak to one degree or another. Seepage may be through the foundation or the embankment, along the foundation-embankment interface, or in any combination of those paths. The seepage volumes may be substantial or barely noticeable. The water may be transporting suspended or dissolved solids. In some cases the seepage may be entirely harmless; in others, it may be extremely serious and immediate treatment becomes imperative.

Whether such leakage can or will lead to serious stability problems or will require expensive repairs depends on many factors, some of which may be related to the original embankment design (for example, inadequate seepage cutoff measures) and some of which may be related to other factors, such as undetected foundation conditions or improper construction procedures or control. Whatever the cause, a developing seepage problem is often quite evident from visual inspection and/or changes in piezometer readings (if such devices have been installed in the dam) and, if not given proper, timely attention, can lead to serious and expensive problems or even catastrophic failure of the dam.

Where treatment is necessary, in the interest of safety, various remedies are available. The choice of remedy is controlled by many factors, including the quantity, path, pattern, and gradient of the seepage flow; the configuration of the dam; its zoning; the characteristics of the foundation formations and geologic structure; the engineering properties of the materials composing the embankment; the foundation treatment and materials placement procedures during construction; and the financial feasibility of comparative costs.

It is not possible and, in many instances, is undesirable to stop the seepage completely. Instead, the objective is to control the forces and actions of the water that would otherwise adversely affect the stability and integrity of the dam-foundation unit. The major forces and actions are softening, saturating, solutioning, pressuring, internal erosion, and transporting. These factors may occur in various combinations to produce abnormal conditions not contemplated in the design of the structure and which cannot be adequately resisted. For example, saturation and pressure may reduce the strength of the materials in an embankment zone sufficiently to cause a slide. Or solutioning and transporting may remove materials from a foundation formation, causing its collapse and disruption of the embankment.

Almost all of these forces and actions are created in some way by the hydraulic gradient of the water as it travels into, through, and out of the dam and foundation. Accordingly, the fundamental achievement of a remedy must be the control of the hydraulic gradient within tolerable limits.

The hydraulic gradient can be controlled by barriers of low permeability, adjacent zones of increasing permeability in the direction of flow, lengthened travel paths that increase friction losses, and forced flow through anisotropic foundation formations in the direction of lesser permeability. All the remedies for seepage employ one or more of these techniques.

Piping, a form of internal embankment erosion, is caused by the progressive movement of soil particles from unprotected exits due to uncontrolled seepage emerging from an abutment or embankment slope. Piping occurrences are a very common cause of dam failures. Areas around and adjacent to conduits are particularly susceptible to piping because of the difficulty in properly compacting fill around these conduits.

A near-failure of Daggs Dam in Arizona in 1973 was attributed to a long-term piping problem associated with a damaged low-level outlet conduit. Repairs included the removal and replacement of the damaged conduit and surrounding embankment material as well as other measures to improve the dam's performance. Piping problems have also developed at dams with many years of satisfactory performance due to solution of soluble materials, such as gypsum or limestone, within the dam foundation. Similar problems in both embankments and foundations have resulted from animal burrows and rotted tree roots.

Embankment cracking due to differential settlement can also provide paths for uncontrolled seepage and progressive internal erosion. In fact, abnormal seepage through either the dam *or* foundation, from whatever cause, even if it does not cause a piping problem, can lead to high hydrostatic uplift pressures or unanticipated uplift pressure distributions. These pressures can, in turn, lead to the formation of boils and springs in, or downstream, of the dam. Also, through the reduction of shearing resistance, these excessive hydrostatic pressures can cause the failure of slopes and abutments. Pervious foundation seams, adversely oriented bedding planes, and open-jointed foundation rock can all provide paths for uncontrolled seepage. If the seepage path is through erodible material or dispersive clays, rapid failure can develop from a small initial seep.

How can seepage problems with potential serious consequences be placed under satisfactory control? There are a number of remedial measures available for such problems, depending on the nature and location of the seepage and on how rapidly the situation is recognized and given the attention it deserves. Examples of successful remedial measures for these more common defects are given below. It must be recognized, however, that in each specific case the details will differ and that remedial construction must be adapted to the actual conditions.

### *Embankment Seepage Corrective Measures*

Treatment for control of hydraulic gradients in the embankment varies with the type of dam. The need for such treatment is rare for rockfill dams because of the mass and high permeability of the downstream zones and resistance to turbulent flows. The treatment for zoned earthfill dams will usually be limited to added zones of higher permeability at the downstream face because the impervious zones are buried and unreachable, although the addition of barriers of low permeability within the upper elevations of an existing impervious zone may sometimes be feasible.

Seepage through so-called homogeneous earth dams, where permeability is relatively high or where leakage may concentrate through anomalous regions or transverse cracks, can be controlled by treatment either, or both, upstream or downstream. Upstream treatment may consist of placing a compacted, more impervious zone on the stripped upstream face of the existing dam. The reservoir must be emptied. If the presence of the impervious blanket on the upstream slope presents a stability problem during reservoir drawdown, the slope can be flattened by adding a pervious zone surmounting the added impervious zone. If the reservoir cannot be emptied, in some cases a layer of bentonite pellets placed underwater may be effective.

If the defect includes excessive seepage through the foundation or along the interface with the dam (often the result of inadequate foundation preparation originally), the new impervious zone can be extended in the form of a cutoff trench or a slurry trench excavated in the bedrock formation across the valley section and into the abutments along the upstream toe of the dam or in the form of an upstream blanket. Time must be allowed for accumulated silt deposits to dry unless excavating by dragline is possible. Partial cutoffs in stratified alluvial formations can reduce the hydraulic gradient by forcing the flow across the strata in the direction of lesser permeability.

If seepage emerges uncontrolled along the toe or over the lower portion of the downstream face, a berm or mildly sloping zone of sand and gravel or cobbles and rock fragments may be added to that face. The grading of the materials positioned immediately against the dam and abutments must be much more pervious than the material upstream and must also prevent movement (piping) of fines from the dam or foundation. If pervious material of the requisite grading is scarce or costly, the main body of the added mass can be comprised of other types of materials, if they are enveloped by pervious materials at all interfaces. The effect of the berm on the stability of the downstream slope must be considered. While it will normally enhance stability, it may not.

If the seepage is largely concentrated along the toe or groins, a drain pipe of clay tile, sewer tile, asbestos-bonded corrugated metal pipe (CMP), or PVC pipe, successively enveloped by gravel and by sand or filter fabric, can be installed in a trench excavated into the foundation along the toe of the dam. If the drain can be safely installed on an alignment upstream of the toe, it will be more effective, especially for slope stability.

Transverse cracking in homogeneous dams can be repaired if the causative forces have stabilized or have attenuated with time. One method was discussed above. When the cracks are limited to the higher elevations in a dam, as they usually are, a narrow trench can be excavated from the top of the dam and backfilled with impervious plastic soils. The reservoir may have to be drawn down or even emptied during repair. The strength of the backfill materials must be adequate; otherwise a critical failure plane may be induced by the backfilled trench. Reinforced plastic fabrics, anchored or buried along their perimeters and placed on a smooth prepared surface on the upstream slope and covered by a protective element, can also be considered.

Excessive leakage caused by disruption of the concrete face elements of a rockfill dam can be reduced or eliminated by selective removal and replacement of damaged panels, if the waterstops from adjacent panels are serviceable. If the embankment is still settling at a significant rate, the repair process will have to be repeated several times. The damaged panels can be covered with courses of redwood tongue and groove planking for increased flexibility during the active settlement period. Anchored butyl rubber sheets have been used successfully on the surface of the panels to waterstop the panel joints.

A leaky rockfill dam can be modified to include an inclined earth core by using the existing dam for the downstream shell and constructing transition zones, filter zones, impervious zones, and shell elements upstream. The opportunity for improved control of foundation seepage, if necessary, is available in such an alteration.

The upper sections of embankments that are riddled with tree roots or rodent holes can be restored by complete removal of the infested portions and by replacement with compacted fill securely bonded to the unaffected portions.

### *Foundation Seepage Corrective Measures*

Seepage through foundations can be controlled by grouting, blanketing, new cutoffs, drainage, and pressure relief wells. Usually the exact nature of the problem will have to be investigated and defined before the most appropriate means of treatment can be identified. Possible remedies follow.

- A grout curtain can be installed beneath the impervious zone of an embankment dam by drilling through the dam. Care must be used to avoid hydraulic fracturing of susceptible fills with the drilling fluid. Injection of grout between the foundation surface and the base of the embankment should be done carefully. Different techniques are available. Portland cement, bentonite, and chemical grout mixes are the three most common types for seepage cutoff applications.
- An impervious blanket of compacted earth or a commercially available liner can be placed on the floor of the reservoir. The blanket must be joined to the impervious element of the dam and to the abutments and must terminate in a satisfactory manner.
- The construction of a new cutoff and an impervious facing was described earlier in this chapter. A new cutoff can also be formed in alluvial deposits with a slurry wall. The wall must be joined to the impervious element of the dam. A horizontal impervious zone (blanket) can sometimes be used. Driven sheet piling cutoffs have been used with limited success.
- Embankment toe drains and drain blankets also were described earlier. The toe drain or part of the blanket drain can also be installed at depth in the foundation for dual service.
- Pressure relief wells or trenches backfilled with drain rock and filter material can be drilled or excavated along or beyond the toe of an embankment dam to control the escape gradients of seepage flowing through the foundation.
- In rock abutment formations, both grouting and drainage curtains can be formed by holes drilled from tunnels or galleries.

### *Use of Synthetic Fabrics*

Geotextiles, or synthetic fabrics, are gaining acceptance in various kinds of construction. These materials may be manufactured from fiberglass, nylon, polyester, polyethylene, polypropylene, or polyvinylchloride and may be woven or nonwoven. Some are available either reinforced or nonreinforced for different exposure and installation or service requirements. The predominant polymers used in civil works are polyester and polypropylene (Koerner and Welsh 1980; Timblin and Frobel 1982).

Although not strictly meeting the definition of a geotextile, closely allied products composed of similar synthetic materials have been developed and are being used for impermeability and for surface protection. Impermeability prevents saturation and uneconomic water loss by controlling seepage and leakage. Surface protection provides erosion control on slopes and bed and bank protection in water conveying channels. These products are the impervious flexible membrane liners and fabric forms manufactured from synthetics, including polyvinylchloride (PVC), chlorinated polyethylene (CPE), and chlorosulfonated polyethylene (CSP). Some are available either reinforced or nonreinforced for different exposure and installation or service requirements.

Current purposes served by geotextiles in construction are (1) filtering, (2) drainage, (3) separation of dissimilar materials or zones, and (4) reinforcement. Filtration arrests the movement of finer soil particles from a protected layer or zone of soil of lesser permeability to one of greater permeability. Drainage promotes the controlled passage of water in the plane of the geotextile either vertically, horizontally, or inclined to an outfall or drain line. Separation prevents the intermixing of adjacent layers or zones of materials of dissimilar grain sizes. Reinforcement helps stabilize a fill or embankment by supplying tensile strength to the mass.

In the improvement of existing dams, these fabrics have a potential for drainage and filter protection that has yet to be developed fully. They can be used as a boundary membrane beneath riprap or gabions to avoid washing of the underlying embankment or natural slope.

In many cases drainage of some kind will be found to be the solution of a deficiency at an earthfill. Sands and gravels have been used commonly to create the necessary filters and drains. Such materials may be in short supply and/or may require expensive washing and screening. Geotextiles may offer some benefits in such cases. Although for various reasons they may not be acceptable as complete substitutes for natural materials, their potential for cost saving can be considered. Since their overall record of performance worldwide is relatively short, the ability of the fabrics to endure and to retain their capabilities for a long time is not yet known. However, especially in remedial work on existing dams, where they may not have to be buried deeply and irretrievably, geotextiles may have suitable applications.

From the descriptions of the functions of geotextiles and these allied products it can be seen that they can have applications in the correction of defects in existing dams and their foundations. Following are two examples.

1. In 1979 a low earthfill dam located in Martin County, Florida, used as a cooling water reservoir, failed from water seepage through the foundation (see Figure 7-1). This foundation defect was corrected by an upstream bentonite slurry wall and by a downstream geotextile drainage system. The slurry wall was formed in a 14-foot-deep trench, 12 inches wide, joined to the impervious dam by an impervious blanket. The drainage system was constructed by first trimming back the downstream face and excavating a trapezoidal trench into the foundation along the toe of the steepened slope.

A layer of sand was spread over the trench surfaces and part way up the slope and was covered with a layer of filter fabric. A 12-inch layer of gravel was spread on the fabric, and a 12-inch perforated bituminous pipe was placed in the trench and covered with additional gravel. A second layer of fabric was placed over the gravel on the slope and in the trench. The entire drain system was then covered with compacted fill to a new, flatter downstream slope. The 12-inch drain pipe is connected to a series of round concrete sumps that collect drain flows for discharge by a float-controlled pump system (Civil Engineering-ASCE 1981).

This example demonstrates the application of several remedial measures for seepage control both in a foundation and in an embankment—the slurry wall and the connecting impervious blanket upstream; the inclined embankment drain and foundation trench downstream. Two basic functions of geotextiles are also demonstrated—filtration for the first layer and separation for the second layer. Also demonstrated is the increased slope stability from the flatter, modified slope in addition to that obtained by the elimination of embankment pore pressures downstream of the inclined drain.

2. The 11,530 acre-foot upper reservoir at Mt. Elbert Forebay, Colorado, (USCOLD 1981) of a large pumped-storage power facility was created by closing the open margins of a topographic depression with a 90-foot-high earthfill dam and a small dike at the power plant intake-outlet located directly above the valley in which the power plant is situated.

The reservoir margin on the valley side is a series of lateral moraines in which there are old landslide scarps. The reservoir bowl was originally lined with a 5-foot-thick compacted earth blanket. The blanket extends to an elevation 3 feet above the maximum reservoir water surface, including the side of the forebay formed by the moraines.

Soon after the first partial filling of the forebay, water levels in the piezometers and observation wells placed along the valley side began rising significantly. From a study of the situation it appeared that enough water might possibly seep through the blanket into the morainal formations and adversely affect the stability of the valley side.

A flexible membrane liner was installed over the entire 290 acres of the reservoir bowl. A 45-millimeter, reinforced, chlorinated polyethylene liner

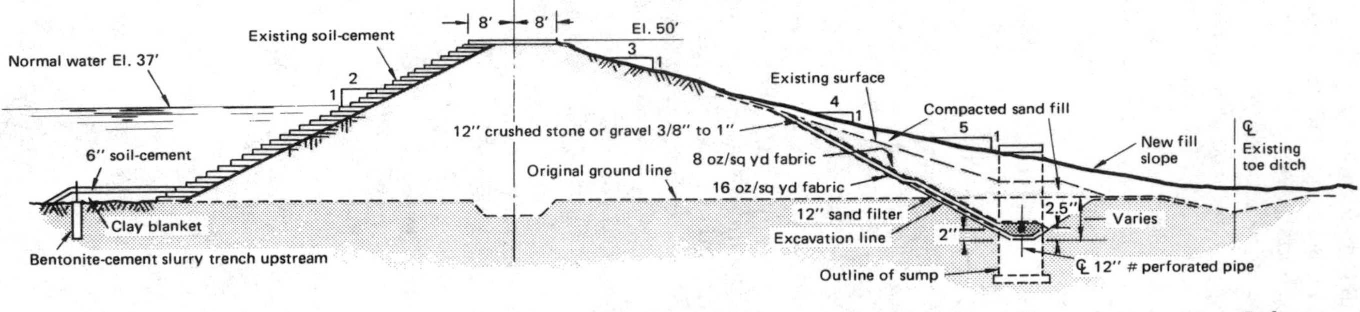


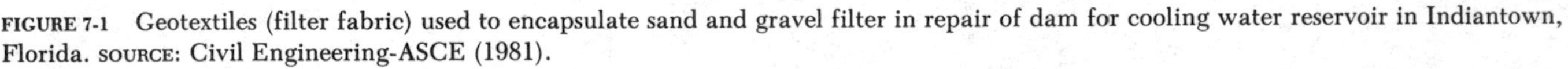
FIGURE 7-1 Geotextiles (filter fabric) used to encapsulate sand and gravel filter in repair of dam for cooling water reservoir in Indiantown, Florida. SOURCE: Civil Engineering-ASCE (1981).

was used. Riprap and other slope protection material were first removed from the earth blanket. The top 2 feet of the blanket were excavated and screened to obtain minus 1-inch material for the membrane subgrade and earth cover. The reservoir slopes were trimmed to 3:1 or flatter to facilitate placement of the earth cover over the liner and for stability.

The bedding for the liner was placed and compacted to obtain a smooth surface with two passes of a pneumatic-tired roller followed by two passes of a vibratory steel roller. The liner was anchored by planting it in a 1-foot-wide by 2.5-foot-deep trench at the top of the reservoir side slopes. The liner was unfolded and positioned manually by work crews. All field seams were overlapped, chemically cleaned, and then sealed with solvent adhesives. The liner is protected from weathering, vandalism, animal traffic, and ice action with an 18-inch earth cover. The earth cover is protected from erosion on the slide slopes by coarse gravel and bedded riprap. It was unnecessary to line the upstream face of the earthfill dam so the membrane liner was terminated by planting it in an anchor trench in the dam embankment along the toe of the slope (USCOLD 1981).

### *Slurry Walls*

Excessive seepage in a dam may be remedied by installing a trench or slot filled with an impervious material along the dam's axis, working from or near the top. In some instances it may be desirable and feasible to extend the seepage barrier into the foundation to remedy foundation seepage problems.

Caution must be used in employing a slurry wall or membrane for controlling seepage in a dam. The wall obviously *must not* be placed downstream from the centerline, since uplift pressures will increase upstream from the wall. Furthermore, the wall may introduce a definite plane of weakness through the dam that could result in structural failure of the dam, especially if the phreatic surface is not significantly lowered by the barrier. Another consideration is cracking (and, possibly, significant movement) of the soils along the wall as the transition from at-rest to active earth pressures occurs during and after wall installation. Cracking of the wall itself can occur if the backfill is not sufficiently plastic. The reservoir should be drained prior to wall installation. Embankment and foundation slurry wall installations must be closely monitored, refilling the reservoir should be done slowly, and piezometers should be installed to monitor seepage pressures.

Three basic techniques exist for installing slurry walls.

1. The "trench" method uses a backhoe, dragline, clamshell, or similar equipment for excavating a relatively wide (1.5- to 2-foot) trench. During

excavation the trench is kept filled with a bentonite-cement slurry to support the walls of the trench, so the excavation is done "in the wet." After excavation the slurry is left in place to form a relatively impermeable seepage barrier. This wall type is relatively simple to construct. Problems can include difficulties with quality control, controlling depth of excavation, segregation of materials in the slurry, and the necessity of disposing of excavated materials that may be semiliquid in form (due to mixing with the slurry).

A variation of the bentonite-filled slurry trench method consists of excavating panels in a bentonite slurry, using piles spaced a few feet apart as guides for the excavating tools, then tremieing concrete into each panel to construct a concrete diaphragm wall.

2. The "ditch" method uses a rapid-acting ditching machine that has multiple rotating buckets (normally 10 to 12 inches wide) that feed the excavated material onto a conveyor belt for loading and disposal. This method also employs a bentonite-cement slurry to support the excavating process and to form the relatively impermeable barrier. Quality control and disposal problems may be reduced by this method as compared with the trench method.

3. The "injection" method uses pressure jetting a bentonite-cement or other impervious slurry through ports in a beam. Multiple contiguous insertions of the jetting beam are made along a line to form a relatively impermeable thin membrane. A successful development of this concept uses a vibrating pile driver and an H-beam with jetting pipes welded inside the flanges; each insertion of the vibrating and jetting H-beam partially overlaps the previous insertion to ensure continuity and alignment of the seepage cutoff wall. Injection is continuous during insertion and extraction. Quality control is exercised at the slurry batch plant and by monitoring the pump pressures during jetting. The width of the membrane is sufficient to control seepage. Barriers of this type have been installed to depths of over 100 feet, and some have been installed through relatively dense and gravelly materials. Where the seepage water is chemically reactive with bentonite-cement (such as storage lagoons for some chemical-processing facilities), chemically resistant slurries have been developed and used. This technique would not be suitable where large boulders exist or where hard rock layers have to be penetrated.

Following are several examples of slurry walls in dams.

- Trenching Method

1. Razaza Dam, Iraq. A slurry trench wall was installed in 1969 in this 30-foot-high dam and through about 30 feet of aeolian and fluvial foundation soils to remedy excessive seepage problems. The trench was excavated

with a kelly mounted hydraulic grab bucket to form a wall about 20 inches wide, up to 80 feet deep, and 8,200 feet long. The slurry used consisted of cement (C/W = 0.10), bentonite, and local sands and silts, mixed to a paste with a water content of about 35 to 40%. After slurry wall installation and reservoir refilling, piezometers indicated a 90% seepage head loss across the slurry wall (Japan Dam Foundation 1977).

Other examples of successful remedial use of bentonite-cement slurry trenches are at the Eberlaste Dam, Austria; Kranji Dam, Singapore; Laguna Dam and others in Mexico (Japan Dam Foundation 1977).

2. Wolf Creek Dam, Kentucky. Construction was completed in 1951 on the 259-foot-high, 3,940-foot-long homogeneous earth embankment portion of this dam (flanking a concrete gravity overflow section). The foundation includes up to 40 feet of alluvial soils overlying limestone. In 1967 muddy seepage discharge and sinkholes developed, caused by piping of cavity-filling soils in the limestone foundation and collapse of overlying fill. In 1968-1970 a remedial limestone grouting program beneath a 250-foot-long section adjacent to the concrete gravity section was successful in treating that immediate area, but it was decided that a positive cutoff in the foundation was needed for an additional 2,000 feet beneath the embankment section. A 2-foot wide, 260-foot-deep concrete panel wall was installed by a modified slurry trench technique. The lower 100 feet of the wall had to penetrate cavernous limestone. The wall was constructed by sinking steel casings on 4.5-foot centers, excavating between the casings in panels using a specially designed clamshell bucket working in a bentonite slurry mix, then tremieing portland cement concrete (3/4-inch maximum aggregate size, 6- to 8-inch slump, 3,000 psi compressive strength) into the panels. A sequence of five stepped casing sizes (26 to 51 inches) had to be used because of the depth and rock conditions. Rock excavation was accomplished with percussion chisels, expandable biconcave chisels, and clamshell buckets (*USCOLD Newsletter* 1977).

This remedial work was completed in 1979 and is reported to be performing satisfactorily (Fetzer 1979).

Similar concrete panel construction is being done at the Walter F. George Dam in Alabama, Clemson Lower Diversion Dam in South Carolina, and at other locations. At the Clemson project, panels 20 feet long and up to 85 feet deep are being employed to intercept seepage in alluvial sands and gravel deposits in the foundation. At the Walter F. George dam the panels are extended up to 100 feet deep, into pervious limestones underlying the dam.

- Martin County Power Plant Cooling Pond Dike, Florida. A piping failure of a portion of this 20-mile-long dam in 1979 was caused by excessive foundation seepage. Remedial work included installation of a 10-inch-wide bentonite-cement partial cutoff by using a rapid ditching machine to excavate a slurry wall about 20 miles long. The slurry ditch was installed at the upstream toe of the dike, 14 feet into permeable foundation sands, down to a porous shell layer, and was tied into an upstream blanket. The slurry ditch is thought to be effective in reducing foundation seepage problems.

- Martin-Marietta Sodyeco Chemical Plant Waste Lagoon Dike, North Carolina. A 3- to 6-inch-wide vibrated beam bentonite-cement slurry wall about 2,500 feet long was installed to depths of 25 feet through permeable alluvial foundation sands and gravels to intercept anticipated foundation seepage. The vibrating beam also penetrated underlying residual micaceous sandy silts having standard penetration resistances of 20 to over 100 blows per foot. Penetration refusal depths for the jetting vibrating beam closely corresponded to power auger refusal depths of the site investigation test borings. At the time of this report, insufficient head has been put on this wall to determine the effectiveness of the cutoff.

Vibrated beam slurry walls have been put in numerous dam foundations in the United States and in Europe, prior to dam construction, with good performance. At chemical waste lagoons near Romulus, Michigan, and Richmond, California, chemically resistant asphaltic slurry walls have been successfully installed using this technique.

### *Repair of Timber Facings*

A number of older rockfill dams were originally faced with tongue and grooved dimensioned lumber secured on timber sleepers set in vertical chases in the upstream slope of the rockfill that was usually hand or derrick placed. These facings gradually rotted or sometimes were quickly destroyed by fire when the reservoir stages were low.

The water retention capability has been restored by removal of any facing remnants, except the sleepers, followed by the application of a thin (3 to 4 inches) gunite membrane facing. The membrane is reinforced for temperature with steel mesh or by an orthogonal system of reinforcing bars. The membrane is locally thickened and more heavily reinforced as a beam horizontally across the sleeper locations. The membrane is joined to the existing concrete of the foundation cutoff by shooting the gunite into the cleaned-out grooves in which the original timber facing had been installed. If the existing foundation cutoff is unservicable or ineffective, a new cutoff slab doweled to bedrock might be feasible.

Of course the reservoir must be fully emptied or locally cofferdammed to expose the lowest elevations of the top of the cutoff. If the reservoir stage can be suitably controlled for the inflows anticipated during the selected construction season, the facing remnants can be removed and the new facing installed by working from barges or rafts floating on the falling and rising reservoir stages.

Because the embankment has already undergone most of the time-dependent settlement due to its own weight and the settlement from initial water-load application, any residual settlement will be very small. Consequently, the facing will not be significantly stressed except by temperature and possibly earthquake.

Several dams in California owned by the Pacific Gas and Electric Company and by the Nevada Irrigation District have been successfully restored in this manner and are serving satisfactorily many years later, one for as long as 52 years.

## Overtopping

In the engineering of permanent embankment dams, overtopping has been strictly avoided, for sound reasons. While the advantages of earthfills are well known, their vulnerability to erosion unless properly safeguarded is basic. There is also ample record of the dislodging of rockfills subjected to overspill. An almost inviolable rule in the design of these structures has been to keep them free of superimposed conveyance structures, whether they be spillways, fish ladders, or pipelines. Any such facility that would obscure the embankment from inspection or conceal underlying adverse conditions has been regarded as objectionable.

These rules are well founded, and any deviation from them should not be taken lightly. However, confronted with the current reality of numerous dams of this type that are inadequately protected from flood damage and considering the insufficiency of funds for immediate full-scale increase in spillway capacity by traditional methods, the engineer may be obligated to weigh the merits of passing water over the embankment, at least as a temporary expedient.

### *Remedial Measures*

Various ways have been proposed to protect embankments during overtopping. Principal among these have been either armoring or reinforcing the embankment. Also, the use and/or contribution of parapets have been proposed to prevent the overtopping.

*Protective Armor.* Pravidets and Slissky (1981) have discussed the comparative effectiveness of riprap, packed edge-to-edge concrete cubes, and precast reinforced concrete slabs. Their work on a test spillway chute at the

Dnieper power station and their experience at a full-scale facility at the Dniester power development have demonstrated that important cost savings are obtainable by installation of precast slabs. These are reinforced wedge-shaped concrete members with drain holes, laid on the slope in shingle fashion.

The Dnieper test section was 46 feet wide and 118 feet long, installed in the spillway at the dam on a 6.5:1 slope. Dimensions of an individual slab or panel were 9.8 by 6.6 by 2.3 feet. Tests reportedly were conducted with a drop of 115 feet, unit discharge of about 640 cfs per foot of crest, and mean velocities as high as 75 feet per second. At the Dniester project the crest and downstream face of an embankment cofferdam 23 feet high and 820 feet long were covered with concrete slabs fastened together, each of these having dimensions of about 33 feet by 15 feet by 20 inches thick. This overflow facility was reported to have passed flood peaks with unit discharges as high as 140 cfs per foot. Both of these structures performed without significant problems.

The stability of this protective armor is enhanced by its multistepped profile and the favorable hydrodynamic pressure on the concrete surface. The drain holes and an underlying filter drain contribute substantially to stability and seepage control. A stepped toe is provided at the toe of the overflow structure for erosion protection and to ensure proper entry of the jet into the tailwater, with an unsubmerged hydraulic jump.

Advocates of this method of embankment protection believe that it is amenable to construction under a range of site and weather conditions. The slabs can be repaired readily.

*Reinforcement.* The experience already accumulated in the steel reinforcement of rockfill diversion dams shows the way to possible applications to other embankment barriers. Certainly, even in the case of comparatively resistant rockfills, extreme care must be exercised in designing for overtopping. Most critical is the protection of the downstream slope and especially its toe, which will be exposed to the potentially destructive velocities of both surface and seepage flow. In projects where the effectiveness of reinforced rockfill diversion dams has been demonstrated, essential elements have been (1) a membrane or zone to ensure relative impermeability, (2) safeguards for erosion protection of the top of dam and the upper part of the upstream slope, (3) containment of the rockfill on the downstream slope, and (4) anchorage of the steel reinforcement to prevent bursting of the contained fill by seepage pressures.

Various designs have been adopted for such reinforcement. The most successful are those that provide complete enclosure of the surface rockfill units by steel mesh that is deeply anchored in the dam, commonly by horizontal steel bars tied to anchor plates. Side anchorage into the abutment is also important. The gabions, or gabion-like elements, may be of large size and may therefore require internal rib-bars and tie-bars to maintain the geometry of the steel-mesh cage. In some cases, also, lean concrete has been placed in the toe of each armoring unit for shape retention. The rock material enclosed in the cage may have a range of gradations (e.g., 3 to 6 inches) depending on what is produced by rock excavations at the site. Obviously, the confined stones must be large enough so that they will not be washed through the mesh. The use of roller-compacted concrete might also be a solution.

*Parapets.* Although, in the most widely accepted practice, papapets on embankment dams are not intended to have water stored against them, such encroachment has happened on many projects and has in some cases become an approved part of the operating regimen during extreme floods. The addition of parapets and the changes in their purpose over a period of years may be part of the evolution of a project as conditions and demands change. Among the important changes may be the updating of hydrologic data or the development or application of different techniques for hydrologic analysis since the project was placed into operation.

Parapets originally designed as ornamental features or for residual freeboard may have only marginal structural capability. This could include (1) comparatively thin, sometimes dry masonry, walls with only a modest layer of mortar added to the upstream face; (2) thin gunite walls placed on steel mesh against a single vertical form; or (3) timber walls. Any of these must be analyzed carefully for structural adequacy against the surcharge water load. This may not be easy if knowledge is incomplete regarding the parapet's materials and details, including bonding, anchorage, reinforcement, and tying to the adjoining structural elements. Not infrequently strengthening is found to be necessary.

The importance of a competent parapet cannot be overemphasized if it is depended upon to support stored water, even temporarily, and if the top and downstream slope of the embankment are vulnerable to erosion.

## Malfunctioning Drains

Adequate embankment and foundation drainage is one of the most important aspects of maintaining a stable embankment dam. If a proper, well-functioning drainage system is originally incorporated into an embankment dam, it is quite possible that this system will continue to perform adequately throughout the entire service life of the structure. However, it is also possible that a drainage system that performed satisfactorily during the early life of a dam may become plugged or broken, or otherwise malfunction, thereby creating excessive uplift (internal hydrostatic pressures) within the dam and foundation. It is also possible that an outward appearance of drain malfunction can be created not by changes in the drains themselves but by deterioration of the dam embankment or foundation, permitting increased seepage that the drainage system was not designed to handle. On the other hand, silt deposits in the reservoir may reduce the seepage appearing at foundation drain outlets. It is extremely important, therefore, that any situation that appears to involve a deterioration of drainage capacity be thoroughly evaluated by an experienced engineer before corrective measures are defined and implemented.

One excellent means of inexpensively providing information that can be highly useful for diagnosing a potentially deteriorating drainage system is to channel to one or two locations and regularly measure visible seepage flows emerging from the dam toe area and any installed drains. This should be done over as wide a range in reservoir levels as possible, so that a relationship can be established between reservoir level and anticipated drainage flow rate. Any significant changes in the flows defined by this relationship may be cause for further investigation. Drain water samples can also be taken for chemical testing if it is suspected that chemical or bacterial reactions are involved in changing drain flows. If flow *reductions* are found to be occurring, backflushing of the drains can sometimes alleviate the problem. If flow *increases* are occurring, the water chemistry testing might indicate foundation solution activity.

Other corrective measures for malfunctioning drains or inadequate drainage can assume a number of forms. Excessive uplift resulting from inadequate control of seepage can be reduced to acceptable levels. If there are foundation drains and formed drains in the dam that have become plugged with chemical deposits, they can sometimes be reamed and their effectiveness restored if they are accessible from drain galleries or from the top of the dam. New foundation drains, both vertical and horizontal, can be drilled. If water losses are excessive, the foundation can be regrouted from galleries, if they exist, or from the top of the dam, but usually the more effective way to reduce uplift is by the addition of drainage.

### Foundation-Embankment Interface

The foundation-embankment interface is a critical area from the principal standpoints of both overall stability and seepage prevention and control. A poor bond between embankment and foundation can lead to piping by creating a favored seepage path along the contact. Improper or incomplete treatment of foundation joints and fractures, alone or coupled with an inadequate filter relationship between the embankment and the joints, can also lead to embankment piping and eventual collapse from internal erosion. Inadequate stripping of loose or otherwise undesirable foundation materials can create a weak plane along which major embankment instability could occur.

Where one or more grout curtains are constructed beneath a dam, it is imperative that continuity of the impervious dam zone and the grout curtain be maintained. This cannot be achieved without proper treatment of the foundation-embankment interface.

Once a dam embankment is constructed, it is obviously rather difficult, and sometimes impossible, to correct construction deficiencies along the foundation-embankment interface without first removing the embankment material from the problem area. As discussed earlier, problems such as incomplete treatment of foundation fractures and joints can sometimes be remedied by grouting through the dam embankment if the deficient areas are discovered and their extent defined, and they can be corrected before a major failure occurs.

The physical features of a defect at the foundation-embankment interface usually are not directly observable because they are hidden by the dam. The presence of the defect characteristics must, therefore, be deduced from indirect as well as direct evidence, obtained instrumentally or from drilling cores and logs and a study of visual manifestations, such as dissolved solids in seepage water or movements in the dam itself. For this type of problem, evaluation by an experienced engineer is essential, but even if the problem is properly defined, the cost of its solution may be very high.

## Trees and Brush

Trees and brush are frequently allowed to grow on the slopes and tops of embankment dams. These forms of vegetation should be removed, especially for small dams, for the following reasons:

- Potential for loss of freeboard and breaching if trees on the top are blown over during high-water conditions.
- Potential dangerous loss of dam cross section if trees on or near the slopes are blown over.
- Potential initiation of leakage by piping if trees die and root systems rot to become channels for flow.
- Obstruction of visibility and access to hamper observation and maintenance of embankments.

Each of these potential problems is considered in the following paragraphs and recommended general criteria for removal are included.

As a general rule, trees and brush should be removed from, and their growth prevented on, dam and dike embankments. Vegetation on slopes should consist of grass and should be cut at least annually so that there can be effective monitoring for animal burrows and seepage. Trees and brush adjacent to embankment slopes should be cut back at least far enough to permit such observation and to allow access to the toes of the slopes by maintenance equipment.

Tree root systems will vary with soil type and groundwater conditions. The following general comments, subject to further consideration at each site, are offered for general guidance:

- Root systems of usual tree types do not grow into the zone of saturation, i.e., below the steady-state phreatic surface. One result is that trees in swamp areas are shallow rooted and easily blown over.
- The spread of root systems is generally comparable to the spread of the branches but will vary with tree type and soil conditions.
- Root system penetration tends to be as follows:

  Pine: typical mat depth of 1 to 2 feet, maximum of 2.5 to 3 feet.

  Softwoods: generally shallow rooted.

  Oak: both deep and shallow rooted, typically 2 to 5 feet maximum mat depth (in glacial till likely to be 1 to 3 feet, more typical in loose-to-medium-compact, fine sand).

  Maple: 10 to 20% shallower than oak, typically 1 to 2.5 feet for major part of mat.

  Ash: relatively deep rooted but less dense mat than oak.

  Birch: relatively shallow rooted, typically 1 to 2 feet maximum mat.

Criteria for removal of individual trees and stumps should also consider the potential for damage due to the root systems. Living or dead trees whose uprooted root systems could endanger a dam or dike should be cut. This would include trees that could damage upstream slope protection and trees on a crest where uprooting could leave less than a 10- or 12-foot width of undamaged embankment. Trees on or near a downstream slope should be removed if their root systems can penetrate significantly into the minimum necessary embankment cross section.

Stump and major root removal should also be based on potential for damage. Major root systems in the top of the dam or within a minimum embankment section offer some potential for embankment damage by decay and should probably be removed. However, the decay will generally be to humus rather than to a void. There is some possibility of inside root decay with attendant potential for water flow, more so with hardwoods, and it would be prudent to remove major roots that traverse the top of a dam. Stumps and root systems that are not in critical locations can be left to rot in place unless removal is necessary for slope grading. The soil will gradually close in.

In connection with tree work on dam and dike embankments, the following additional comments are offered:

- The integrity of trees that remain in place should be considered—their root systems may be damaged by adjacent work, or they may be more exposed to wind.
- Cutting shallow roots to limit growth will not work without a barrier. Cutting will stimulate additional growth.
- Vegetation must be reestablished in work areas to protect embankments from erosion.
- Backfill material after stump or root removal should have characteristics similar to the embankment material at that location.

### Rodents and Other Burrowing Animals

The burrowing of holes in earthfill dams by rodents is a widespread maintenance problem. This problem is known or suspected to have caused several failures of small dams. The animals that have caused the most problems are beavers, muskrats, groundhogs, foxes, and moles. Beavers and muskrats cause the largest problem because they operate below the water level. They sometimes burrow holes below the water from the lakeside all the way through the dam.

Frequent visual inspections of the earthfill embankment should be made to detect the presence of animals or the holes they have made. If the presence of these animals is detected in the vicinity of the dam area, the animals should be eradicated by either trapping, shooting, or with poison. If they have made holes that are carrying or could carry water through the dam, these holes should be immediately repaired by excavating and recompaction or by filling with a thick slurry grout.

## STABILITY ANALYSES

The stability of an embankment dam, in conjunction with its foundation, must be evaluated from a number of different standpoints, as can be appreciated from the preceding discussions concerning the many potential defects that can create unstable conditions within the structure-foundation system. Among the various methods of stability analyses available to the engineer is the conventional analysis of slope stability, which, although it is a valuable tool in the assessment of embankment adequacy, must be performed and used with experienced judgment or it can produce completely misleading results that could lead to erroneous, even disastrous, conclusions regarding the safety of a dam. This is true because, as with most other types of numerical analysis, the final results are only as valid as the data used as input to the computations.

In the case of earth dams and their foundations, the input data themselves are often subject to fairly wide ranges of interpretation simply because the engineer is working with native materials that have been altered in varying degrees by the forces of nature.

Especially in the case of the analysis of stability of existing embankment dams, the exploration, sampling, laboratory testing, and materials properties evaluation program always has physical and economic constraints that limit the extent of knowledge that can be gained about the important physical properties of any given structure. For this reason the physical properties data that must be input to a numerical stability analysis are always subject to varying uncertainties that must be put in their proper perspective for each individual case. It is in this critical area that experience, as well as engineering judgment, are critical to the performance of the numerical analysis and evaluation of results. Even when the results of an analysis appear favorable, they cannot be viewed in a vacuum but must be integrated with all the other information available on the safety of the particular structure, thus becoming an important part, but still only a part, of the overall stability evaluation.

### Methods of Slope Stability Analysis

Various methods of slope and foundation stability analyses are available. The more common ones are two-dimensional and are based on limiting equilibrium. These analyses are known by a variety of titles, including slip circle, Swedish circle, Fellenius method, method of slices, and sliding block. There are differences in assumptions and force resolutions in the different methods. When forces representing earthquake effects are included, the analysis is often termed pseudostatic.

An analysis is made by assuming some form and location of failure surface, such as a circular arc, compound curved surface, or a series of connected plane surfaces. The configuration and positioning of the surface depend on the kind of embankment dam, the internal zoning, and the foundation's geologic structure. For example, connected plane surfaces are often used for an inclined or sloping core rockfill dam. Also, the trial failure surfaces are positioned judgmentally to pass through weaker or more highly stressed regions. Thus, a plane surface may be positioned in a confined fluvial foundation susceptible to high pore pressure. The most critical surface is defined as the one having the least computed factor of safety. The factor of safety is considered to be the ratio of forces or moments resisting the movement of the mass above the surface being considered to the forces or moments tending to cause movements. Both embankment slopes are analyzed for the specific service conditions expected.

Allied analyses are used during stability studies to determine seepage patterns and amounts, pore pressures, uplift forces, hydraulic gradients, and escape gradients in the embankment zones and the foundation by the application of the principles of flow through porous media and the graphical or mathematical modeling of flow nets (Cedergren 1967).

It is beyond the scope of this report to present the details of the many numerical methods available to analyze the stability of an embankment dam foundation system. These methods are discussed in great detail, with examples, in university textbooks for fundamentals; professional engineering society publications, such as the journals of the American Society of Civil Engineers, which consider practical, specific applications; and design manuals, monographs, handbooks, and standards of federal and state agencies engaged in the design of earth dams, such as the U.S. Army Corps of Engineers and the U.S. Bureau of Reclamation. The reader is referred to the references in this chapter for details of the subjects discussed in the following sections.

## Loading Conditions

Dams that have been stable for a period of time may become unstable when subjected to more severe loading conditions. Conditions that should be considered in the stability analyses of dams are listed in Table 7-2. These include (1) steady seepage with the highest pool level that may persist for a significant period, (2) rapid drawdown from normal pool to lower pool elevations, and (3) earthquake loading conditions. If a dam has not been subjected to the most severe loading conditions expected, its safety can be evaluated by measuring the strengths of the materials of which it is built and by performing analyses to compare these strengths to the stresses in the dam.

For modern dams, where factors of safety as shown in Table 7-2 were evaluated during design, sufficient information may already be available such that only a review is needed to establish the adequacy of the embankment and its foundation. As-built drawings, construction records, tests and

**TABLE 7-2** Loading Conditions, Required Factors of Safety, and Shear Strength for Evaluations for Embankment Dams

| Case | Loading Condition | Required Factor of Safety[a] | Shear Strength for Evaluation[b] |
|---|---|---|---|
| 1 | Steady seepage at high pool level | 1.5 | S strength |
| 2 | Rapid drawdown from pool level | 1.2 | Minimum composite of R and S |
| 3 | Earthquake reservoir at high pool for downstream slope; reservoir at intermediate pool for upstream slope | 1.0 | R tests with cyclic loading during shear |

[a]Ratio of available shear strength to shear stress, required for stable equilibrium.
[b]Terminology from U.S. Army Corps of Engineers. R = total stress shear strength from consolidated-undrained shear tests; S = effective stress shear strength from drained or consolidated undrained shear tests.

SOURCE: U.S. Army Corps of Engineers.

record samples, and performance records of piezometric levels and movements can be used to confirm or refute the suitability of design assumptions and evaluations.

For older dams, where factors of safety as shown in Table 7-2 have not been calculated, evaluating safety will require collecting data to estimate the strength properties of the embankment and the pore pressures within it, and performing stability analyses.

### *Steady Seepage Conditions*

The highest reservoir level that may persist over a significant period of time constitutes the most severe conditions of steady seepage, resulting in the lowest factor of safety for the downstream slope. A knowledge of water pressures within the various zones in a dam and its foundation is essential for a stability analysis. Field data may be obtained from observation wells and piezometers, as described in Chapter 10. Thereafter, pore water pressure throughout the embankment can be predicted from seepage analyses, providing the information needed for an effective stress analysis of slope stability.

### *Rapid Drawdown Condition*

Rapid drawdown subjects the upstream slopes of dams to severe loading by quickly reducing the stabilizing effect of the water acting against the slope without significant reduction of the pore water pressures within the soils forming the upstream embankment. Rapid drawdown slides occur in soils of low permeability, which do not drain freely. Generally, they are shallow slides within the upstream slope that pose no significant threat of loss of impoundment. In some cases (notably the slide at San Luis Dam) rapid drawdown slides extend into the foundation, and such deeper slides may pose a hazard for loss of impoundment if the slide cuts through the top of the dam.

### *Earthquake Condition*

Earthquake accelerations impose forces on embankments and their foundations; these forces are superimposed on the static forces. As a result, embankment dams may suffer a number of kinds of damage during earthquakes. According to Seed et al. (1977), these include disruption by fault movement, loss of freeboard owing to fault movement, slope failures induced by ground shaking, liquefaction of the foundation or embankment material, loss of freeboard due to slope failures, loss of freeboard owing to compaction of embankment materials, sliding of the dam on weak foundation soils or rock, piping through cracks induced by shaking, overtopping by earthquake-generated waves in the reservoir, and overtopping by waves caused by earthquake-induced landslides or rockfalls into the reservoir. The type of damage is highly dependent on the type of soil acted on by the earthquake forces. For example, embankment fill material that contains significant amounts of clay is able to withstand short-lived increases in load without a catastrophic failure; however, such embankments may suffer some slumping and permanent deformation. Cohesionless soils that are saturated may suffer dramatic loss of shearing resistance when subjected to cyclic loading. In the extreme case, saturated cohesionless materials may assume the properties of a dense viscous liquid. This liquefied state may persist for several minutes under the earthquake motion and cause the embankment fill and/or the foundation to flow as a liquid. The most severe failures of embankment dams during earthquakes have occurred as a result of this liquefaction of loose sandy soils.

To evaluate the possible effects of earthquakes on embankment dams, two possibilities must be considered: (1) the fault motion in the foundation can disrupt the embankment or cause loss of freeboard and (2) there may be some form of damage caused by the ground shaking. In the first case the dam must be able to absorb cracks and shears without suffering damaging piping or erosion; it must have an adequate amount of freeboard prior to the earthquake; and the dam designer must accurately estimate the potential magnitude, location, and direction of the fault movement during the earthquake. The second effect can be sliding within the embankment or the foundation material, settlement due to compaction of the soil materials, and cracking and/or subsequent erosion of the embankment materials. To survive this type of earthquake motion the dam must be designed with adequate density of soils, the construction process must have had adequate quality control, and finally the intensity of the earthquake shaking must have been properly estimated by the designer. Experience with embankment dams during earthquakes has shown a marked difference in performance dependent on the type of material of which the dam is built and the quality of construction. The performance of nearly 150 dams during earthquakes has shown that hydraulic-fill dams and dams built on loose materials frequently suffered severe damage. On the other hand, dams built of clay soils on stable foundations performed very well, although many were subjected to very strong shaking (Seed et al. 1977).

## Shear Strength Evaluation

Soil strengths for stability analyses are most often evaluated through laboratory triaxial or direct shear tests. To provide useful information, the tests must be performed under conditions corresponding to those in the field (drained or undrained, static or cyclic loading), and the samples must be representative of the soils in the field with respect to density and water content.

For most cohesive soils it is possible to obtain "undisturbed" samples for testing that retain essentially the same properties as in the field. It is possible to sample cohesionless soils only by very expensive and elaborate procedures requiring highly sophisticated equipment and procedures, and it is common to estimate the in situ relative densities of such soils based on the results of static or dynamic penetration tests. The shearing resistance of cohesionless soils may be evaluated by performing laboratory tests on samples compacted to the in situ relative density, by correlations between shearing resistance and relative density for similar soils, or by large-scale field direct shear tests.

Large triaxial shear testing equipment developed in the past 30 years has enabled more accurate determination of strengths of rockfills. Friction angles vary widely, depending on characteristics of the rock in the fill. Confining pressure is an important parameter (Leps 1970).

## Seismic Analyses

Pseudostatic methods of analysis, in which dynamic earthquake loads are represented by static loads, can be used to assess the stability of dams built of cohesive soils on stable foundations (Makdisi and Seed 1977). Pseudo-static analyses do not provide a suitable means for evaluating stability of dams built of or built on loose cohesionless materials, because they do not provide a means for including the potential these materials have for strength loss under cyclic loading. To evaluate the stability of loose cohesionless materials, more realistic dynamic analyses should be used, in conjunction with special laboratory tests to evaluate soil strength under cyclic loading.

Although they generally perform well during earthquakes, dams of cohesive soils on stable foundations may suffer some permanent deformation and loss of freeboard due to earthquake shaking. These deformations may be estimated using a simplified procedure suggested by Makdisi and Seed (1977), or they may be analyzed in greater detail through dynamic finite element analyses of embankment and foundation response to seismic loading.

### Factors of Safety

Typical factors of safety for the loading conditions discussed previously are shown in Table 7-2. These are the minimum values required for dams under the jurisdiction of the U.S. Army Corps of Engineers and thus represent standards of practice that find wide application, even though they may not be universally accepted by all agencies and for all circumstances.

## REFERENCES

ASCE/USCOLD (1975) *Lessons from Dam Incidents, USA*, American Society of Civil Engineers, New York.

Cedergren, H. R. (1967) *Seepage, Drainage, and Flow Nets*, John Wiley & Sons, New York.

Civil Engineering-ASCE (1981) *Dike Safety Upgraded with Millions of Square Feet of Fabric*, January.

Davis, C. V., and Sorenson, K. E. (1969) *Handbook of Applied Hydraulics*. Section 18 by John Lowe III, Embankment Dams, and Section 19 by I. C. Steele and J. B. Cooke, Concrete-Face Rock-Fill Dams.

Fetzer, C. A. (1979) *Wolf Creek Dam, Remedial Work Engineering Concepts, Actions and Results*, Transactions of ICOLD Congress, New Delhi.

Gordon, B. B., Dayton, D. J., and Sadigh, K. (1973) *Seismic Stability of Upper San Leandro Dam*, ASCE, San Francisco.

Jansen, R. B., Dukleth, G. W., and Barrett, K. G. (1976) *Problems of Hydraulic Fill Dams*, Transactions of ICOLD Congress, Mexico.

Japan Dam Foundation, Tokyo (1977) "Use of Slurry Trench Cut-Off Walls in Construction and Repair of Earth Dams," in *World Dam Today '77*, 576 pp.

Koerner, R. M., and Welsh, J. P. (1980) *Construction and Geotechnical Engineering Using Synthetic Fabrics*, John Wiley & Sons, New York.

Lamberton, B. (1980) "Fabric Forms for Erosion Control and Pile Jacketing," *Concrete Construction Magazine*, May.

Leps, T. M., Strassburger, A. G., and Meehan, R. L. (1978) "Seismic Stability of Hydraulic Fill Dams," *Water Power and Dam Construction*, October/November.

Leps, T. M. (1970) "Review of Shearing Strength of Rockfill," *Journal, ASCE Soil Mechanics and Foundations Division*, July.

Makdisi, F. I., and Seed, H. B. (1977) *A Simplified Procedure for Estimating Earthquake-Induced Deformations in Dams and Embankments*, EERC, University of California.

Pravidets, Y. P., and Slissky, S. M. (1981) "Passing Floodwaters Over Embankment Dams," *Water Power and Dam Construction*, July.

Proceedings of Engineering Foundation Conference (1974) "Foundation for Dams," Asilomar, Calif.

Seed, H. B., Makdisi, F. I., and DeAlba, P. (1977) *The Performance of Earth Dams During Earthquakes*, Report No. UCB/EERC-77/20, Earthquake Engineering Research Center, University of California, Berkeley.

Sowers, G. F. (1962) *Earth and Rockfill Dam Engineering*, Asia Publishing House.

Timblin, L. O., Jr., and Frobel, R. K. (1982) Geotextiles—A State-of-the-Art Review, paper presented to USCOLD annual meeting.

U.S. Army Corps of Engineers (1982) *National Program for Inspection of Nonfederal Dams. Final Report to Congress, May 1982* (contains ER 1110-2-106, September 26, 1979).

U.S. Bureau of Reclamation (1974) *Design of Small Dams*, Water Resources, Technical Publication, Government Printing Office, Washington, D.C.

United States Committee on Large Dams (1981) "Mt. Elbert Forebay Reservoir," *USCOLD Newsletter*, March.

Wilson, S. D,. and Marsal, R. J. (1979) *Current Trends in Design and Construction of Embankment Dams*, ICOLD/ASCE, New York.

## RECOMMENDED READING

ASCE, Soil Mechanics and Foundation Division (1969) *Stability and Performance of Slopes and Embankments*, New York.

Cortright, C. J. (1970) "Reevaluation and Reconstruction of California Dams," *Journal of the Power Division*, ASCE, January.

D'Appolonia, D. J. (1980) "Soil-Bentonite Slurry Trench Cutoffs," *Journal of the Geotechnical Engineering Division*, ASCE, April.

Golze, A. R., et al. (1977) *Handbook of Dam Engineering*, Van Nostrand Reinhold Company, New York.

Hollingworth, F., and Druyts, F. H. W. M. (1982) *Filter Cloth Partially Replaces and Supplements Filter Materials for Protection of Poor Quality Core Material in Rockfill Dam*, Transactions, ICOLD.

*USCOLD Newsletter* (1977) "Diaphram Wall for Wolf Creek Dam," July.

# 8
# Appurtenant Structures

## INTRODUCTION

Appurtenant structures are other structures around a dam that are necessary to the operation of the dam project. These include spillways, outlet works, power plants, penstocks, gates, valves, trash racks, diversion works, and switchyards. Generally, these are smaller structures than the dam, but they can be of considerable importance to the project because they control the flow of water and power.

Incidents of failure or near-failure of all types of dams that were attributed to defects in the appurtenant structures have been well documented in the literature. Often, the defects in appurtenant structures, when identified in the early stages, can be corrected by taking preventive maintenance measures without endangering the integrity of the dam. In cases where more extensive repair work is required, it may be necessary to lower the reservoir level to provide a sufficient factor of safety during repairs. In extreme instances, defects in appurtenant structures can be of such magnitude that they lead to complete failure and subsequent abandonment of the dam.

This chapter describes some problems common to appurtenant structures, together with suggested solutions. Table 8-1 summarizes defects, causes, effects, and remedies.

## DEFECTIVE SPILLWAYS

The main appurtenant structure of a dam is usually the spillway. The primary defect most often indicated is inadequate discharge capacity. Inade-

**TABLE 8-1** Evaluation Matrix of Appurtenant Structures

| Type of Defect | Causes | Effects | Remedies |
|---|---|---|---|
| Defective spillways | Insufficient analysis | Overtopping[a] Erosion or washout on downstream side | Reevaluate spillway capacity using present-day hydrologic techniques |
| | Design error | | Use watershed model simulation and prototype studies in design |
| | New criteria established | Erosion along and around spillway chute | Institute major repairs: Increase spillway capacity; Construction of auxiliary or emergency |
| | Major or unpredicted events | Breach | Alternate methods:[d] Revise reservoir operating procedures; Restrict reservoir utilization; Require attendance of dam personnel during flood events; Establish well-defined emergency procedures |
| Obstruction to spillways and outlet works | Excess trash[b] burden | Overtopping | Install log booms or trash racks based on use of reservoir, anticipated trash burden, etc. |
| | | Erosion | Perform maintenance as required to remove excess trash buildup |
| | | Damage to trash racks | |
| Defective gates and hoists | Mechanical breakdown | Upsets normal operation characteristics of dam | Perform regular maintenance on mechanical equipment |
| | Inadequate gate seals | Vibrations; Fatigue cracking | Check bottom gate seals for damage; Provide for sharp clean flow breakoff |
| | Cavitation around gate guides | Damage to gate frames and operating shaft | Repair cavitated areas[c] with steel liners; check that all gate frames are securely mounted |

**TABLE 8-1** Evaluation Matrix of Appurtenant Structures (*continued*)

| Type of Defect | Causes | Effects | Remedies |
|---|---|---|---|
| | Differential foundation settlement | Gates becoming inoperable<br>Gate frames crack | Repair foundation |
| | Trash and debris | Vibration<br>Trash can knock gates from frames | Install trash racks |
| | Galvanic corrosion and/or mineral deposits | Corrode moveable parts; makes gates inoperable | Provide cathodic protection<br>Exercise gate to prevent formation of deposits |
| | Poor design and/or inadequate operational procedures | Vibration | Revise operating procedures |
| | | Unbalanced flow (can cause other problems to occur, such as buckling of steel liners and concrete erosion) | Provide adequate air vents |
| Defective conduits | Surface irregularities (offset joints, voids, transverse grooves, roughness) | Cavitation erosion<br>Piping | Grinding surface to smoothness that will prevent cavitation erosion<br>Air vents at irregularities<br>Require close construction tolerances<br>Provide aeration grooves to draw air into flowing water |
| | Sealing in conduit | Unsteady flow conditions<br>Structural vibrations | Perform prototype studies and modify<br>Adequate air vents |
| | Unsymmetrical flow | Cavitation | Repair concrete<br>Install guide vanes |
| | | Erosion in stilling basins | Baffle blocks at terminal structure<br>Adequate air vents |
| | Settlement of foundations | Joint separation<br>Structural cracking<br>Piping | Stabilize foundations<br>Replace joint collars<br>Replace joint seals |
| | Corrosion | Piping of embankment material through holes | Replace or repair conduit |

**TABLE 8-1** Evaluation Matrix of Appurtenant Structures (*continued*)

| Type of Defect | Causes | Effects | Remedies |
|---|---|---|---|
| Defective drainage system | Inadequate design<br>Improper installation | Uncontrolled seepage<br>Piping<br>Boils | Investigate and modify<br>Install new or improve existing drain field<br>Provide relief wells<br>Reduce reservoir pool level |
| | Inadequate filter layer | Saturated conditions<br>Seepage of fines from foundation | Improve filter layer |
| | Mineral deposition | Clogging | Ream drains. Drill supplemental drains |
| Erosion | Inadequate design of spillways and stilling basins | Fluctuating positive to negative or uplift pressures can develop on spillways and stilling basins (can cause cracking of concrete slabs in stilling basins and subsequent removal of embankment material); this fluctuation of pressure can demolish a spillway or stilling basin | Increase thickness of concrete slabs<br>Impose tailwater elevation that will force hydraulic jump<br>Provide floor drain openings in locations to avoid subjecting them to fluctuating pressures |
| | Structural cracks in concrete slabs of spillways and stilling basins | Water seepage through slab and eroding of embankment materials | Pressure grout cracks in slab<br>Replace with thicker slab |
| | | Development of voids under slab<br>Loss of slab support<br>Breakup of slab | Evaluate effectiveness of energy dissipators and replace if necessary<br>Fill voids under concrete slabs<br>Anchor invert |
| | Unsymmetrical operation of outlet gates | Unsymmetrical loading of spillway<br>Scour actions in discharge area | Operate gates symmetrically<br>Repair with erosion resistant aggregate and high-strength concrete |

**TABLE 8-1** Evaluation Matrix of Appurtenant Structures (*continued*)

| Type of Defect | Causes | Effects | Remedies |
|---|---|---|---|
| | Excessive discharges<br>Abrasive objects in stilling basin (rocks, construction debris, etc.) | Abrasion and cavitation erosion of concrete in spillway and stilling basins | Repair with special concretes and steel plates<br>Line dissipators with steel plates |
| | | Damage to chute blocks and energy dissipators<br>Breakup of slabs and destruction of spillway | Install rip-rap |

[a]Overtopping is more critical on earth or rockfilled dams. Concrete dams can stand a limited amount of overtopping.
[b]Large trash, such as logs, etc., can damage spillways, stilling basins, and energy-dissipating blocks as it is carried over the spillway.
[c]New techniques for repair: polymer-impregnated concrete has been used to repair cavitation in concrete tunnels and stilling basins.
[d]New technique for repair: for spillway repair, rollcrete has been used as an alternative repair method.

quate capacity can lead to overtopping of the dam, which is particularly critical in earth or rockfill dams because overtopping can cause failure. In the evaluation of older dams a determination of inadequate spillway capacity is generally the result of new criteria and updated hydrological procedures and records rather than design or construction faults.

The subjects of spillway design floods; the ability of the spillway and reservoir acting together to control safely the design flood; and the general types of mitigating measures where that ability is lacking are discussed at length in Chapter 4. Some of the specific defects and poor hydraulic behavior that have been observed and remedies that have been used are discussed below.

Siphon spillways have been constructed at a few earthfill dams, usually as towers or imbedded risers in combination with the outlet works conduit. In some instances subsequent performance has demonstrated that the discharge capacity is much less than what was theoretically predicted. Where topography permits, a supplemental open channel spillway can be constructed beyond one end of the dam with the control elevation above that of the siphon. Vegetated linings may suffice depending on the frequency and duration of discharge in excess of that which can be handled by the siphon spillway (Cortright 1970).

Some spillway stilling basins originally constructed with the basin invert elevation incorrectly set in relation to tailwater and the conjugate depth of design discharge have been destroyed or severely damaged. It is sometimes possible to terminate the spillway discharge channel as a bucket at an elevation above tailwater. The bucket is supported on a foundation level at or below the expected depth of the eroding plunge basin. The support can be provided by cast-in-place piling in drilled or cased holes in granular formations or by a reinforced concrete substructure on a hard rock formation.

Nonsuperelevated horizontal curves have been built in spillway discharge channels where flow velocity is supercritical. As a result, flows have overtopped the outer wall with consequent erosion and structural damage. It may be possible to raise the outer wall and accept the transverse slope of the water surface provided the erratic wave pattern created is safely contained beyond the curve. In one instance the curved portion of the channel was compartmented with several vertical walls so that the outer rise in the transversely sloping water surface was diminished sufficiently for containment within the available freeboard.

Spillways with converging training walls are often susceptible to having an actual capacity less than theoretical. This is caused by water piling up along the converging walls and overtopping, often with serious results. Hydraulic model testing is often the optimum and only effective way of determining actual capacity.

In recent years diversion facilities during construction of some rockfill dams (both impervious core and faced) have included crest and downstream face reinforcing. Floods have been successfully passed over the top and down the slope of the uncompleted embankment with minimal damage. This suggests the possibility of a less costly way of increasing the spilling capability at an existing rockfill dam where the spillway capacity is too small. In some cases such treatment would be a temporary betterment until a permanent solution could be financed. In other cases the treatment might be justifiably considered permanent, for example, where the required spilling capability was expected to operate rarely, if ever, during the project life.

A decision to adopt this remedy would depend on full consideration of the quality and size of the rock in the top and face layers of the dam, the character of the foundation rock along the toe, a limiting dam height, the length of dam to be so treated, anticipated depth and duration of overtopping, river channel characteristics immediately downstream, and a nearby source of additional rock (original abutment or downstream quarries, perhaps).

The face reinforcing commonly consists of a heavy square steel mesh retained by an orthogonal pattern of spaced horizontal and sloping reinforcing bars. The bar network is retained against the face with bent anchor bars embedded in the rock mass. The horizontal and sloping bars are anchored to bedrock along the toe. During original construction the anchor bars are embedded as the rock lifts are being placed. At an existing dam the embedment would have to be made in a sliver fill of rock placed against the existing face.

Spillway capacities can be increased by constructing vertical concrete parapet walls on the tops of dam embankments. This can be feasible both for rockfill and earthfill dams. Usually the parapet wall is considered to provide residual freeboard only and the effective head on the spillway is thus measured from dam top elevation (Cortright 1970).

## OBSTRUCTIONS IN SPILLWAYS AND OUTLETS

In conjunction with spillway capacity, obstructions in the spillways and outlet works also can affect the stability and desired operating characteristics of dams. These obstructions can be caused by faulty design, structural defects, excessive reservoir trash burden, siltation, landsliding, or a combination of these factors. One documented incident that illustrates the results of obstructions occurred at the Nacimiento Dam near Bradley, California (ASCE/USCOLD 1975). After several intense storms the high-level outlet slide gate clogged with trash and failed.

Where trash racks are used, their proper design and placement plus regularly scheduled maintenance and cleaning of debris from the racks can help prevent such incidents. The design of trash racks generally must consider such factors as the intended use of the reservoir (recreation, water supply, flood, etc.), types of gates, and maintenance requirements (U.S. Bureau of Reclamation 1974). Where log booms are used to prevent obstructions to spillways and intakes, accumulated debris should be continuously removed and inspection made for damaged, corroded, or inadequate log booms.

A frequent deficiency in the outlet works of embankment dams relates to the elevation at which the intake structure was placed when originally constructed. If inadequate dead storage capacity was provided, the intake structure may be in danger of becoming obstructed by a mixture of waterlogged trash, sediment, and debris. Loss of withdrawal capability is of great concern when a more serious dam defect appears. Permanently submerged outlet gates in particular can be a problem because they are difficult to inspect or to maintain. They are assumed to function until sometime when they no longer do so. By then it may be too late to take corrective measures short of emergency measures.

Where it is still possible to empty the reservoir, a vertical freestanding riser or possibly a sloping riser laid on the abutment or possibly the face of the dam can be constructed to a new and higher intake elevation. The existing gate control can be set at the new entrance to the structure, or it may be desirable to modify the type of control at the same time. Vertical risers or stub towers have been successfully installed on several dams owned by the Santa Clara Valley Water District in California.

Where the reservoir cannot be drained and where extensive work by divers is impractical, a prefabricated riser can be added underwater. By making the riser watertight and by fitting it with connections for hoses through which ballast water or compressed air can be pumped, it is possible to tip up and position the riser vertically over the existing intake structure. A bottom end cover temporarily held in position by bolts and clamps is removed after the riser is in the vertical position and then blown off by compressed air. With the bottom cover off, the riser can be made to rise or sink by adding or releasing compressed air. The riser can then be joined to the existing intake structure and totally flooded. The top cover can then be removed, and a prefabricated trash rack arrangement can be installed by divers. This remedy was successfully made at Santa Felicia Dam in California by the United Water Conservation District (Bengry and Caltrider 1978).

## DEFECTIVE CONDUITS

Surface irregularities such as offset joints, voids, and roughness create turbulence within a conduit, which can cause cavitation, leakage, and piping. Grinding surfaces to a smooth finish, applying a smooth coat of epoxy, and providing air vents and/or aeration grooves to draw air into the flowing water are some solutions for turbulence problems.

Sealing or the transition from a free surface flow to full pipe flow in horizontal or inclined conduits can result in structural vibrations because of unsteady flow conditions and in an undesirable variation of the water flow (ASCE 1978). Sealing can be mitigated by providing adequate air vents in the conduit. In any case it is a condition to avoid through adequate design of the conduit and the use of prototype studies.

Unsymmetric flow conditions through conduits caused by bends or irregular gate operation can result in cavitation in the conduit and erosion in the stilling basin. Guide vanes installed in the conduit and adequate air vents can help streamline the flow. Baffle blocks and energy dissipators at the terminal structure can help control erosion.

Settlement of conduit foundations can cause joint separation and structural cracking that can lead to leaking and piping. In such cases the foundations need to be stabilized and joint collars and seals repaired or replaced. In corroded metal conduits, embankment material can be piped through the corroded holes and may require total lining/grouting or replacement of the conduit.

Bare metal conduits are frequently found in and beneath embankment dams especially in smaller, older, privately owned dams storing water for farm use and recreation. The conduits consist of welded or riveted steel or corrugated metal pipe. The transverse joints are welded, riveted, banded, or even slip-jointed. Bell and spigot cast iron pipe have been used. The conduits were installed by bedding them either on embankment or granular foundation surfaces and surrounding them with the materials of the overlying embankment zones. Little or no compaction was achieved beneath the overhanging portions of the pipe sections. The overlying embankments themselves may not have been placed with controlled moisture and compaction procedures. The outlet discharge is often controlled only by downstream gates or valves, and the conduits are subjected to full reservoir head when the outlet is closed.

The steel pipes are corroded by electrolysis and/or chemical action externally and are pitted internally. The rivets are no longer in intimate contact with the surrounding plate material. The banded joints are loose and rusted. Slip joints and bell and spigot joints have been opened by the base spreading forces of the embankment.

These defects are reason for great concern and have caused a number of dam failures. An outlet conduit is subjected to full hydrostatic reservoir pressure when closed downstream and transmits that pressure directly to all portions of the embankment and foundation along its entire length. The conduit is subjected to the lesser pressure of the hydraulic grade line when flowing. If the point of free discharge is far beyond the downstream end of the conduit the pressure can approach reservoir head. Any leakage under pressure from the conduit into the surrounding embankment or foundation can cause failure by internal erosion.

The existence of the defect can be determined by physical examination and reference to any reliable construction records. The deterioration of the conduit interior can be examined by closed-circuit television if the conduit is too small for entry. Inspectors can be pulled on wheeled dollies through dewatered conduits as small as 30 inches in diameter. Precautions should be taken to provide adequate air supply to the inspectors. Leakage appearing about the periphery of the conduit at the downstream face may have its source in the conduit. Temperature and chemical comparisons of the water may verify that source. By varying downstream gate settings, corresponding changes in the leakage rate may help identify the source.

If the conduit is of sufficient diameter and its full discharge capacity is not needed, a smaller-diameter pipe can be inserted and centered in the conduit, the annular space bulkheaded at both ends, and the space pressure filled with a sand-cement mixture. If the existing gate or valve was installed on the downstream end, it should be relocated to the upstream end of the conduit (Cortright 1970).

Conduits can be taken out of service by filling them with a sand-cement mixture under pressure. A drain and filter system can be installed around the exterior of the conduit near the downstream end to protect the embankment against piping from any leakage that may tend to flow along the exterior surfaces of the conduit. If drawdown limitations allow, a new, shorter outlet works can be constructed and founded on an abutment by removal and replacement of a portion of the embankment.

Although not favored as a permanent solution, an interim siphon outlet can be installed over the top of the dam until a permanent gravity-flow outlet works can be financed. If downstream releases are not required and the project has no other defects, the defective outlet can be taken out of service as described earlier and its replacement deferred temporarily until it can be financed.

## DEFECTIVE GATES AND HOISTS

Defective gates and hoists, especially those under high head, can cause unexpected problems and threaten dam stability when malfunctions occur. Being mechanical devices, these gates and hoists are subject to breakdown. Generally, two types of gates can be found on dams, depending on the design: (1) spillway gates used to control flow releases over the spillway if reservoir storage above the spillway crest is desired and (2) gates that function as regulating and guard gates in conduits.

One major problem that can occur with gates is induced vibrations from hydraulic forces during opening and closing. The problem has been most acute with radial gates on spillways. In Japan in 1967 (*Journal of Fluids Engineering* 1977), oscillations due to fluid-induced structural loadings caused the collapse of a radial gate. This resulted in a sudden rise in the water level downstream, with a subsequent loss of human lives. In the United States it was reported that vibrations on spillway gates on the Arkansas River were severe enough to cause fatigue cracks (ASCE 1972). An investigation determined that to eliminate the vibrations a sharp, clean flow break-off point was required. Soft rubber seals should not be used on the bottom of the gates. They should be rubber bar seals rigidly attached.

Differential foundation settlement of the gate structure can crack gate frames or skew the frames, so that the gates become inoperable. Cavitation erosion of concrete around gate frames can weaken the supports and cause subsequent failures. Cavitated areas can be repaired by using steel plates.

Gate operation can be stopped by formation of ice in the guideways or by reservoir ice. Galvanic action and/or mineral deposits can corrode movable parts on gates, rendering them inoperable. Cathodic protection and regular exercising of gates can help prevent formation of deposits and eliminate the problem. If trash racks are damaged or not provided, large trash and debris also can knock gates from frames.

Poor gate design and/or inadequate operational procedures can cause vibration and unbalanced flow. An unbalanced flow condition can lead to other problems such as cavitation and abrasive erosion and damage to gate frames. In such cases a revision of operating procedures may be all that is required to solve the problem. Providing adequate air ventilation behind the gate also will help mitigate such problems. Basic hydraulic design guides and criteria have been established by the U.S. Army Corps of Engineers and U.S. Bureau of Reclamation (ASCE 1973).

Other problems related to gate structures have been caused by downslope movement of riprap due to frost action/creep, causing the gate structure to tilt or crack. Where service bridges are used to gain access to the gate structure, any movement of the gate structure or any displacement of the bridge support foundation may induce stress or buckling of the structural elements of the bridge.

## DEFECTIVE DRAINAGE SYSTEMS

Defective drainage systems are often a source of problems. Clogged or plugged drains can lead to saturated conditions and create uplift pressures on spillways. Inadequate filter systems also can cause saturated conditions and allow piping of fines from the foundation. When such conditions occur the problems can be temporarily lessened by a reduction of the reservoir pool level. Long-term solutions can include providing a new design with an improved drain field or installation of new field and/or relief wells.

## EROSION

Erosion in and around dams is sometimes associated with defective appurtenant structures. Erosion can play a dual role in that it can be both the cause and the effect of defects and, if left untreated, can lead to dam failure. Erosion is far more evident with spillways and stilling basins because the tremendous force of moving water makes the effects of erosion highly visible.

An inadequate spillway or stilling basin design can lead to erosion with subsequent undermining of the dam itself. If uplift pressures on the spillway are not adequately controlled, the fluctuating positive to negative pressure can lead to cracking of the concrete slabs and removal of foundation materials (ASCE 1972). Structural cracks in spillways and stilling basins or poorly constructed joints allow seepage through the slab; this can cause piping of embankment material with a subsequent loss of slab support. If corrective measures are not taken, complete breakup of the slabs may result.

Unsymmetrical operation of outlet gates and unsymmetrical loading of the spillway can result in scouring action in the discharge area. Excessive discharges due to major storm events likewise can cause abrasion, erosion, and cavitation of concrete on spillways and in stilling basins. Chute blocks and energy dissipators can be damaged. Remedial measures for erosion vary extensively depending on the size of the dam and severity of the erosion.

Corrective action as simple as the placement of riprap in the discharge area may be all that is required to solve minor problems. For more complex problems a redesign and reconstruction of the spillway and stilling basin may be necessary. In cases where uplift pressures are determined to be the cause of erosion, slab thickness can be increased or rock anchors used to tie down the slabs to underlying rock. In addition, spillway and floor drains can be installed to relieve the excess pressure. Voids under spillway and stilling basin slabs need to be filled and the cracks grouted. In such grouting, care must be exercised to avoid having grout fill underlying filters and uplift the slab. If slabs are to be replaced, the use of erosion resistant aggregate and high-strength concrete is advantageous. Eroded concrete slabs and energy dissipators can be repaired and strengthened by use of steel plates. Nonstructural solutions to erosion can include better reservoir management through the controlled releases of water.

## EARTHQUAKES

After an earthquake it is essential that appurtenant structures continue to function in order to keep water and power flow under control. The uncontrolled water flow could result in damage to all or part of the dam and the surrounding terrain. The location of these structures is significant in engineering studies because much more shearing energy may be transmitted to the base of one of these structures if it is located on a ridge than if it were located on firm, level ground. Likewise, its foundation may indicate the susceptibility to damage of the structure, e.g., a structure on alluvium may suffer considerably more damage than one located on bedrock.

Freestanding structures, such as intake towers, tend to magnify ground motions. If they are submerged in the reservoir, the motion of such a tower causes a certain mass of the lake to move with it. This increases the apparent mass of the structure and, consequently, its response to the earthquake.

For all these reasons the appurtenant structures must be studied for their response to earthquake. They must be structurally stable, i.e., the earthquake-induced stresses in the concrete and steel must be within acceptable limits. Also a check must be made to see if the machinery will continue to function at the conclusion of the seismic incident.

## REFERENCES

ASCE (1978) *Size Determination of Partly Full Conduits.* Proceedings 104 (HY 7 No. 13862).

ASCE (1973) *High Head Gates and Valves in the U.S.* Proceedings 99:1727-75, October.

ASCE (1972) *Spillway Gate Vibrations on Arkansas River Dams.* Proceedings (HY) 99:219-238, January.

ASCE/USCOLD (1975) *Lessons from Dam Incidents, USA*, American Society of Civil Engineers, New York, pp. 259-61.

Bengry, E. O., and Caltrider, W. T. (1978) "Reservoir Outlet Extended Above Silt to Prevent Clogging," Civil Engineering-ASCE, September.

Cortright, C. J. (1970) "Reevaluation and Reconstruction of California Dams," *Journal of the Power Division*, American Society of Civil Engineers, January, pp. 63, 65.

*Journal of Fluids Engineering* (1977) "Instability of Elastically Suspended Tainter-gate System Caused by Surface Waves on the Reservoir of a Dam," Vol. 99, December, pp. 699-708.

U.S. Bureau of Reclamation (1974) *Design of Small Dams*, Government Printing Office, Washington, D.C.

## RECOMMENDED READING

ASCE (1973) *Reevaluation Spillway Adequacy of Existing Dams*, Proceedings 99 (HY), pp. 337-382, February.

ASCE Proceedings 98 (1972) *Damage to Kannafuli Dam Spillway*, (HY 12 No. 9452), December, pp. 2155-2170.

Chopra, A. K., and Liaw, C-Y. (1975) "Earthquake Resistant Design of Intake-Outlet Towers," *Journal of the Structural Division*, ASCE, Vol. 101, No. 577 (July).

Engineering News Record 200:11 (1978) Brazil Blames Earth Dam Collapses on Failure to Open Spillway Gates, February 2.

U.S. Army Corps of Engineers (1964) *Structural Design of Spillways and Outlet Works*, EM 1110-2-2400.

U.S. Bureau of Reclamation (1974) *Safety of Dams*, ASCE Proceedings 100 (HY), February, pp. 267-277.

# 9
# Reservoir Problems

## INTRODUCTION

The need to extend dam safety investigation to include the reservoir rim was dramatically illustrated by the 1963 landslide on the valley wall of the Vaiont Reservoir in Italy (Kiersch 1964). In that case the wave caused by the slide of rock (some 200 million cubic meters) into the reservoir overtopped the dam by about 125 meters. Although the dam withstood the impact with only minor damage, the wave continued downstream into the town of Longarone killing an estimated 2,000 people. This chapter discusses the reservoir problems of slope instability, induced earthquake, excessive seepage, backwater flooding, and ice.

## SLOPE INSTABILITY

### Slides

Of concern to dam safety is the possible movement of large masses of rock or soil into the reservoir. The Vaiont reservoir slide is a spectacular and catastrophic example (see Figure 9-1). These movements can be initiated by changes in the piezometric conditions within the mass as a result of reservoir loading, saturation of soil and weak rock materials with a resulting lowering of the internal coefficient of friction, erosion and subsequent undercutting of large earth masses by wave action and reservoir operations (raising and lowering), earthquake stresses on potentially unstable masses of earth, or increased erosion as a result of the removal of protective vegetation.

FIGURE 9-1 Vaiont reservoir slide (aerial view). SOURCE: Courtesy, L. Mueller.

Major concerns from reservoir bank instabilities are (1) the *sudden* release of large masses of material may generate reservoir waves that overtop the dam; (2) large masses dropping or slowly sliding into a reservoir severely reduce reservoir capacity; and (3) highways, railroads, or developed land adjacent to the reservoir may be undercut or displaced by such movements.

## Sedimentation

Slides can increase sedimentation in the reservoir. Another consideration is that the reservoir water may cause adverse chemical or mechanical alteration of the materials composing the reservoir banks. This would result in increased erosion and possible landslides. Increases in the siltation of a reservoir, whether from slide materials or normal sedimentation processes, can reduce the reservoir capacity to store floods. As a result overtopping can occur if such reduced storage capacity has not been considered in the dam and spillway design. Sedimentation also can block or inhibit flow through low-level gates and emergency outlets.

In stability analyses of concrete dams, consideration must be given to the expected silt load on the structure. An approximation of this load can be obtained by including a fluid pressure of 85 pcf for the horizontal pressure from the silt and 120 pcf for the vertical pressure.

### Bank Storage

Another possible undesirable situation could result from high-volume bank storage. The latter occurs when the rock or soil in the reservoir is cavernous or highly permeable. If such storage occurs and a reservoir is lowered rapidly, the drainage to the reservoir from the reservoir bank may be too slow to allow rapid dissipation of pore pressure within the rock or soil mass. This can produce instability in these masses.

## INDUCED EARTHQUAKES

There is a question of whether a reservoir may induce earthquakes. To date, there is no universally accepted proof that this can occur, but it is a possibility that should be given consideration. The causative factors still are uncertain, e.g., the weight of the reservoir may increase stresses at epicentral depths sufficiently to trigger fault activity (there is little to support this thesis), the downward movement of reservoir water may increase hydrostatic pressures in the rock masses (this increase presumably would increase pressure on a fault plane that may be under a critical state of strain), or the addition of water to a critically stressed fault plane may decrease the shear friction values of the fault sufficiently to trigger an earthquake. A recent study (Meade 1982) indicates that the possibility of reservoir-induced earthquakes is very limited and probably should not be considered except for extremely large and deep reservoirs.

## EXCESSIVE SEEPAGE

Excessive seepage through sinkholes, rock formation, or any pervious soil formation is a principal type of defect in reservoirs. The effect of this defect is normally obvious in that the reservoir will lose water. Reservoir leakage of this nature would not be expected to cause any loss of basin integrity or catastrophic release of storage, except where it might occur in the immediate proximity of the dam or thin natural barriers around the periphery of the reservoir. Seepage rates can be estimated by keeping a log of water levels, rainfall, and spillway discharges and then calculating total inflow and total water removal, including evaporation and transpiration. The difference should be seepage.

After it is determined that there is excess seepage, the problem is to find where and how the water is escaping. By keeping a log of water levels and rainfall, one may sometimes obtain a good indication of the elevation at which the excessive seepage is occurring. For example, if a reservoir rises rapidly to a certain elevation but rises slowly after reaching that elevation, this may be a good indicator of excessive seepage at that elevation. Excessive seepage locations can sometimes be detected by visual inspection. It is better to make a visual inspection on a clear, calm day by walking around the rim looking for a vortex, any movement in the water surface, or differences in the turbidity. These visual inspections should especially be made during the initial filling of any new reservoir. If visual inspection fails to identify the problem, a qualified geotechnical firm probably should be employed to investigate and recommend a solution. A reservoir that leaks over a large area probably cannot be sealed economically.

Excessive seepage may be corrected by using one or more of the following corrective measures:

- Grouting.
- Installation of a layer of clay or impervious soil over the problem area.
- Use of bentonite.
- Mixing soda ash with the pervious soil in the problem area.
- Use of liquid chemical soil sealants.
- Installation of a polyvinylchloride (PVC) liner.

Which corrective measure should be used can only be decided after one has located and quantified the seepage loss and learned all that practicably can be learned about the surrounding soil and rock formations. Core drilling may be used in the localized problem area to learn more about the way the water is leaving the reservoir and to help in planning the corrective program.

## BACKWATER FLOODING

Any dam and reservoir owner should own all of the land that would be flooded in the event of a maximum flood or should have the permission of the other land owner(s) involved to flood their land.

## ICE

The development of an ice layer on a reservoir surface can cause structural damage and produce maintenance difficulties. The main damage is commonly from the impact of ice against thin-walled structures such as parapets on masonry dams. Vertical and horizontal motion of ice against a structure (because of reservoir operations or deep waves) can induce high thrusts and cause considerable damage to concrete. Freezing and thawing of the ice also can damage structures, particularly if poor drainage allows such action underneath or behind concrete slabs and walls.

Ice can adhere to structures, and, when reservoirs are raised or lowered, there can be a concurrent drag of the ice upon the structure. This drag force can induce uplift or increase compression loads on such structures as intake gates and trash racks. There are many cases of such damage. Ice around intake gates and/or towers can impede the passage of water into the intake and interfere with the operation of gates and valves.

Ice problems adjacent to intake gates and valves have been mitigated by the continuous introduction of compressed air bubbles along the water side of the structure. This inhibits formation of ice directly on metal or concrete structures. The establishment of minimum operating levels above such facilities also is a possible solution.

During a thaw the floating ice can block intake areas or be driven by wind against man-made structures with resultant damage. In general, thermal expansion and wind loads are the major cause of structural damage from ice. Ice can also block spillway control structures, particularly gated ones, thereby reducing the spillway capacity, raising the water level, or resulting in a sudden water surge when the ice finally breaks loose.

A number of empirical formulas have been proposed for calculating ice loads on a dam. Parameters to consider are the slope of the upstream face and the slope and roughness of the valley walls. In addition, wind blowing

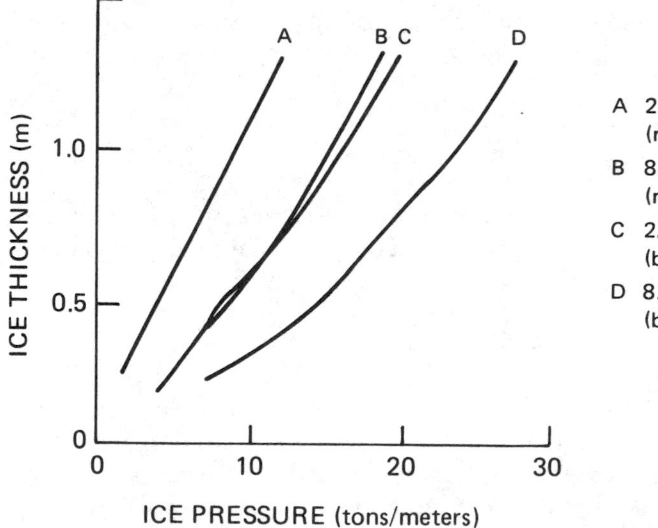


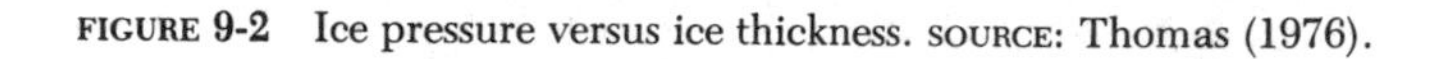
FIGURE 9-2 Ice pressure versus ice thickness. SOURCE: Thomas (1976).

down a reservoir can increase ice loading by 4 to 5 tons per meter on an exposed face. One study showed that, theoretically, thrusts on the order of 7.5 to 30 tons per meter were possible under North American climatic conditions, and in Canada a figure of 15 tons per meter is commonly used. At one time Norway used 45 tons per meter but now has decreased this to 5 tons per meter for dams with sloping upstream faces. Some tests by the U.S. Bureau of Reclamation near Denver, Colorado, indicated the highest thrusts were in the range of 20 to 30 tons per meter, depending on ice temperature. The latter is important because the rate of temperature rise and its duration can determine the pressure likely to be exerted. Figure 9-2 compares ice pressure with ice thickness and considers whether the reservoir banks act as a restraint. It was developed in Japan in 1970 (Thomas 1976).

## REFERENCES

Kiersch, G. A. (1964) "Vaiont Reservoir Disaster," *Civil Engineering*, March.

Meade, R. B. (1982) *State of the Art for Assessing Earthquake Hazards in the U.S.—The Evidence for Reservoir Induced Macro-Earthquakes*, U.S. Army Corps of Engineers, Waterways Experiment Station, Miscellaneous Paper S-73-1.

Thomas, H. H. (1976) *The Engineering of Large Dams*, Vol. 1, John Wiley & Sons, New York, p. 57.

## RECOMMENDED READING

Daly, W., Judd, W., and Meade, R. (1977) *Evaluation of Seismicity at U.S. Reservoirs*, USCOLD Committee on Earthquakes, Panel on Evaluation of Seismicity.

Monfore, G. E., and Taylor, F. W. (1948) *The Problem of an Expanding Ice Sheet*, U.S. Bureau of Reclamation Memorandum, March 18.

Muller, L. (1968) "New Considerations on the Vaiont Slide," *Rock Mechanics and Engineering Geology*, Vol. 6/1-2.

Rothe, J. P. (1969) *Earthquake and Reservoir Loadings*, Proceedings of Fourth World Conference on Earthquake Engineering, International Association for Earthquake Engineering.

# 10
# Instrumentation

## INTRODUCTION

The safety of an existing dam can be improved and its life lengthened by a carefully planned and implemented surveillance program. A key part of such a program is a visual examination of the structure, the reservoir, and the appurtenant works (as discussed in Chapter 2). However, surveillance must be more than visual observations. Settlements may go undetected without proper measurements of the dam. Comparison of seepage quantities from one inspection to another and over the years is difficult by visual observation and estimation. There are also conditions within a dam that cannot be seen but that can be measured by instrumentation. Thus, even for a simple structure, some type of instrumentation may be needed to improve and supplement the visual examination.

The purpose of instrumentation in an existing dam is to furnish data to determine if the completed structure is functioning as intended and "to provide a continuing surveillance of the structure to warn of any developments which endanger its safety" (ICOLD 1969).

The means and methods available to monitor geotechnical phenomena that can lead to a dam failure extend over a wide spectrum of instrumentation devices, consisting of very simple to very complex ones. The program for dam safety instrumentation requires detailed design that is consistent with all other project components; it must be based on prevailing geotechnical conditions of the dam and impoundment site and on the hydrologic and hydraulic factors prevalent both before and after the project was in operation.

A proposal for instrumentation for monitoring potential deficiencies at existing dams must take into account the threat to life or property that the project presents. The extent and nature of the instrumentation depends on the complexity of the dam, the size of the impoundment, and the potential for loss of life and property damage downstream. The program should incorporate instrumentation and evaluation methods that are as simple and straightforward as the location for installation and monitoring will allow. The owner(s) must make a definite commitment to a continuing monitoring program because any installation of instrumentation devices is wasted without a continuing program.

A primary factor of any system is the involvement of qualified personnel at all times, especially during the installation of instrumentation devices. The preparation of comprehensive installation reports is a necessary adjunct to future data evaluation.

While instrumentation can be tied to automatic warning systems, the experience of the committee indicates that no computer or automatic warning system can replace engineering judgment. Instrumentation data must be carefully reviewed periodically by an engineer experienced in the field.

This chapter discusses general deficiencies that may be noted or suspected during an examination of the dam and describes the instrumentation that will monitor a deficiency. Increased knowledge of the potential deficiency through such monitoring may provide sufficient data to determine the cause and prescribe the necessary treatment. Table 10-1 summarizes the deficiencies and the instrumentation to monitor that deficiency. Various types of instrumentation and their manufacturers or suppliers are listed, and methods of installation are discussed. After the instrumentation is in place, the data collection and analysis provide the owner and engineer with the information to define the problem more clearly. This will be discussed in the section Data Collection and Analysis.

## MONITORING OF CONCRETE AND MASONRY DAMS

Concrete and masonry dams must be inspected and monitored on a continuous basis, following a carefully planned monitoring program. To aid in these inspections and in the analysis of the condition of the dam, a number of monitoring methods and devices are used. Where these devices are installed, they should be maintained in good condition, and the data obtained should be regularly recorded and evaluated. Changes in the magnitude of the measurements recorded are the significant factors to be

**TABLE 10-1** Causes of Deficient Behavior, Means of Detection

| Causes of Deficient Behavior | Means of Detection, Measurement and Observations | | | | | | | | | | | |
|---|---|---|---|---|---|---|---|---|---|---|---|---|
| | Direct Observations | Relative Movements | Vertical Displacements | Angular Displacements | Horizontal Displacements | Uplift and Pore Pressure | Seismic Measurements | Rainfall Measurements | Flows | Turbidity | Sound Investigations | Crack and Joint Measurements |
| Causes of Deficient Behavior | | | | | | | | | | | | |
| *Concrete and Masonry Dams* | | | | | | | | | | | | |
| Due to foundation | X | X | X | X | X | X | | X | X | X | X | X |
| Due to concrete | X | | X | X | X | | | X | X | | X | X |
| Due to unforeseen action[a] | X | | X | X | X | X | X | | | | | |
| Due to structural behavior of arch and multiple arch dams | X | | X | X | X | | | | | | X | X |
| Due to structural behavior of gravity and buttress dams | X | | X | X | X | | | | | | X | X |

| | | | | | | | | | | | | |
|---|---|---|---|---|---|---|---|---|---|---|---|---|
| Due to maintenance[b] | X | | | | | | | | X | X | | |
| *Earth and Rockfill Dams* | | | | | | | | | | | | |
| Due to foundation | X | X | X | X | X | X | | X | X | X | X | |
| Due to embankment materials[c] | | | | X | | X | | | | | | |
| Due to unforeseen actions[a] | X | | | X | | X | X | | | | | |
| Due to structural behavior[d] | X | X | X | X | X | X | | X | X | X | X | X |
| Due to maintenance[b] | X | | | | | | | | X | X | X | |
| *Reservoirs* | | | | | | | | | | | | |
| Slope sliding | X | | X | X | X | | | X | | | X | |
| Movement of rock blocks | X | | X | X | X | | | | | | X | |
| Permeability | X | | | | | X | | | X | X | | |

[a]Or actions of exceptional magnitude, such as uplift, earthquakes, external or internal temperature variation, moisture variation, freezing, and thawing.
[b]Includes periodic inspections, cleaning of drains, control of seepage, deterioration of instrumentation, maintenance of slope protection, burrowing animals.
[c]Includes method of construction; excludes filters and drains.
[d]Includes filters and drains.

SOURCE: ICOLD (1981).

observed and evaluated by a trained observer. However, the devices must be properly maintained to ensure that the readings and measurements obtained are appropriate.

### Drainage Systems

In many concrete and masonry dams a foundation drainage system is installed to reduce uplift pressures on the dam. These systems are usually installed during construction but can be installed or supplemented at any time. They consist of holes drilled through the base of the dam into the foundation and may contain pipes where the foundation formation will not remain open. Also, monolith joint drains and face drains are commonly installed to intercept seepage along monolith and lift joints. It is very important to maintain the drain in usable condition; drains should be cleaned and periodically checked to maintain free flow conditions.

Water levels in or flow from the individual drains should be routinely measured. The flows from all drains or groups of drains should be collected and measured at weir installations. The water should be checked for chemical and suspended sediment content to aid in evaluation of solution or erosion that may be taking place. The elevation of the reservoir and tailwater elevations should be recorded at the time of drainage measurements so that relationships between these parameters can be developed.

### Seepage and Leakage

Seepage and leakage from the abutments, foundation, and joints or cracks in a dam should be collected and measured on a routine basis. It is important to review such flows for changes in magnitude and material, both dissolved and suspended, transported by these flows. Increases in these items are early warning indicators of potential problems. Weirs and venturi flumes with upstream stilling basins are frequently used to measure seepage and leakage. Flow measurements in the downstream discharge channels can add information on the amount of seepage and leakage that is not observed at surface leaks or seeps. On critical and or remote structures it is sometimes desirable to telemeter the flow information to another location.

### Uplift Pressures

Uplift pressures in the foundation and in the dam should be measured routinely as indicators of stability or instability. Changes in pressure should be looked for; increases may result in instability. Uplift pressures are measured by piezometers inserted in holes drilled into the foundation of the dam. In some cases foundation drainage holes can serve as piezometers if packers are inserted temporarily in the top of the drain hole. Packers can also be used within the holes with connected gages to isolate the interval in which the measurement of water pressure is desired. Observations of the reservoir and tailwater elevations should be recorded when uplift pressures are measured.

### Movements

Movement of concrete and masonry dams and their abutments can be expected during and after construction. These movements will occur as the reservoir is filled and may periodically cycle as it is emptied and filled during succeeding seasons. Small movements are of little concern, but increases in the magnitude of the movement or direction of movement should be immediately evaluated as to their potentially adverse impact on the structure.

Movements are measured by surveying the location of monuments located at various points on or adjacent to the dam. The benchmark or starting location for surveys should be located outside of the influence of the dam or reservoir if possible. Special measurement techniques may be used where very precise measurements are desired.

Measurements of the locations of the monuments should be such that changes in vertical, horizontal (both longitudinal and transverse to the dam axis), and angular locations are measured. The number of monuments surveyed depends on the size and type of the structure. The locations are tailored to the structure and might include locations to measure movement between blocks, displacement at joints and cracks, deflections of various parts of the structure, settlement of the foundation, and movement of the abutments.

The locations of the monuments should be recorded at relatively short intervals in the initial years of the life of the structure and less frequently as the satisfactory history of the dam lengthens. They should be more frequent if any tendency toward weakness or unsatisfactory performance is indicated. The data collected should be carefully recorded and should include observations on the relative water levels in the reservoir and downstream. The records accumulated should be plotted to provide graphic displays of the locations of the monuments and displacements between monuments. Computers can be used to create and display three-dimensional time-lapse sequences of the structure. This allows the normal seasonal movement cycles to be differentiated from changes that may be indicators of potential problems.

### Deterioration

Routine visual inspection of concrete and masonry dams can be of great value in determining the integrity of the structure. Descriptions of concrete conditions should conform with the appendix to "Guide for Making a Condition Survey of Concrete in Service" (American Concrete Institute 1968). Comparative photographs can aid the trained observer in distinguishing changes that might otherwise be more difficult to identify. Where deterioration of concrete, mortar, or masonry appears to be taking place, cores and samples can be taken and tested in the laboratory to provide absolute strength values. A routine schedule of nondestructive testing, such as ultrasonic velocity measurements, can be useful in determining trends of changes in strength. These types of tests should be carried out whenever deterioration appears or is suspected to be taking place.

### Seismic Instrument Program

A seismic instrument program is an essential part of evaluating existing dams in areas of high potential for seismic activity. Devices to measure ground motions and dam response can facilitate rational design decisions for repair and strengthening of a structure if damage has occurred as a result of an earthquake. These records are also desirable to compare the performance of the structure with design expectations and to estimate the structure's performance during other, larger shocks. However, the type of seismic instrument installation, or whether there even should be one, depends on the size and location of the dam. Such installations are desirable on larger structures, dams of unique design, and dams with large downstream hazard potential.

## MONITORING OF EMBANKMENT DAMS

A visual examination by a trained professional of an embankment dam is a reliable way to detect potential malfunctions or deteriorations of a structure. Surveillance can be aided by devices that measure seepage and leakage through and around the embankment, movements of the embankment and foundation, and water levels and pressures within the embankment and the foundation. Adequate records of such measurement devices, along with the visual observations, should be maintained. To be effective, these records should be continuous and periodically reviewed by a professional engineer versed in the design and vulnerability of embankment structures. These reviewers should be able to distinguish the important indicators from the unimportant. A tendency toward change in behavior of the dam should signal a need for further review and analyses. The record review must focus on the anomalies as opposed to the norm. Obviously this requires a continuous data base reflecting measurements made and records kept over a period of time.

### Seepage and Leakage

Seepage and leakage through the embankment, abutments, or foundation can be measured by various types of weirs. It is important to be able to check changes in amounts of seepage or leakage and in the material transported by these flows. Water from seeps can be collected by various drain structures into a weir box and measured by flow over the weir. The weir box would serve as a settling basin for some materials that may be carried from the embankment or abutment. Therefore, the weir box should be watched for deposition of materials. When measuring the flow at the weir plate the water should be observed for turbidity and changes in color, and samples should be taken for analyses of dissolved minerals if the abutment or foundation contains soluble solids. Springs and stream flow downstream of the embankment should be periodically monitored since changes in flow could indicate that piping or solution may be taking place.

### Movements

Considerable movement of embankment dams can be anticipated during and immediately after construction. Much of the movement may be attributed to foundation settlement under the loading of the embankment. The embankment will also move as the reservoir is filled for the first time and may periodically cycle movements as the reservoir is emptied and filled in succeeding seasons. Movements are determined by periodic measurements of monuments placed in or on the structure and abutments during construction or located on the structure and the abutments after construction. For existing dams monumentation to measure movements is usually limited to the crest and downstream slopes. The monuments usually consist of steel rods or surveyor's markers imbedded in concrete placed in excavations on the embankment and abutments. Differences in elevation and location of the monuments are measured by transit and level surveys of the monuments.

Measurements of the locations of the monuments on the surface of the embankment should be such that changes in both vertical and horizontal locations are measured. The measurements should be reduced to graphical displays of changes in vertical location, changes in longitudinal location along the axis of the embankment, and changes in horizontal location transverse to the axis of the embankment (upstream and downstream). Relationships to the water surface elevation in the reservoir at the time of measurement of the monuments are important and should be recorded along with the monument location data. Whenever possible the monuments should be tied to a benchmark that is outside the influence of the dam and reservoir. Monuments should be located such that they are not damaged by normal traffic or operations.

The number of monuments used depends on the size of the structure. The interval between measurements would depend on the history of the embankment. The interval should be relatively short during the initial years of the embankment life and may be extended as the satisfactory history of the embankment lengthens. If the structure shows any tendency toward weakness or unsatisfactory performance, the time interval between measurements should be shortened appropriately to provide analytic data that can warn of impending problems.

Inclinometers can be used to measure internal movements within embankments, in abutments, and in the reservoir rim. An inclinometer is a vertical tube placed in the embankment after construction. An electronic device is lowered within the inclinometer tube to detect any change in the location of the tube since the last measurement. The tube has vertical furrows or ridges that control the location of the pendulum device used to detect movement. The use of an inclinometer can give a continuous record of movement from the surface to the bottom of the inclinometer allowing differentials to be calculated at any elevation. This information has been useful in plotting the movements of slide masses and can provide valuable information on movements within embankment dams. They are not cheap to install and are relatively expensive to monitor. Therefore their use in existing embankments may be limited to the more important structures and to those with obvious deficiencies.

Another device to measure movement is the extensometer. It is generally used to measure strain in rock masses but can also be used in soil. The best use of an extensometer is to study relaxation or movement in rock excavations, such as tunnels or mines. The extensometer assembly can be sensed either mechanically or electrically.

## Piezometric Pressures

A primary indicator of the performance of an embankment is the water pressure distribution within the structure and its foundation. Water pressures in the embankments are measured by piezometers. There are basically three types of piezometers in common usage: (1) a hydraulic piezometer in which the water pressure is obtained directly by measuring the elevation of water standing in a pipe or vertical tube, (2) an electronic piezometer in which the water pressure deflects a calibrated membrane and the deflection is measured electronically to give the water pressure, and (3) a gas pressure unit in which the water pressure is measured by balancing it with a pressurized gas in a calibrated unit. The electronic and gas piezometers are usually installed during construction of the embankment, whereas the standpipe type can be installed at any time and is commonly used on existing structures.

One of the simplest piezometers is a performance type installed vertically in the embankment. These may be driven or augered into place or installed in holes drilled specifically for the purpose. Care should be taken during drilling to prevent hydraulic fracturing within the embankment. If there is a chance that embankment material might move through the perforations in the piezometer tube, a graded filter should be placed around the piezometer pipe in the drilled hole, and the annular space above the piezometer tip location should be backfilled with material of low permeability. The surface area around the piezometer should be sealed to prevent the entry of surface water along the casing. There are many variations of hydraulic piezometer units designed for special applications and to provide various levels of accuracy and ease of measurement.

Piezometers should be installed in an embankment structure so that the location of the free water surface or phreatic line can be determined. The line of piezometers would be perpendicular to the longitudinal axis of the embankment. In large structures there may be several lines of piezometers, while in smaller structures and existing dams perhaps one line would be adequate. The time interval between measurements of the water levels or pressures in piezometers depends on the age and condition of the structure. More frequent measurements are appropriate for relatively new structures and those with apparent or suspected defects. The elevation of the water surface in the reservoir and other conditions should be noted at the time of measurement of water levels and pressures in piezometers. If the piezometer is of the vertical standpipe type it should be kept capped at all times when measurements are not being made in order to prevent entry of material that would render future measurements impossible.

## RESERVOIR RIM

The construction of a dam and the subsequent impoundment cause more interference with natural conditions than do almost any other works of the civil engineer (Legget 1967). The groundwater level along the reservoir valley will be directly affected by the rise in water level, generally for a considerable distance away from the actual shore line of the reservoir. Materials in the sphere of influence of the reservoir water may fail to retain their former stability and landslides may result. Leakage from the reservoir is another potential source of trouble.

In the publication *General Considerations on Reservoir Instrumentation*, by the Committee on Measurements (ICOLD 1969), Komie discusses several consequences of altering natural groundwater regime relative to the reservoir rim. Seepage into adjoining basins is a potential problem, seepage from solid waste disposal reservoirs can cause deterioration of the regional groundwater, and seepage may contribute to local subsidence. Preconstruction and postconstruction hydrogeologic studies and instrumentation are recommended. The instrumentation can primarily be done by installing and monitoring observation wells, weirs, and piezometers. Komie suggests that postconstruction monitoring should be continued for several cycles of reservoir operation to document cyclic changes. Periodic observation of natural springs and comparison of pre- and postreservoir water temperature are also often helpful. Figures 10-1 through 10-4 are typical instrumentation installations from Komie's paper.

Instability of reservoir slopes caused by saturation of overburden or uplift water pressure in bedrock is a potential danger. Tockstein (USCOLD 1979/1981) states that whenever the in situ conditions of an area are disturbed by the impoundment of a reservoir a potential for ground movements is created along the reservoir slopes and the slopes adjacent to, but outside of, the reservoir rim. The damage potential can be reduced by identifying existing and potential landslide areas, monitoring these areas, and implementing a preplanned course of action in the event that excessive movement of an unstable area is imminent.

In planning a monitoring program it is necessary to remember that a landslide will occur when the driving forces exceed the resisting forces. Thus, the parameters that affect or measure these forces on a given slope should be identified. The parameters most commonly used are pore water pressure and displacement, both at the surface and at various depths within the slope, which indicate if the resisting forces are being exceeded and at what location. By correlating changes in these parameters with each other and with external influences on the slope, the cause of the movement can be identified and appropriate remedial action initiated. A systematic approach to planning a monitoring program is presented by Dunnicliff (1981). It is important that both the person installing the instruments and the person making the readings and maintaining the instruments understand the purpose of the instrumentation. The reduced data must be reviewed periodically by a professional engineer or engineering geologist with expertise in slope stability. Another excellent reference on instrumentation to monitor ground movements relative to slope instability is by

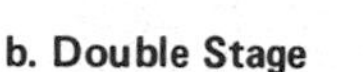


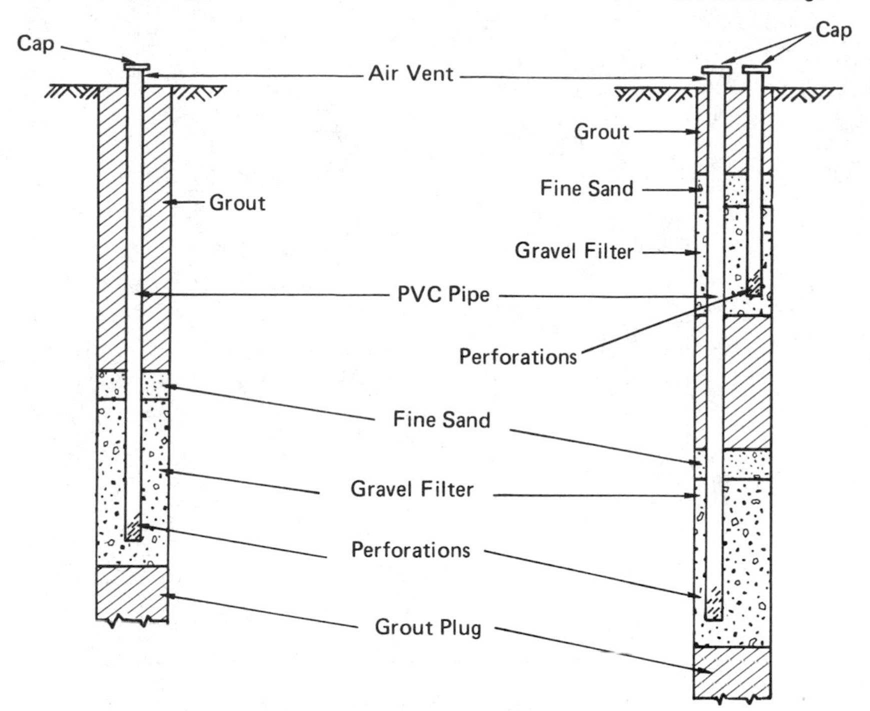


FIGURE 10-1 Typical observation well installations. SOURCE: ICOLD (1969, 1981).

Wilson (1970). Figure 10-5 is an example of slope instrumentation and data plots from Wilson's paper.

## INDUCED SEISMICITY

As discussed in Chapter 9, controversy exists over whether there is a significant increase in seismic activity associated with impoundment of some large reservoirs. Increased instrumentation for new and old projects can assist in solving this controversy.

Bolt and Hudson (1975) discuss the need for seismic instrumentation and recommend the minimum instrumentation for recording basic earthquake data. They recommend that where there may be potential for determining if earthquakes are induced from reservoir loading a network of seismographs should be in operation prior to impounding the reservoir. They recommend simple and reliable instruments. The seismographs should be

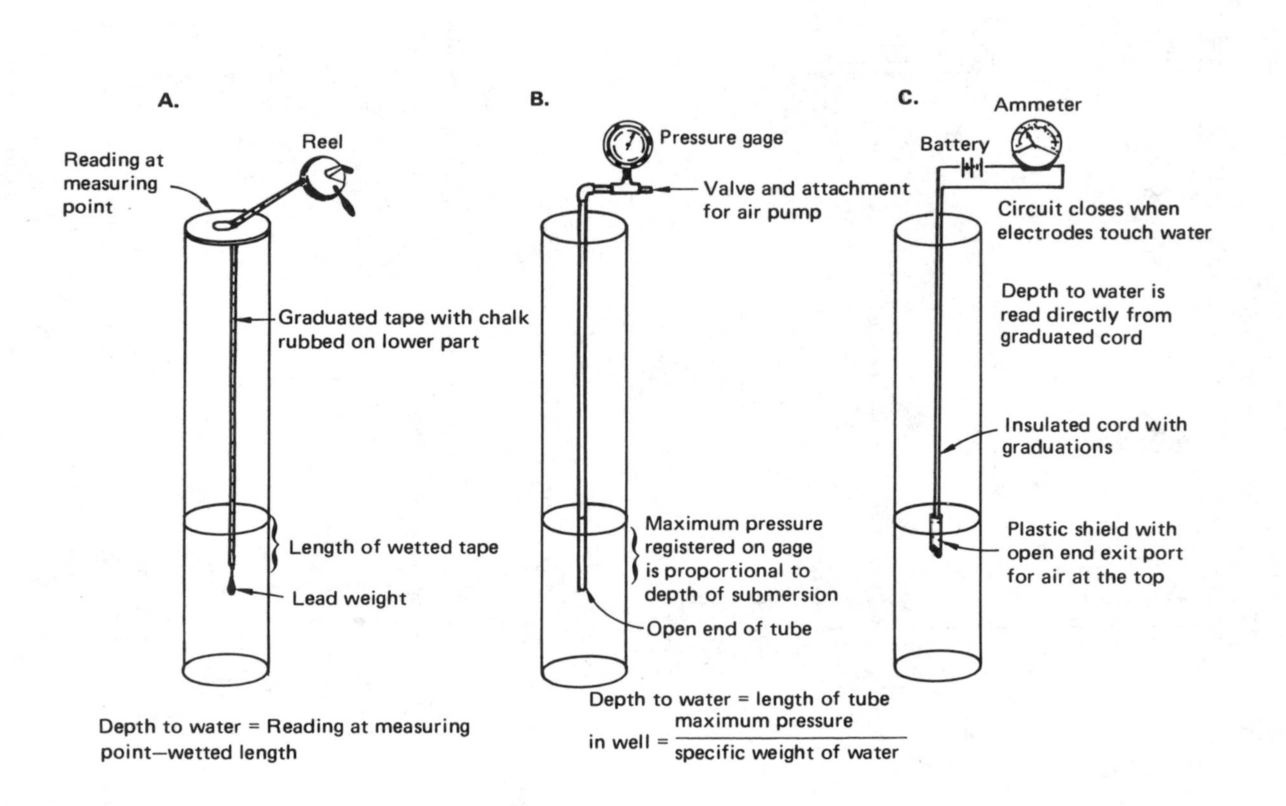
A.
Reel
Reading at measuring point
Graduated tape with chalk rubbed on lower part
Length of wetted tape
Lead weight
Depth to water = Reading at measuring point—wetted length
B.
Pressure gage
Valve and attachment for air pump
Maximum pressure registered on gage is proportional to depth of submersion
Open end of tube
Depth to water = length of tube
in well = maximum pressure / specific weight of water
C.
Ammeter
Battery
Circuit closes when electrodes touch water
Depth to water is read directly from graduated cord
Insulated cord with graduations
Plastic shield with open end exit port for air at the top

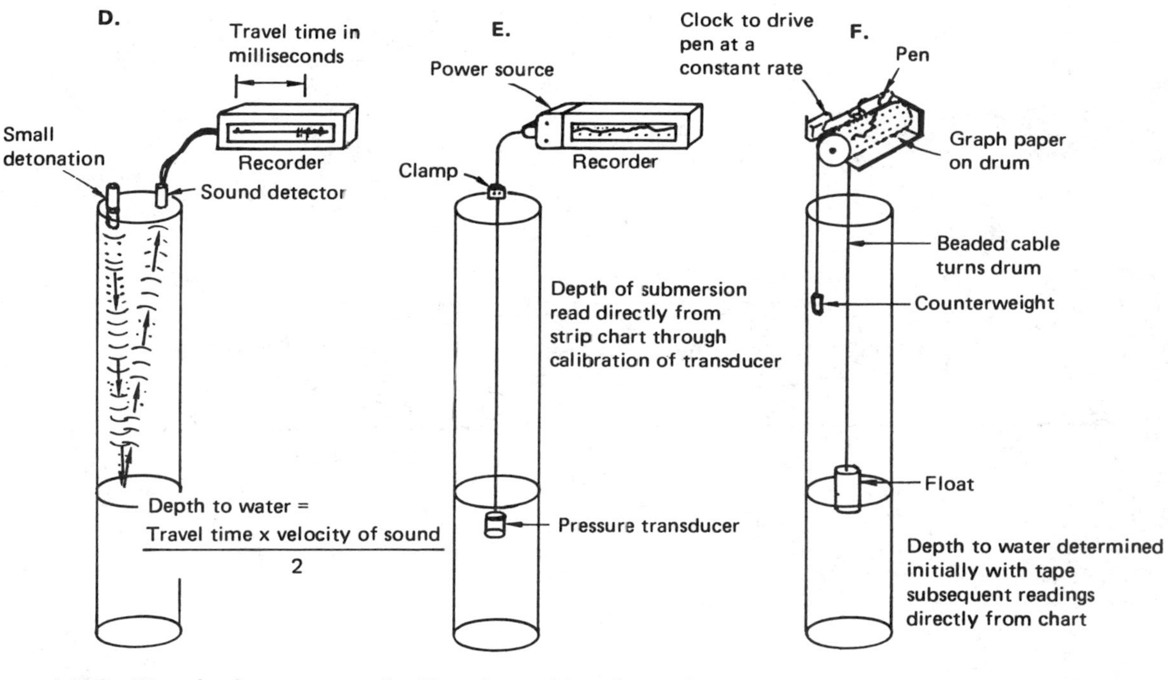


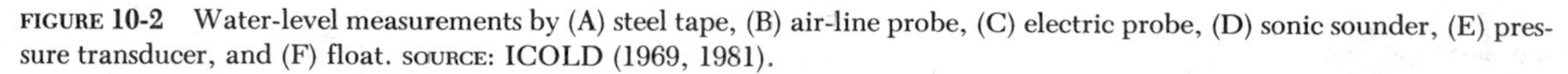

FIGURE 10-2 Water-level measurements by (A) steel tape, (B) air-line probe, (C) electric probe, (D) sonic sounder, (E) pressure transducer, and (F) float. SOURCE: ICOLD (1969, 1981).

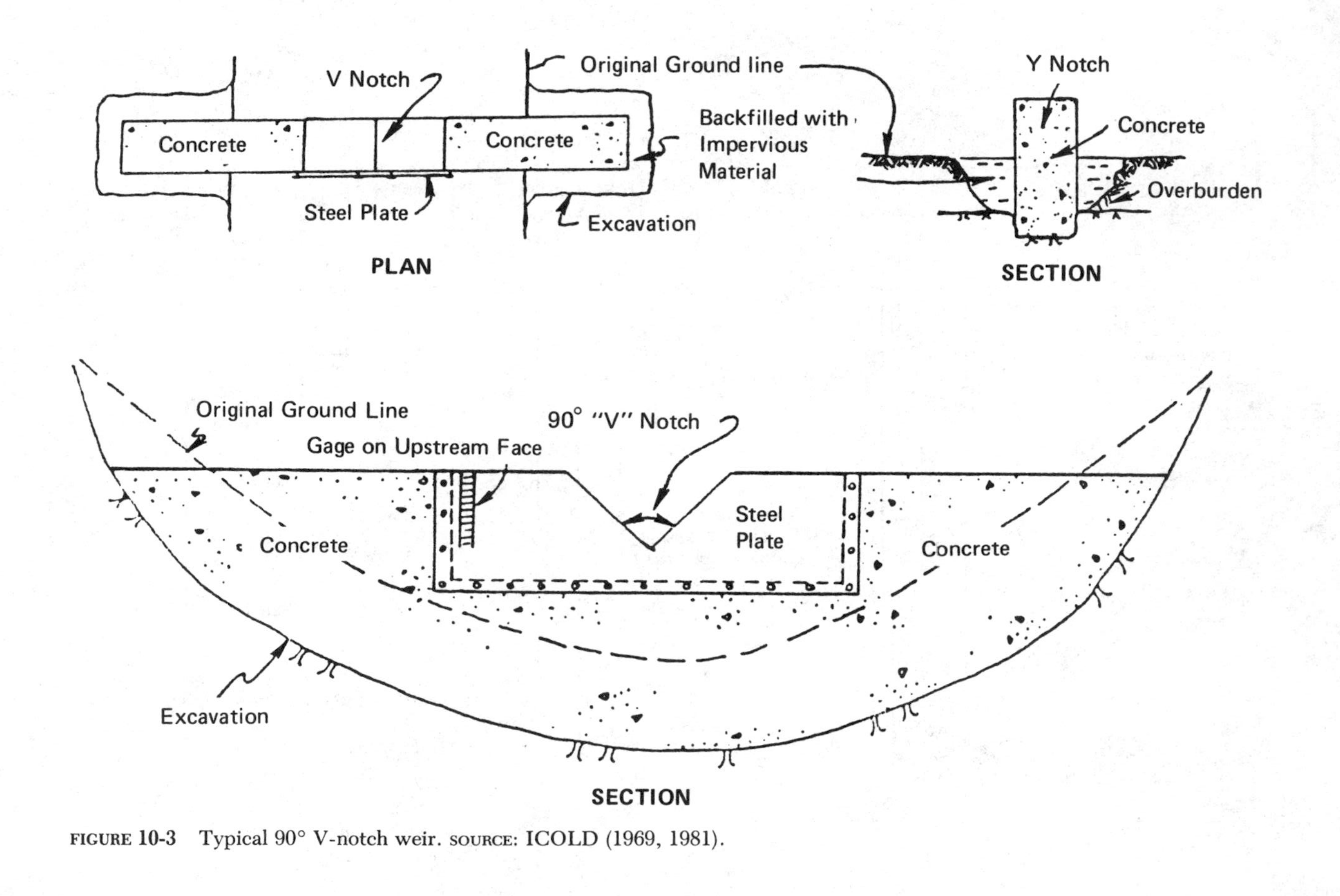


FIGURE 10-3 Typical 90° V-notch weir. SOURCE: ICOLD (1969, 1981).

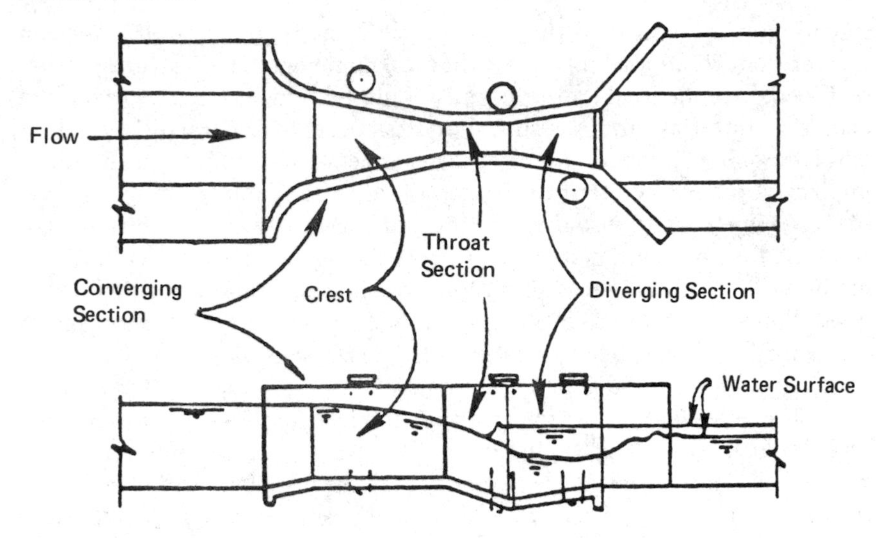


FIGURE 10-4 Typical parshall flume. SOURCE: ICOLD (1969, 1981).

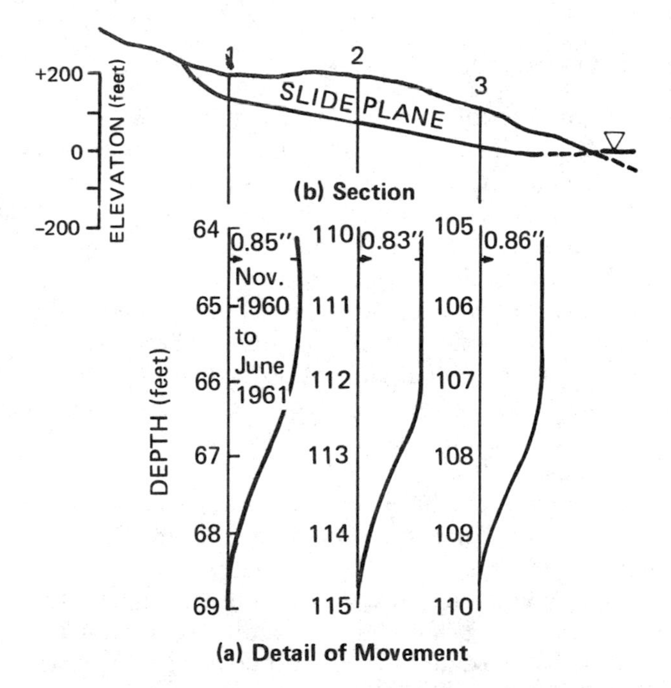


FIGURE 10-5 Point Loma landslide, California. SOURCE: Wilson (1970).

spread in azimuth around the reservoir with the interstation distance not greater than 30 kilometers or less than 5 kilometers. The instruments are best located on bedrock outcrop and should be remote from construction activities, quarries, and streams. For seismographic characteristics, Bolt and Hudson recommend two alternative schemes that will meet the minimum requirement and that have been field tested. One system uses portable seismometers and visual recording units; the network stations are not connected. The second system telemeters the signals from individual seismometers to a central recording room, often using commercial telephone lines. This system is more expensive but has the advantage of recording at one central location. The operation of either system does not require an instrumentation specialist or seismologist.

Another excellent paper on reservoir-induced seismicity is by Sharma and Raphael (USCOLD 1979/1981). They stress that instrumentation to study local reservoir-induced seismicity must be installed prior to reservoir filling to establish the extent to which local seismicity is a consequence of the reservoir or part of a more general seismic pattern. These authors discuss objectives of seismic instrumentation, type of measurements, instruments, schedule for making measurements and data analysis.

## TYPES OF INSTRUMENTS

The need and purpose for instrumentation was described earlier in this chapter. This section describes various types of instruments which may be used to monitor dams. Table 10-1 compares causes of deficient behavior with the types of measurements or observations used to monitor that behavior. A brief inventory of the instruments and the factors measured and monitored is listed in Table 10-2.

Table 10-3 lists the various commercially available instrument types with the name of the manufacturer. The information on addresses of the U.S. or Canadian suppliers was current at the time of preparation of the report by Dunnicliff (1981). Installation of instruments presupposes the observation (measurements), reduction, and evaluation of the data. In a later section of this chapter, Monitoring of Concrete Dams, a proposed observation schedule is provided, including tables on the frequency of readings (see Tables 10-4 and 10-5). These data are described fully in U.S. Bureau of Reclamation (1974).

The procurement and timely evaluation of instrumentation data are primary prerequisites for determining the conditions of dams. The physical conditions existing at the dam should be observed and noted when making observations. Such information supplements periodic inspections. In some instances, such as where dams are not readily accessible, data acquisition

**TABLE 10-2** Inventory of Geotechnical Instruments

| Phenomena Measure | Instrument | Suppliers' Numbers Refer to Table 10-3 |
|---|---|---|
| Pore water and ground measurement | Piezometer, closed system and open system, observation wells | 3, 6, 8, 11, 16, 18, 22, 25, 29, 30, 31, 32, 36, 37, 41, 42, 44, 45, 49, 50, 53 |
| Earth pressure measurements | Earth pressure cells, wholly embedded in soil; at contact plane between soil and structure | 6, 16, 18, 19, 27, 36, 42, 44, 49, 50 |
| Deformation measurements; horizontal and vertical | Survey equipment transits theodoloites, electronic distance measurement equipments (EDME), and levels | 1, 6, 9, 16, 18, 19, 22, 23, 24, 26, 27, 28, 29, 33, 35, 36, 41, 42, 43, 44, 45, 46, 47, 48, 49, 50, 53, and numerous others |
| Internal deformation, rotational and tilting | Extensometers, inclinometers, tiltmeters | 4, 8, 9, 10, 15, 16, 19, 20, 22, 26, 27, 32, 41, 42, 43, 44, 47, 48, 53 |
| Load and strain measurement: Surface installation or embedded | Strain meters<br>Load cells<br>Concrete stress cells | 2, 5, 7, 8, 12, 13, 14, 15, 16, 18, 19, 22, 26, 27, 32, 33, 38, 39, 40, 44, 45, 47, 48, 49, 51, 52 |
| Temperature measurement | Temperature sensors | 8, 16, 27, 32, 42, 44, 48, 49, and numerous others |
| Seepage | Weirs, flow meters, and flumes | Local |

SOURCE: Adapted from Dunnicliff (1981).

by automated means may be appropriate. A method to achieve these prerequisites, including automated data evaluation and plotting, is found in a paper by Lytle (ICOLD 1972).

## METHODS OF INSTALLATION

Proposed methods and procedures suggested for installation of the various instrumentation types are presented in the U.S. Army Corps of Engineers (1971, 1976, 1980) and U.S. Bureau of Reclamation (1974). A timely reference is *Geotechnical Instrumentation for Monitoring Field Performance* (National Research Council 1982). Chapter 5 of that report presents numerous instrumentation types and installation procedures. A caveat is expressed, however, concerning the application of the recommendations

**TABLE 10-3** Names and Addresses of Manufacturers and North American Suppliers

1. A & S Co., 9 Ferguson Street, Milford, MA 01757
2. Ailtech, 19535 E. Walnut Drive, City of Industry, CA 91748
3. Apparatus Specialties Co., Box 122, Saddle River, NJ 07458
4. Bison Instruments, Inc., 5708 West 36th Street, Minneapolis, MN 55429
5. BLH Electronics, 42 Fourth Avenue, Waltham, MA 02254
6. *Borros Co., Ltd., Box 3063, S-17103 Solna 3, SWEDEN (NA supplier: Roctest, Ltd.)
7. Brewer Engineering Lab., P.O. Box 288, Marion, MA 02738
8. Carlson Instruments, 1190-C Dell Avenue, Campbell, CA 95008
9. *Coyne et Bellier, 5 Rue d'Heliopolis, 75017 Paris, FRANCE (NA supplier: Roctest, Ltd.)
10. Eastman Whipstock, Inc., P.O. Box 14609, Houston, TX 77021
11. Engineering Laboratory Equipment, Inc. (ELE), 10606 Hempstead, Suite 112, Houston, TX 77092
12. Evergreen Weight, Inc., 15125 Highway 99, Lynnwood, WA 98036
13. *Gage Technique, Ltd., P.O. Box 30, Trowbridge, Wilts, England (NA supplier: Terrametrics, Inc.)
14. Gentran, Inc., 1290 Hammerwood Avenue, Sunnyvale, CA 94086
15. Geokon, Inc., 7 Central Avenue, West Lebanon, NH 03784
16. *Geonor, Grini Molle, P.O. Box 99, Roa, Oslo 7, Norway (NA suppliers: ELE, Roctest Ltd., Slope Indicator Co., and Terrametrics)
17. Geotechniques International, Inc., P.O. Box E, Middleton, MA 01949
18. *Franz Gloetzl, D-7501 Forchheim, Baumesstechnik, West Germany (NA suppliers: Terrametrics, Inc., and Roctest Ltd.)
19. Hall, Inc., 1050 Northgage Drive, San Rafael, CA 94903
20. Hamlin, Inc., Lake and Grove Streets, Lake Mills, WI 53551
21. Hitec Corporation, Nardone Industrial Park, Westford, MA 01886
22. *Huggenberger AG Zurich, Hohlstrasse 176, CH-8040, Zurich, Switzerland (NA supplier: Slope Indicator Co.)
23. Hewlett-Packard: contact local office
24. *Dr. Ing. Heinz Idel, Potthoffs Borde 15, 43 Essen, West Germany (NA supplier: Terrametrics, Inc.)
25. *Ingenjorsfirman Geotech AB, Varslevagen 39, S-43600, Askin, Sweden (NA supplier: Roctest, Ltd.)
26. *Interfels, Zweigniederlassung, Bentheim, West Germany (NA supplier: Roctest, Ltd.)
27. Irad Gage, Etna Road, Lebanon, NH 03766
28. Kern Instruments, Inc.: contact local office
29. Keuffel & Esser Co.: contact local office
30. Landtest Ltd., 43 Baywood Road, Rexdale, Ontario, Canada M9V 3Y8
31. *Linden-Alimak AB, S-93103 Skelleftea, Sweden (NA supplier: Burcan Industries, 1255 Laird Boulevard, Montreal, Quebec, Canada H3P 2T1)
32. *H. Maihak, 2000 Hamburg, 39 Semper Street, Hamburg, West Germany (NA suppliers: Ampower Corporation, 1 Marine Plaza, North Bergen, NY 07047 and Roctest, Ltd.)
33. W. H. Mayes & Sons, Ltd., Vansittart Estate, Arthur Road, Windsor, Berkshire, England
34. Micro-Measurements, Box 306, 38905 Chase Road, Romulus, MI 48174
35. Walter Nold Company, 24 Birch Road, Natick, MA 01760
36. Petur Instrument Company, Inc., 11300 25th Avenue NE, Seattle, WA 98125
37. Piezometer R & D, Inc., 33 Magee Avenue, Stamford, CT 06902
38. Prewitt Associates, Dawson Building, 1634 N. Broadway, Lexington, KY 40505

39. Proceq, SA, Riesbachstrasse 57, CH-8034, Zurich 8, Switzerland
40. Remote Systems, Inc., P.O. Box 12914, Pittsburgh, PA 15241
41. Roctest Ltd., 665 Pine, St. Lambert (Montreal), Quebec, Canada J4P 2P4; *also* Roctest, Inc., 7 Pond Street, Plattsburgh, NY 12901
42. Slope Indicator Co., 3668 Albion Place North, Seattle, WA 98103
43. *Peter Smith Instrumentation Ltd., Gosforth Industrial Estate, Newcastle Upon Tyne NE3 1XF, Gosforth, England (NA supplier: Roctest, Ltd.)
44. *Soil Instruments, Ltd., Bell Lane, Uckfield, East Sussex, TN22, 10L, England (NA supplier: Solinst Canada, Ltd., 5-2440 Industrial Street, Burlington, Ontario, Canada L7P 1A5)
45. Soiltest, Inc., 2205 Lee Street, Evanston, IL 60202
46. Spectra-Physics, Inc., 1250 W. Middlefield Rd., Mountain View, CA 94042
47. Structural Behavior Engineering Laboratories (SBEL), P.O. Box 23167, Phoenix, AZ 85063
48. *Telemac, 17 Rue Alfred Roll, 75 Paris 17eme, France (NA Supplier: Roctest, Ltd.)
49. Terrametrics, Inc., 16027 West 5th Avenue, Golden, CO 80401
50. Terra Technology Corporation, 3860 148th Avenue, N.E., Redmond, WA 98052
51. Texas Measurements, Inc., P.O. Box 2618, College Station, TX 77840
52. Transducers, Inc., 14030 Bolsa Lane, Cerritos, CA 90701
53. Westbay Instruments, Ltd., Suite 1B, 265-25th Street, West Vancouver, British Columbia, Canada V7V 4H9
54. Wild-Heerbrugg Ltd: contact local office

---

*Contact North American (NA) supplier, not manufacturer.

SOURCE: Dunnicliff (1981).

without proper guidance by a competent engineer for an evaluation of the materials in which the instruments will be installed and the purpose and period for which monitoring is required.

The following listed chapters from the referenced material cover details for installation of the instrumentation most basic for monitoring dams:

- U.S. Corps of Engineers Manual, EM 1110-2 1908, Part 1 of 2, 31 Aug 71: Instrumentation of Earth and Rockfill Dams (Groundwater and Pore Pressure Observations); Chapter 5, Installation, Maintenance of Piezometers, and Observations.
- U.S. Corps of Engineers Manual, EM 1110-2-1908, Part 2 of 2, 19 Nov. 76: Instrumentation of Earth and Rockfill Dams (Earth Movement and Pressure Measuring Devices); Chapter 2, Movement Devices for Embankments and Foundations.
- U.S. Corps of Engineers Manual, EM 1110-2-4300, 15 Sept. 1980: Instrumentation for Concrete Structures, Chapter 3, Uplift and Leakage, and Chapter 4, Plumbing Instruments and Tilt Measuring Devices.

• National Research Council, Geotechnical Instrumentation for Monitoring Field Performance, April 1982; Chapter 5, Details of Instrumentation.

Adding instrumentation to existing dams will require specialized equipment and drilling techniques for boreholes in which the instruments are to be installed or in fastening to existing structures; therefore, it is recommended that firms having subsurface exploration experience and expertise should be obtained for making the installations. During the drilling of boreholes samples of material and logs of the borings should be obtained. These data will be of significant value for subsequent data evaluation and prognostication concerning the ongoing safe operating conditions. The drilling methods and operation procedures must be specified and carefully monitored to prevent damage from hydraulic fracturing of earth embankments.

Portable instruments also have been useful for investigation of deficiencies at dams. For example, borehole cameras and periscopes have been used successfully in examining concrete structures and foundations. Television cameras have important applications in underwater work, such as in surveys of conduits and the submerged faces of dams. Sonic devices have been used effectively to locate leaks in dams. They have been particularly useful in measuring leakage through the concrete facings of rockfill embankments where disruption of the slab joints or cracks in the panels were primary sources of leakage. A basic instrument used for this purpose is a hydrophone, which essentially is a waterproof microphone that can be lowered on a cable. The test involves the comparison of background sound intensity with intensity measured in the vicinity of leaks. The microphone may be lowered from a boat located over the points of suspected leakage. Although areas of high sound intensity can be found by using such equipment, the size of the leak does not necessarily correlate with the indicated intensity. A small leak can produce a high sound level if there is a sharp disturbance at the entrance. On the other hand, a large leak with a smooth entrance could produce a much lower reading.

Two effective methods of visual inspection of a dam face involve the employment of divers and closed-circuit television. Divers can be effective in water depths up to about 150 feet without very expensive equipment. At greater depths the long decompression time required for each dive may become prohibitive. An effective way to operate at relatively shallow depths is to have the diver carry a television camera with him so that leaks can be viewed by an engineer at the surface or recorded on videotape. When inspecting with closed-circuit television, a hydrophone and a "pigtail" mounted on the camera facilitate location of the leak. The sonic level will increase and the pigtail will be drawn in the direction of flow as the camera is moved into the region of higher velocity (Jansen 1968).

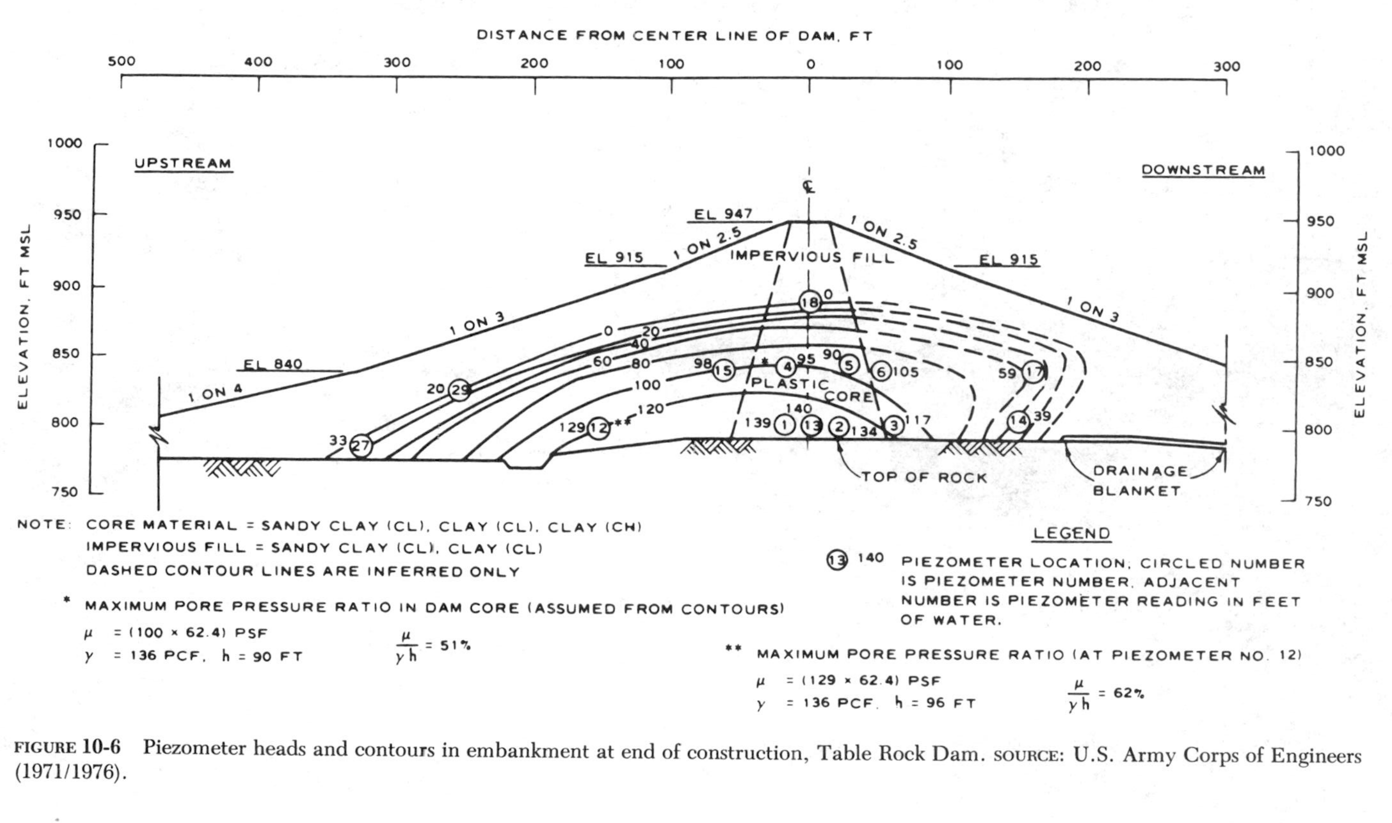


FIGURE 10-6 Piezometer heads and contours in embankment at end of construction, Table Rock Dam. SOURCE: U.S. Army Corps of Engineers (1971/1976).

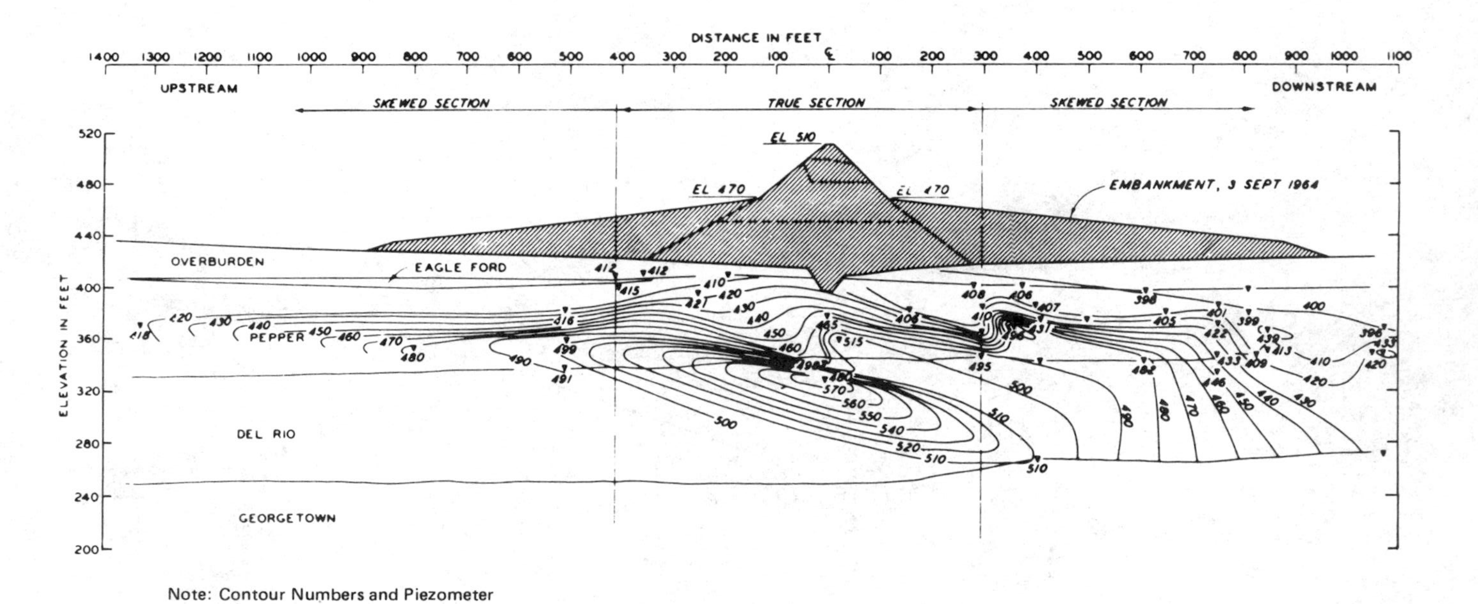


Note: Contour Numbers and Piezometer
Readings are in Feet of Water.
V Piezometer Location

**PIEZOMETRIC HEADS AND CONTOURS, MIDFAULT, SECTION**

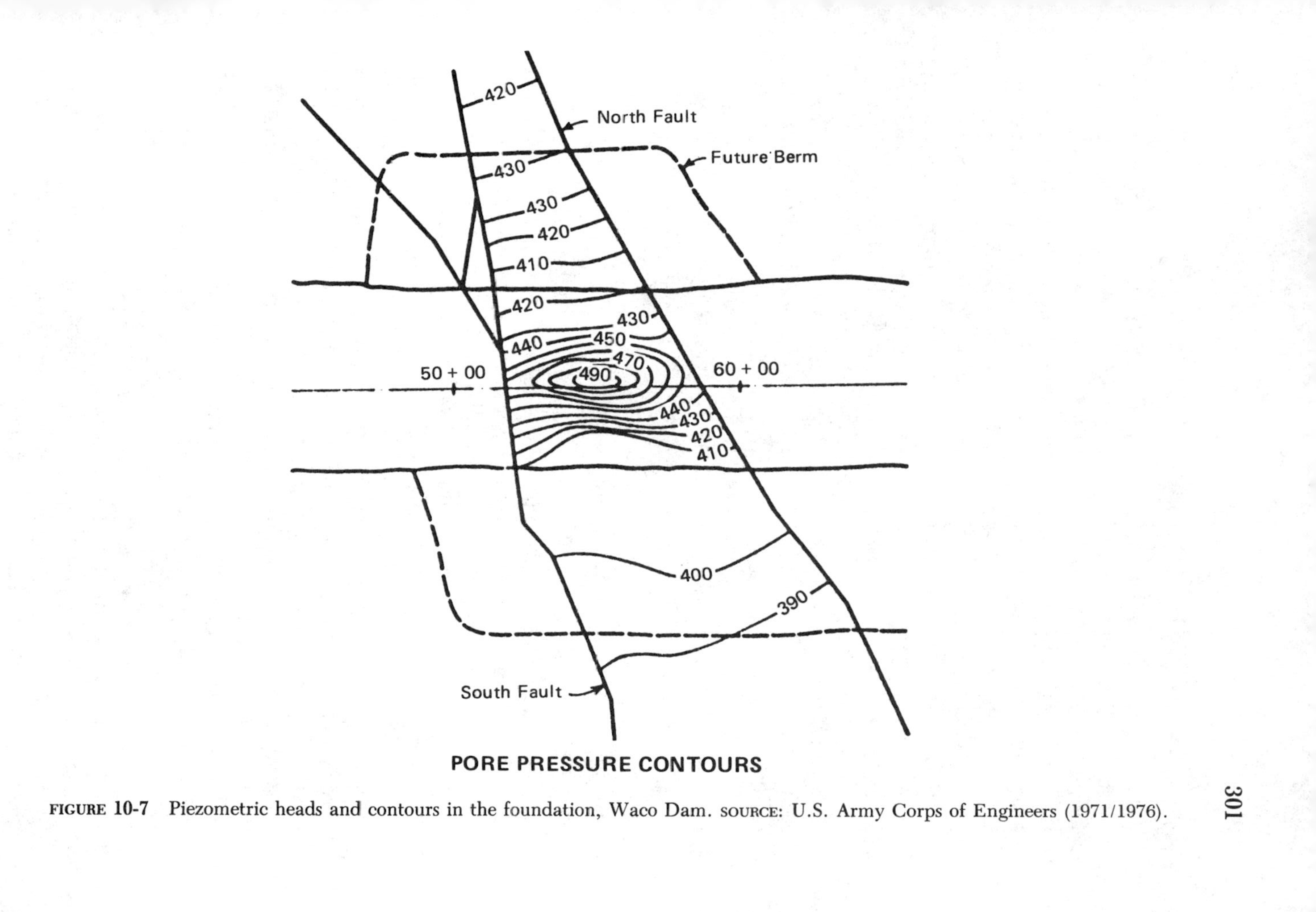


FIGURE 10-7 Piezometric heads and contours in the foundation, Waco Dam. SOURCE: U.S. Army Corps of Engineers (1971/1976).

## DATA COLLECTION AND ANALYSIS

The reason for installing instruments in dams is to monitor them during construction and operation. One of the specific applications of measurements is to furnish data to determine if the completed structure will continue to function as intended. The processing of large masses of raw data can be efficiently handled by computer methods. The interpretation of the data requires careful examination of measurements as well as other influencing effects, such as reservoir operation, air temperature, precipitation, drain flow and leakage around the structure, contraction joint grouting, concrete placement schedule, seasonal shutdown during construction, concrete testing data, and periodic instrument evaluations. The display of data should be both tabular and graphical and should be simple and readily understood. The data should be reviewed periodically by a professional engineer versed in the design, construction, and operation of embankment and/or concrete dams.

Because of the various types of instrumentation used in the different kinds of dams, data collection and analysis will be discussed for embankment dams and for concrete gravity and concrete arch dams separately.

### Monitoring of Embankment Dams

The U.S. Corps of Engineers' *Engineer Manual* (1976) discusses collecting, recording, and analyzing data and makes recommendations on frequency of reading, recording, and reporting measurements and analysis. Profiles of piezometric levels in the foundation and contours of pore water pressures in the embankment and foundation should be made periodically to evaluate the data. Examples of plotting the data are shown in Figures 10-6 and 10-7. After construction and during reservoir filling, embankment and foundation piezometers should be read weekly or once for every 5-foot rise in the water level. Readings after reservoir filling should be continued on a quarterly or semiannual basis depending on the dissipation of pore water pressures. It is also recommended in the Corps' *Manual* that additional observations be made at times of varying heads.

The movement measuring devices should be read immediately after installation, since all subsequent readings will be referred to the initial reading. The field data should be reduced to a reportable form promptly and prepared in graphical form to evaluate relations and trends. Figure 10-8 shows a typical presentation of vertical and horizontal movement data.

Wilson (1973) states that it is customary to install surface monuments on the top of a dam after completion and to observe postconstruction settlement for several years. Internal movement instruments should be installed inside a dam before the embankment is topped out to provide a history of

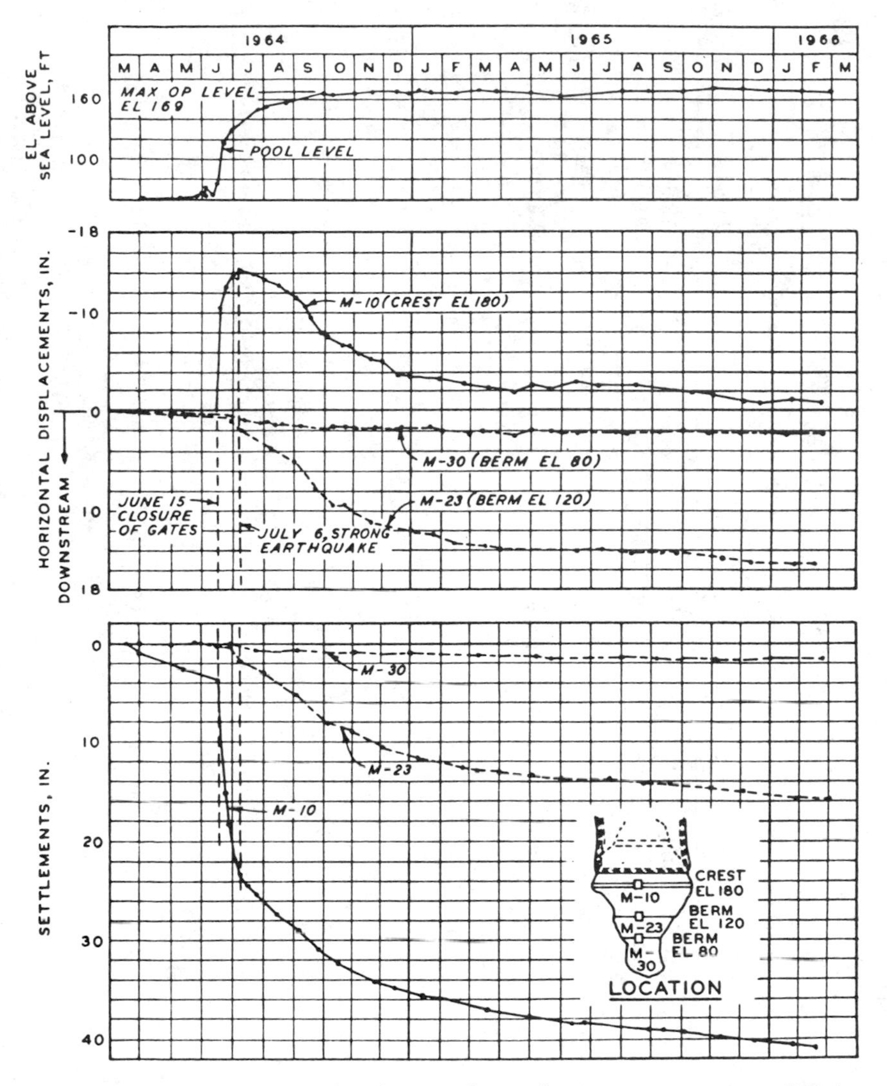


FIGURE 10-8 Displacements and settlements of central points at crest. SOURCE: U.S. Army Corps of Engineers (1971/1976).

deformations. The reading of these instruments should continue for years to establish performance during operation. Figure 10-9 shows a plot of horizontal displacements versus pool levels at El Infiernillo Dam (Marsal and de Arellano 1972). The observations of the well-instrumented El Infiernillo Dam led the authors to conclude that the unexpected behavior that developed two years after the first filling of the reservoir was the result of the

**TABLE 10-4** Frequency of Readings for Earth Dam Instrumentation

| | Progress Report During Construction | | Periodic Report Operation | |
|---|---|---|---|---|
| | Frequency of Readings | | Frequency of Readings | |
| | Construction | Shutdown | First Year | Regular |
| Piezometer readings (separate gages) | Twice monthly | Monthly | Monthly | Monthly |
| Piezometer readings (master gage) | Monthly | Alternate months | Approximately 6 months after completion of dam | Annually on same date as a set of separate gage reading |
| Porous tube piezometer readings | Twice monthly | Monthly | Monthly | Monthly |
| Internal vertical and horizontal movement readings (crossarm or HMD) | Complete set of readings each time a unit is installed | Monthly | Complete set approximately 6 months after dam is completed | Every 2 years |
| Foundation settlement readings (baseplates) | Complete set of readings each time an extension is added | Monthly | Approximately 6 months after dam is completed | Every 2 years |
| Measurement points—cumulative settlement and deflection readings | Monthly, if required, or when dam is completed | Monthly, if required | Approximately 6 months after dam is completed | Every 2 years |
| Measurement points—cumulative settlement and deflection readings spillway and outlet works | Monthly as portions of structures are completed | Monthly | Approximately 6 months after structure is completed | Every 2 years |
| Measurement points—cumulative settlement readings-spillway floor slabs | Monthly as slabs on structure are completed | Monthly, if required | Approximately 6 months after structure is completed | Every 2 years |

SOURCE: U.S. Bureau of Reclamation (1974).

interaction between the core and the rockfill shells and the wetting of the dry rockfill and that the dam had an acceptable margin of safety.

For earth pressure measuring devices, after a structure is completed, the pressure cells should be read at least annually to evaluate changes in stress with time. Data from earth pressure cells should be reduced and time plots maintained for each cell during and after construction. Analysis of data should include a comparison of observed earth pressure with earth pressure assumed for the design of the structure. The U.S. Bureau of Reclamation (1974) has developed a frequency of readings for earth dam instrumentation. A copy of this illustration is shown in Table 10-4.

## Monitoring of Concrete Dams

The U.S. Bureau of Reclamation (1976, 1977) states that to determine the manner in which a concrete dam and its foundation behave during con-

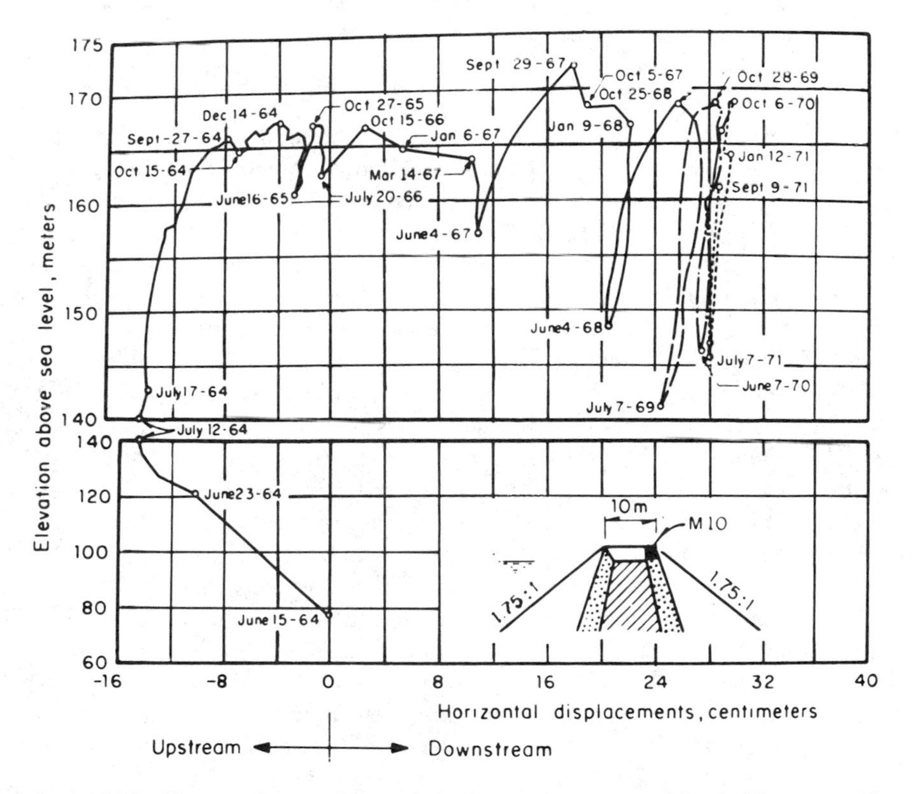


FIGURE 10-9 Horizontal (river direction) displacements versus pool levels, Monument 10. SOURCE: Marsal and de Arellano (1972).

**TABLE 10-5** Frequency of Readings for Concrete Dam Instrumentation

| Type Reading | Frequency |
|---|---|
| Embedded instrument | Seven to 10 days during construction; semimonthly or monthly afterward. More frequently during periods of reservoir filling or rapid drawdown. |
| Deflection measuring devices | Weekly; closer intervals during periods of special interest. |
| Uplift pressure measurement | Monthly, except for initial filling, which should be a 7- to 10-day interval. |
| Target deflection and pier net triangulation measurements | Semiannually during period of minimum and maximum air temperature. Additional measurements during early stages of reservoir filling. |
| Leveling across top of dam and vicinity | Periodically. More frequently in the early stages of operation and less frequently at later stages, depends on conditions encountered. |

SOURCE: U.S. Bureau of Reclamation (1977).

struction and operation, measurements should be made to obtain data on strain, temperature, stress, deflection, and deformation of the foundation. There are two general methods of measurement for obtaining the essential behavioral information. The first method involves several types of instruments that are embedded in the mass concrete of the structure and on the features of the dam. The second method involves several types of precise surveying measurements from targets at various locations on and in the dam. The suggested schedules for collecting data are shown in Table 10-5.

The planned program for measurements should cover a time period that will include a full reservoir plus two cycles of reservoir operation, after which a major portion of the measurement may be suspended. For the remaining measurements the interval between successive readings may be lengthened. The U.S. Corps of Engineers' *Engineering Manual* (1980) discusses instrumentation to measure the structural behavior of concrete gravity dams under five major headings: (1) Carlson-type instruments, (2) uplift and leakage, (3) deflection plum line, (4) precise alignment facilities, and (5) thermocouples. The frequency of collection and evaluation of data is also discussed in detail for each category.

Some excellent illustrations concerning dam foundation uplift pressure histories, methods of showing uplift pressure gradients, typical deflection history, and typical precise alignment marker layout and details can be found in the U.S. Corps of Engineers' *Engineering Manual* (1980).

## REFERENCES


American Concrete Institute (1968) "Guide for Making a Condition Survey of Concrete in Service," *Journal of the American Concrete Institute, Proceedings*, Vol. 65, No. 11, pp. 905-918.

Bolt, B. A., and Hudson, D. E. (1975) "Seismic Instrumentation of Dams," *Journal of the Geotechnical Engineering Division*, ASCE, Vol. 101, No. GT11 (November), pp. 1095-1104.

Dunnicliff, J. (1981) *Measurements Committee Report*, U.S. Committee on Large Dams Section VI, Inventory of Geotechnical Instruments, Manufacturers or Suppliers.

International Commission on Large Dams (ICOLD) (1969) *General Considerations Applicable to Instrumentation for Earth and Rockfill Dams*. Committee on Observations of Dams and Models, Bulletin No. 21, November.

International Commission on Large Dams (ICOLD) (1981) *Automated Observation for Instantaneous Safety Control of Dams and Reservoirs*, Bulletin No. 41, January.

Jansen, R. B. (1968) *A Prescription for Dam Safety—Instrumentation and Surveillance*, Conference of College of Engineering, University of California, Berkeley.

Legget, R. F. (1967) "Reservoirs and Catchment Areas," Chapter 14 in *Geology and Engineering*, 2d ed., McGraw-Hill International Series, New York.

Lytle, J. D. (1982) *Dam Safety Instrumentation; Automation of Data Observations, Processing and Evaluation; Question* 52, in International Commission on Large Dams (ICOLD), Response 14th Congress, Rio de Janeiro.

Marsal, R. J., and de Arellano, L. R. (1972) "Eight Years of Observations at El Infiernillo Dam," Proceedings of the Specialty Conference on Performance of Earth and Earth-Supported Structures, ASCE.

National Research Council (NRC) (1982) *Geotechnical Instrumentation for Monitoring Field Performance*, National Academy Press, Washington, D.C.

Sharma and Raphael (1979/1981) *General Considerations on Reservoir Instrumentation*, Committee on Measurements, USCOLD.

U.S. Army Corps of Engineers (1971 and 1976) *Instrumentation on Earth and Rock-Fill Dams*, Parts 1 and 2, EM 1110-2-1908, August 1971 and November 1976.

U.S. Army Corps of Engineers (1980) *Instrumentation for Measurement of Structural Behavior of Concrete Structures*, EM 1110-2-4300, September.

U.S. Bureau of Reclamation (1974) *Earth Manual*, 2d ed., Government Printing Office, Washington, DC.

U.S. Bureau of Reclamation (1976) "Design of Gravity Dams," Chapter XIII in *Design Manual for Concrete Gravity Dams*.

U.S. Bureau of Reclamation (1977) "Design of Arch Dams," Chapter XIII in *Design Manual for Concrete Arch Dams*.

Wilson, S. D. (1970) "Observational Data on Ground Movements Related to Slope Instability, The Sixth Terzaghi Lecture," *Journal of the Soils Mechanic and Foundation Division*, SM5, September.

Wilson, S. D. (1973) "Deformation of Earth and Rockfill Dams," in *Embankment-Dam Engineering*, Casagrande Volume, John Wiley & Sons, New York.


## RECOMMENDED READING


ASCE (1981) *Conference Proceedings on Recent Developments in Geotechnical Engineering for Hydro Projects*, Fred Kulhawy, ed.

ASCE (1973) *Inspection, Maintenance and Rehabilitation of Old Dams*, New York.

Bannister, J. R., and Pyke, R. (1976) "In-Situ Pore Pressure Measurements at Rio Blanco," *Journal of the Geotechnical Engineering Division*, ASCE, Vol. 102, No. TG10, October, pp. 1073-1091.

De Alba, P., and Seed, H. B. (1976) "Sand Liquefaction in Large-Scale Simple Shear Test," *Journal of the Geotechnical Engineering Division*, ASCE, Vol. 102, No. GT9 (September), pp. 909-974.

ICOLD (1973) *Rock Mechanics and Dam Foundation Design*. International Commission on Large Dams, Boston, Mass.

ICOLD Bulletin No. 23 (1972) *Reports of the Committee on Observation on Dams and Models*, July.

Jansen, R. B. (1980) *Dams and Public Safety*, U.S. Department of the Interior, Bureau of Reclamation (reprinted 1983).

Kovari, K. (1977) *Field Instrumentation in Rock Mechanics*, Vols. I and II, Federal Institute of Technology, Zurich.

O'Rourke, J. E. (1974) "Performance Instrumentation Installed in Oroville Dam," *Journal of the Geotechnical Engineering Division*, ASCE, Vol. 100, No. GTZ (February), pp. 157-174.

Peck, R. B. (1969) "Advantages of Limitations of Observational Method in Applied Soil Mechanics," 9th Rankine Lecture, *Geotechnique*, Vol. 19, No. 2, pp. 171-187.

Raphael, J. M., and Carlson, R. W. (1965) *Measurement of Structural Action in Dams*, 3d ed., James J. Gillick and Co., Berkeley, Calif.

Sarkaria, G. S. (1973) *Proceedings of the Engineering Foundation Conference on Inspection, Maintenance and Rehabilitation of Old Dams, Safety Appraisal of Old Dams: An updated perspective*, September 23-28, ASCE, New York, pp. 405-217.

Sherard, J. L., et al. (1973) *Earth and Earth Rock Dams*, John Wiley & Sons, New York.

Slobodulk, D., and Roof, E. F. (1980) "Monitoring of Dam Movements Using Laser Light," *Transportation Engineering Journal*, November, pp. 829-843.

U.S. Bureau of Reclamation (1980) *Safety Evaluation of Existing Dams (SEED Manual)*, Government Printing Office, Washington, D.C.

Wahler, W. A. (1974) *Evaluation of Mill Tailings Disposal Practices and Potential Dam Stability Problems in Southwestern United States*, Prepared for Bureau of Mines, General Report, Vol. I.

Wilson, S. D., and Handcock, C. W., Jr. (1965) "Instrumentation for Movements Within Rockfill Dams," *Instruments and Apparatus for Soil and Rock Mechanics*, ASTM STP 392, American Society of Testing Materials, pp. 115-130.

# 11
# Recommended Glossary

Many agencies at various levels of government, utilities, professional organizations, and other entities have responsibilities and other interests in the design, construction, and safe operation of dams. Consequently, the nomenclature that has developed over time by the large number of entities and for different purposes in association with dams is not always consistent. Inconsistencies range from subtle to contradictory. These differences in definition have occasionally resulted in confusion—even among professional engineers involved in dam safety. In fact, the National Research Council's Committee on Safety of Nonfederal Dams recommended in its 1982 report* to FEMA that "FEMA, with the help of ICODS, should develop a glossary of common terms for use in dam safety activities." The present committee, based on its review of the various sources of nomenclature and the members' experience, recommends the following glossary of terms to assist FEMA in implementing this recommendation. This glossary is based principally on the International Commission on Large Dams (ICOLD) *Technical Dictionary on Dams*, 1978. Many of the ICOLD terms were modified or supplemented to conform more closely to American usage. Terms, not found in the ICOLD glossary, were supplied by committee members. It is noted that all of these terms are not contained in the body of the present report.

*National Research Council, Committee on Safety of Nonfederal Dams (1982) *Safety of Nonfederal Dams: A Review of the Federal Role*, National Academy Press, Washington, D.C.

ABUTMENT That part of the valley side against which the dam is constructed. An artificial abutment (*see* Block) is sometimes constructed, as a concrete gravity section, to take the thrust of an arch dam where there is no suitable natural abutment. Right and left abutments are those on respective sides of an observer when viewed looking downstream.

ACTIVE STORAGE The volume of the reservoir that is available for use either for power generation, irrigation, flood control, or other purposes. Active storage excludes flood surcharge. It is the reservoir capacity less inactive and dead storages. The terms *useful storage* or *usable storage* or *working storage* are sometimes used instead of active storage but are not recommended.

ADIT Tunnel for exploratory or test purposes; opening in the face of a dam for access to galleries or operating chambers; or access tunnel to a tunnel for construction or maintenance purposes.

AFTERBAY DAM (REREGULATING DAM) A dam constructed to regulate the discharges from an upstream power plant.

AMBURSEN DAM *See* Buttress Dam.

ANCHOR BLOCK *See* Block.

APPURTENANT STRUCTURES Refers to ancillary features of a dam, such as the outlet, spillway, powerhouse, tunnels, etc.

AQUEDUCT An artificial way of conveying water, i.e., by canal, pipe, or tunnel; hence the terms *connecting aqueduct* and *diversion aqueduct*.

ARCH DAM A concrete or masonry dam that is curved in plan so as to transmit the major part of the water load to the abutments.

ARCH CENTERLINE The locus of all midpoints of the thickness of an arch section.

ARCH ELEMENT That portion of a dam bounded by two horizontal planes spaced 1 foot apart.

CANTILEVER ELEMENT That portion of an arch dam that is contained within the vertical planes normal to the extrados and spaced 1 foot apart at the axis.

CONSTANT ANGLE ARCH DAM An arch dam in which the angle subtended by any horizontal section is constant throughout the whole height of the dam.

CONSTANT RADIUS ARCH DAM An arch dam in which every horizontal segment or slice of the dam has approximately the same radius of curvature.

CROWN CANTILEVER That cantilever element located at the point of maximum depth in the canyon.

DOUBLE CURVATURE ARCH DAM An arch dam that is curved vertically as well as horizontally.

FILLET An increase in thickness of a dam beginning near and extending to the abutments of the arches or base of cantilevers. Usually placed at the downstream face.

LENGTH OF ARCH The distance along a curve that is concentric with the extrados and passes through the midpoint of the arch thickness at the crown.

LINE OF CENTERS The loci of centers for circular arcs used to describe a face of a dam or a portion thereof.

ARCH BUTTRESS DAM *See* Buttress Dam.

ARCH GRAVITY DAM *See* Gravity Dam.

AUXILIARY SPILLWAY *See* Spillway.

AXIS OF DAM The plane or curved surface, arbitrarily chosen by a designer, appearing as a line, in plan or in cross section, to which the horizontal dimensions of the dam can be referred.

BACKWATER CURVE The longitudinal profile of the water surface in an open channel where the depth of flow has been increased by an obstruction such as a weir or dam across the channel, or by an increase in channel roughness, by a decrease in channel width, or by a flattening of the bed slope.

BAFFLE BLOCK *See* Block.

BANK STORAGE (GROUND STORAGE) *See* Storage.

BASE WIDTH (BASE THICKNESS) The maximum thickness or width of a dam measured horizontally between upstream and downstream faces and normal to the axis of the dam but excluding projections for outlets, etc.

BATTER Angle of inclination from the vertical.

BERM A horizontal step or bench in the sloping profile of an embankment dam.

BLANKET

DRAINAGE BLANKET A drainage layer placed directly over the foundation material.

GROUT BLANKET *See* Consolidation Grouting.

UPSTREAM BLANKET An impervious layer placed on the reservoir floor upstream of a dam. In the case of an embankment dam the blanket may be connected to the impermeable element in the dam.

BLOCK

BAFFLE BLOCK (IMPACT BLOCK) A block of concrete or concrete and steel constructed in a channel or stilling basin to dissipate the energy of water flowing at high velocity.

CHUTE BLOCK A baffle block constructed in a spillway chute.

THRUST BLOCK (ANCHOR BLOCK) A massive block of concrete built to withstand a thrust or pull from an arch dam.

BULKHEAD GATE *See* Gate.

BUTTRESS DAM A dam consisting of a watertight upstream face supported at intervals on the downstream side by a series of buttresses. Buttress dams can take many forms.

ARCH BUTTRESS DAM (CURVED BUTTRESS DAM) A buttress dam that is curved in plan.

MULTIPLE ARCH DAM A buttress dam the upstream part of which comprises a series of arches.

FLAT SLAB DAM (AMBURSEN DAM) (DECK DAM) A buttress dam in which the upstream part is a relatively thin flat slab usually made of reinforced concrete.

SOLID HEAD BUTTRESS DAM A buttress dam in which the upstream end of each buttress is enlarged to span the gap between buttresses. The terms *round head*, *diamond head*, *tee head* refer to the shape of the upstream enlargement.

CELLULAR GRAVITY DAM *See* Gravity Dam.

COFFERDAM A temporary structure enclosing all or part of the construction area so that construction can proceed in the dry. A diversion cofferdam diverts a river into a pipe, channel, or tunnel.

CONCRETE LIFT In concrete work the vertical distance between successive horizontal construction joints.

CONDUIT A closed channel to convey the discharge through or under a dam.

CONSOLIDATION GROUTING (BLANKET GROUTING) Consolidating a layer of the foundation to achieve greater impermeability and/or strength by injecting grout.

CONSTRUCTION JOINT The interface between two successive placings or pours of concrete where bond, not permanent, separation is intended.

CONTACT GROUTING Filling with cement grout any voids existing at the contact of two zones of different materials, e.g., between a concrete tunnel lining and the surrounding rock. The grouting operation is usually carried out at low pressure.

CORE (IMPERVIOUS CORE) (IMPERVIOUS ZONE) A zone of material of low permeability in an embankment dam; hence the terms *central core*, *inclined core*, *puddle clay core*, and *rolled clay core*.

CORE WALL A wall built of impervious material, usually of concrete or asphaltic concrete in the body of an embankment dam to prevent leakage. *See also* Membrane or Diaphragm.

CREST GATE *See* Gate.

CREST LENGTH The developed length of the top of the dam. This includes the length of spillway, powerhouse, navigation lock, fish pass, etc., where these structures form part of the length of the dam. If detached from the dam these structures should not be included.

CREST OF DAM The crown of an overflow section of the dam. In the United States, the term *crest of dam* is often used when *top of dam* is intended. To avoid confusion, the terms *crest of spillway* and *top of dam* should be used for referring to the overflow section and dam proper, respectively.

CRIB DAM A gravity dam built up of boxes, cribs, crossed timbers, or gabions and filled with earth or rock.

CULVERT (a) A drain or waterway structure built transversely under a road, railway, or embankment. A culvert usually comprises a pipe or a covered channel of box section. (b) A gallery or waterway constructed through any type of dam, which is normally dry but is used occasionally for discharging water; hence the terms *scour culvert*, *drawoff culvert*, and *spillway culvert*.

CURTAIN

GROUT CURTAIN (GROUT CUTOFF) A barrier produced by injecting grout into a vertical zone, usually narrow in horizontal width, in the foundation to reduce seepage under a dam.

DRAINAGE CURTAIN *See* Drainage Wells.

CURVED BUTTRESS DAM (ARCH BUTTRESS DAM) *See* Buttress Dam.

CURVED GRAVITY DAM *See* Gravity Dam.

CUTOFF An impervious construction by means of which seepage is reduced or prevented from passing through foundation material.

CUTOFF TRENCH The excavation later to be filled with impervious material so as to form the cutoff. Sometimes used incorrectly to describe the cutoff itself.

CUTOFF WALL A wall of impervious material (e.g., concrete, asphaltic concrete, steel sheet piling) built into the foundation to reduce seepage under the dam.

DAM A barrier built across a watercourse for impounding or diverting the flow of water.

DEAD STORAGE The storage that lies below the invert of the lowest outlet and that, therefore, cannot be withdrawn from the reservoir.

DESIGN FLOOD *See* Spillway Design Flood.

DIAMOND HEAD BUTTRESS DAM *See* Buttress Dam.

DIAPHRAGM *See* Membrane.

DIKE (LEVEE) A long low embankment. The height is usually less than 4 to 5 meters and the length more than 10 or 15 times the maximum height. Usually applied to embankments or structures built to protect land from flooding. If built of concrete or masonry the structure is usually referred to as a flood wall. Also used to describe embankments that block areas on the reservoir rim that are lower than the top of the main dam and that are quite long.

In the Mississippi River basin, where the old French word *levee* has survived, this now applies to flood protecting embankments whose height can average up to 10 to 15 meters.

DIVERSION CHANNEL, CANAL, OR TUNNEL A waterway used to divert water from its natural course. The term is generally applied to a temporary arrangement, e.g., to bypass water round a dam site during construction. *Channel* is normally used instead of *canal* when the waterway is short. Occasionally the term is applied to a permanent arrangement (diversion canal, diversion tunnel, diversion aqueducts).

DOLOSSE A precast concrete shape, named after the knucklebone of a sheep. Dolosses are placed randomly in an interlocking pattern and are extremely effective in dissipating the energy of waves or flowing water.

DRAINAGE AREA The area that drains naturally to a particular point on a river.

DRAINAGE LAYER OR BLANKET A layer of pervious material in a dam to relieve pore pressures or to facilitate drainage of the fill.

DRAINAGE WELLS (RELIEF WELLS) Vertical wells or boreholes usually downstream of impervious cores, grout curtains, or cutoffs, designed to collect and control seepage through or under a dam so as to reduce uplift pressures under or within a dam. A line of such wells forms a drainage curtain.

DRAWDOWN The resultant lowering of water surface level due to release of water from the reservoir.

EARTH DAM OR EARTHFILL DAM *See* Embankment Dam.

EMBANKMENT Fill material, usually earth or rock, placed with sloping sides.

EMBANKMENT DAM (FILL DAM) Any dam constructed of excavated natural materials or of industrial waste materials.

EARTH DAM (EARTHFILL DAM) An embankment dam in which more than 50% of the total volume is formed of compacted fine-grained material obtained from a borrow area.

HOMOGENEOUS EARTHFILL DAM An embankment dam constructed of similar earth material throughout, except for possible inclusion of internal drains or drainage blankets. Used to differentiate from a zoned earthfill dam.

HYDRAULIC FILL DAM An embankment dam constructed of materials, often dredged, that are conveyed and placed by suspension in flowing water.

ROCKFILL DAM An embankment dam in which more than 50% of the total volume comprises compacted or dumped pervious natural or crushed rock.

ROLLED FILL DAM An embankment dam of earth or rock in which the material is placed in layers and compacted by using rollers or rolling equipment.

ZONED EMBANKMENT DAM An embankment dam the thickness of which is composed of zones of selected materials having different degrees of porosity, permeability, and density.

EMERGENCY ACTION PLAN A predetermined plan of action to be taken to reduce the potential for property damage and loss of lives in an area affected by a dam break.

EMERGENCY GATE A standby or reserve gate used only when the normal means of water control is not available.

EMERGENCY SPILLWAY *See* Spillway.

ENERGY DISSIPATOR Any device constructed in a waterway to reduce or destroy the energy of fast-flowing water.

ENERGY DISSIPATING VALVE A generic term used to describe those regulating valves that are designed to dissipate as much energy as possible through the valve. Included are Howell-Bunger valves and Hollow Jet valves.

EPICENTER That point on the earth's surface that is directly above the focus of an earthquake.

EXTRADOS The curved upstream surface of an arch dam.

FACE With reference to a structure, the external surface that limits the structure, e.g., the face of a wall or dam.

FACING With reference to a wall or concrete dam, a coating of a different material, masonry or brick, for architectural or protection purposes, e.g., stonework facing, brickwork facing. With reference to an embankment dam, an impervious coating or face on the upstream slope of the dam.

FAILURE An incident resulting in the uncontrolled release of water from an operating dam.

FETCH The straight line distance between a dam and the farthest reservoir shore. The fetch is one of the factors used in calculating wave heights in a reservoir.

FILTER (FILTER ZONE) A band or zone of granular material that is incorporated in a dam and is graded (either naturally or by selection) so as to allow seepage to flow across or down the filter without causing the migration of material from zones adjacent to the filter.

FINGER DRAINS A series of parallel drains of narrow width (instead of a continuous drainage blanket) draining to the downstream toe of the embankment dam.

FIXED CREST WEIR *See* Weir.

FLASHBOARDS Lengths of timber, concrete, or steel placed on the crest of a spillway to raise the retention water level but that may be quickly removed in the event of a flood either by a tripping device or by deliberately designed failure of the flashboards or their supports.

FLAT SLAB DAM *See* Buttress Dam.

FLIP BUCKET (SKIJUMP SPILLWAY) The downstream end of a spillway shaped such that water flowing at high velocity is deflected upward in a trajectory away from the end of the spillway.

FLOOD PLAIN An area adjoining a body of water or natural stream that has been or may be covered by flood water.

FLOOD PLAIN MANAGEMENT A management program to reduce the consequences of flooding, either by natural runoff or by dam failure, to properties in a flood plain, both existing and future.

FLOOD ROUTING The determination of the attenuating effect of storage on a flood passing through a valley, channel, or reservoir.

FLOOD STORAGE *See* Storage.

FLOOD SURCHARGE The volume or space in a reservoir between the controlled retention water level and the maximum water level. Flood surcharge cannot be retained in the reservoir but will flow over the spillway until the controlled retention water level is reached. (The term *wet freeboard* for describing the depth of flood surcharge is not recommended; *see* Freeboard.)

FLOOD WALL A concrete wall constructed adjacent to a stream for the purpose of preventing flooding of property on the landside of the wall; normally constructed in lieu of or to supplement a levee where the land required for levee construction is more expensive or not available.

FOCUS (HYPOCENTER) The point within the earth that is the center of an earthquake and the origin of its elastic waves.

FOUNDATION OF DAM The natural material on which the dam structure is placed.

FREEBOARD The vertical distance between a stated water level and the top of a dam. NET FREEBOARD, DRY FREEBOARD, FLOOD FREEBOARD, or RESIDUAL FREEBOARD is the vertical distance between the estimated maximum water level and the top of a dam. GROSS FREEBOARD or TOTAL FREEBOARD is the vertical distance between the maximum planned controlled retention water level and the top of a dam. (That part of the GROSS FREEBOARD attributable to the depth of flood surcharge is sometimes referred to as the WET FREEBOARD, but this term is not recommended as it is preferable that freeboard be stated with reference to the top of a dam.)

FUSE PLUG SPILLWAY *See* Spillway.

GABION A prefabricated basket of rock within a wire cage that is free draining and capable of being stacked.

GABION DAM Name given to a crib dam when built of gabions.

GALLERY (a) A passageway within the body of a dam or abutment; hence the terms *grouting gallery*, *inspection gallery*, and *drainage gallery*. (b) A long and rather narrow hall; hence the following terms for a power plant: *valve gallery*, *transformer gallery*, and *busbar gallery*.

GATE In general, a device in which a leaf or member is moved across the waterway from an external position to control or stop the flow.

BASCULE GATE *See* Flap Gate.

BULKHEAD GATE A gate used either for temporary closure of a channel or conduit before dewatering it for inspection or maintenance or for closure against flowing water when the head difference is small, e.g., for diversion tunnel closure. Although a bulkhead gate is usually opened and closed under nearly balanced pressures, it nevertheless may be capable of withstanding a high differential head when in the closed position.

CREST GATE (SPILLWAY GATE) A gate on the crest of a spillway to control overflow or reservoir water level.

DRUM GATE A type of spillway gate consisting of a long hollow drum. The drum may be held in its raised position by the water pressure in a flotation chamber beneath the drum.

EMERGENCY GATE A standby or reserve gate used only when the normal means of water control is not available.

FIXED WHEEL GATE (FIXED ROLLER GATE) (FIXED AXLE GATE) A gate having wheels or rollers mounted on the end posts of the gate. The wheels bear against rails fixed in side grooves or gate guides.

FLAP GATE A gate hinged along one edge, usually either the top or bottom edge. Examples of bottom-hinged flap gates are tilting gates and fish belly gates so called from their shape in cross section.

FLOOD GATE A gate to control flood release from a reservoir.

GUARD GATE (GUARD VALVE) Gate or valve that operates fully open or closed. May function as a secondary device for shutting off the flow of water in case the primary closure device becomes inoperable. Usually operated under balanced pressure no-flow conditions, except for closure in emergencies.

OUTLET GATE A gate controlling the outflow of water from a reservoir.

RADIAL GATE (TAINTER GATE) A gate with a curved upstream plate and radial arms hinged to piers or other supporting structure.

REGULATING GATE (REGULATING VALVE) A gate or valve that operates under full pressure and flow conditions to throttle and vary the rate of discharge.

ROLLER DRUM GATE A crest gate for dam spillways comprising a long horizontal cylinder spanning between piers. The cylinder is fitted with a toothed rim at each end and rotates as it is moved up and down on inclined racks fixed to the piers.

ROLLER GATE (STONEY GATE) A gate for large openings that bears on an intermediate train of rollers in each gate guide.

SKIMMER GATE A gate at the dam crest whose prime purpose is to control the release of debris and logs with a limited amount of water. It is usually a flap or Bascule gate.

SLIDE GATE (SLUICE GATE) A gate that can be opened or closed by sliding in supporting guides.

GEOPHYSICAL METHODS Methods of studying soil and rock properties and geologic structure without taking samples.

GRAVITY DAM A dam constructed of concrete and/or masonry that relies on its weight for stability.

ARCH GRAVITY DAM An arch dam where part of the water thrust is transmitted to the abutments by horizontal thrust and part to the foundation by cantilever action.

CURVED GRAVITY DAM A gravity dam that is curved in plan.

HOLLOW GRAVITY DAM (CELLULAR GRAVITY DAM) A dam that has the outward appearance of a gravity dam but that is of hollow construction.

GROIN That area along the contact (or intersection) of the face of a dam with the abutments.

GROSS STORAGE (RESERVOIR CAPACITY) (GROSS CAPACITY OF RESERVOIR) The gross capacity of a reservoir from the river bed up to maximum controlled retention water level. It includes active, inactive, and dead storage.

GROUT BLANKET *See* Blanket.

GROUT CAP A concrete pad or wall constructed to facilitate subsequent pressure grouting of the grout curtain beneath the grout cap.

GROUT CURTAIN *See* Curtain.

HAZARD A source of danger. In other words, something that has the potential for creating adverse consequences.

HEADRACE A free-flow tunnel or open channel that conveys water to the upper end of a penstock; hence the terms *headrace tunnel* and *headrace canal.*

HEADWATER LEVEL The level of the water in the reservoir or in the headrace at the nearest free surface to the turbine.

HEEL OF DAM The junction of the upstream face of a gravity or arch dam with the foundation surface. In the case of an embankment dam the junction is referred to as the upstream toe of the dam.

HEIGHT ABOVE GROUND LEVEL The maximum height from natural ground surface to the top of a dam.

HEIGHT ABOVE LOWEST FOUNDATION The maximum height from the lowest point of the general foundation to the top of a dam.

HYDRAULIC HEIGHT Height to which the water rises behind a dam and the difference between the lowest point in the original streambed at the axis of the dam and the maximum controllable water surface.

HYDROGRAPH A graphical representation of discharge, stage, or other hydraulic property with respect to time for a particular point on a stream. (At times the term is applied to the phenomenon the graphical representation describes; hence a *flood hydrograph* is the passage of a flood discharge past the observation point.)

IMPERVIOUS CORE *See* Core.

INACTIVE STORAGE The storage volume of a reservoir measured between the invert level of the lowest outlet and minimum operating level.

INCLINOMETER (INCLOMETER) An instrument, usually comprising a metal or plastic tube inserted in a drill hole, and a sensitized monitor either lowered into the tube or fixed within the tube. This measures at different points the tube's inclination to the vertical. By integration, the lateral position at different levels of the tube may be found relative to a point, usually the top or bottom of the tube, assumed to be fixed. The system may be adapted to measure settlement.

INTAKE Any structure in a reservoir, dam, or river through which water can be drawn into an aqueduct.

INTENSITY SCALE An arbitrary scale to describe the degree of shaking at a particular place. The scale is not based on measurement but on a descriptive scale by an experienced observer. Several scales are used (e.g, the Modified Mercalli scale, the MSK scale) all with grades indicated by Roman numerals from I to XII.

INTERNAL EROSION *See* Piping.

INTRADOS The curved downstream surface of an arch dam.

INUNDATION MAP A map delineating the area that would be inundated in the event of a dam failure.

LEAKAGE Uncontrolled loss of water by flow through a hole or crack.

LENGTH OF RESERVOIR The distance along the thalweg of the valley forming the reservoir from the dam to the farthest point where the principal river or a tributary enters the reservoir.

LEVEE *See* Dike.

LINING With reference to a canal, tunnel, shaft, or reservoir, a coating of asphaltic concrete, reinforced or unreinforced concrete, shotcrete, rubber or plastic to provide watertightness, prevent erosion, reduce friction, or support the periphery of the structure. May also refer to the lining, such as steel or concrete, of an outlet pipe or conduit.

LIVE STORAGE The sum of active and inactive storage volumes. When there is no inactive storage, e.g., in some irrigation reservoirs, live storage and active storage describe the same storage that is generally termed *live storage*.

LOADING CONDITIONS Events to which the dam is exposed, e.g., earthquake, flood, gravity loading.

LOWEST POINT OF FOUNDATION The lowest point of the dam foundation excluding cutoff trenches less than 10 meters wide and isolated pockets of excavation.

LOW-LEVEL OUTLET (BOTTOM OUTLET) An opening at a low level from the reservoir generally used for emptying or for scouring sediment and sometimes also for irrigation releases.

MAGNITUDE (*see also* RICHTER SCALE) A rating of a given earthquake independant of the place of observation. It is calculated from measurements on seismographs and is properly expressed in ordinary numbers and decimals based on a logarithmic scale. Each higher number expresses an amount of earthquake energy that is 10 times greater than expressed by the preceding lower number, e.g., a magnitude 6 earthquake will have 10 times more energy than a magnitude 5.

MASONRY DAM Any dam constructed mainly of stone, brick, or concrete blocks that may or may not be joined with mortar. A dam having only a masonry facing should not be referred to as a masonry dam.

MAXIMUM CREDIBLE EARTHQUAKE (MCE) The severest earthquake that is believed to be possible at the site on the basis of geologic and seismological evidence. It is determined by regional and local studies that include a complete review of all historic earthquake data of events sufficiently nearby to influence the project, all faults in the area, and attenuations from causative faults to the site.

MAXIMUM CROSS SECTION OF DAM Cross section of a dam at the point where the height of the dam is a maximum.

MAXIMUM WATER LEVEL The maximum water level, including the flood surcharge the dam is designed to withstand.

MEMBRANE (DIAPHRAGM) A sheet or thin zone or facing made of a flexible impervious material such as asphaltic concrete, plastic concrete, steel, wood, copper, plastic, etc. A *cutoff wall*, or *core wall*, if thin and flexible is sometimes referred to as a *diaphragm wall* or *diaphragm*.

MINIMUM OPERATING LEVEL The lowest level to which the reservoir is drawn down under normal operating conditions. The lower limit of active storage.

MORNING GLORY SPILLWAY *See* Spillway.

MULTIPLE ARCH DAM *See* Buttress Dam.

NAPPE The overfalling stream from a weir or spillway.

NONOVERFLOW DAM (NONSPILL DAM) A dam or section of dam that is not designed to be overtopped.

NORMAL WATER LEVEL For a reservoir with a fixed overflow sill it is the lowest crest level of that sill. For a reservoir the outflow from which is controlled wholly or partly by movable gates, syphons or by other means, it is the maximum level at the dam to which water may rise under normal operating conditions, exclusive of any provision for flood surcharge.

OGEE SPILLWAY *See* Spillway.

OPERATING BASIS EARTHQUAKE More moderate than the MCE and may be selected on a probabilistic basis from regional and local geology and seismology studies as being likely to occur during the life of the project. Generally, it is at least as large as earthquakes that have occurred in the seismotectonic province in which the site is located.

ONE-HUNDRED YEAR (100-YEAR) EXCEEDANCE INTERVAL The flood magnitude expected to be equalled or exceeded on the average of once in 100 years. It may also be expressed as an exceedance frequency with a 1% chance of being exceeded in any given year.

OUTLET An opening through which water can be freely discharged for a particular purpose from a reservoir.

OVERBURDEN All earth materials that naturally overlie rock.

OVERFLOW DAM (OVERTOPPABLE DAM) A dam designed to be overtopped.

PARAPET WALL A solid wall built along the top of a dam for ornament, for the safety of vehicles and pedestrians, or to prevent overtopping.

PEAK FLOW The maximum instantaneous discharge that occurs during a flood. It is coincident with the peak of a flood hydrograph.

PENSTOCK A pipeline or pressure shaft leading from the headrace or reservoir to the turbines.

PERVIOUS ZONE A part of the cross section of an embankment dam comprising material of high permeability.

PHREATIC SURFACE The free surface of groundwater at atmospheric pressure.

PIEZOMETER An instrument for measuring pore water pressure within soil, rock, or concrete.

PIPING The progressive development of internal erosion by seepage, appearing downstream as a hole or seam discharging water that contains soil particles.

PLUNGE BASIN (PLUNGE POOL) A natural or sometimes artificially created pool that dissipates the energy of free-falling water. The basin is located at a safe distance downstream of the structure from which water is being released.

PORE PRESSURE The interstitial pressure of water within a mass of soil, rock, or concrete.

POWER TUNNEL A tunnel carrying water to a hydropower plant.

PRECAST DAM A dam constructed mainly of large precast concrete blocks or sections.

PRESSURE CELL An instrument for measuring pressure within a mass of soil, rock, or concrete or at an interface between one and the other.

PRESSURE RELIEF PIPES Pipes used to relieve uplift or pore water pressure in a dam foundation or in the dam structure.

PRESTRESSED DAM A dam the stability of which depends in part on the tension in steel wires, cables, or rods that pass through the dam and that are anchored into the foundation rock.

PROBABILITY The likelihood of an event occurring.

PROBABLE MAXIMUM FLOOD (PMF) The flood that may be expected from the most severe combination of critical meteorologic and hydrologic conditions that are possible in the region.

ONE-HALF PMF That flood with a peak flow equal to one-half of the peak flow of a probable maximum flood.

PROBABLE MAXIMUM PRECIPITATION (PMP) The maximum amount and duration of precipitation that can be expected to occur on a drainage basin.

PUMPED STORAGE RESERVOIR A reservoir filled entirely or mainly with water pumped from outside its natural drainage area.

RANDOM FILL Earth or rockfill the grading of which is not specified and that is placed without treatment just as it comes from the excavation.

REGULATING DAM A dam impounding a reservoir from which water is released to regulate the flow in a river.

RELIEF WELLS *See* Drainage Wells.

REREGULATING DAM *See* Afterbay Dam.

RESERVOIR (MAN-MADE LAKE) An artificial lake, basin, or tank in which water can be stored.

RESERVOIR AREA The surface area of a reservoir when filled to controlled retention water level.

RESERVOIR ROUTING The computation by which the interrelated effects of the inflow hydrograph, reservoir storage, and discharge from the reservoir are evaluated.

RESERVOIR SURFACE The surface of a reservoir at any level.

RICHTER SCALE A scale proposed by C. F. Richter to describe the magnitude of an earthquake by measurements made in well-defined conditions and with a given type of seismograph. The zero of the scale is fixed arbitrarily to fit the smallest recorded earthquakes. The largest recorded earthquake magnitudes are near 8.7 and are the result of observations and not an arbitrary upper limit like that of the intensity scale (see Table 5-4).

RIPRAP A layer of large uncoursed stones, broken rock, or precast blocks placed in random fashion on the upstream slope of an embankment dam, on a reservoir shore, or on the sides of a channel as a protection against wave and ice action. Very large riprap sometimes is referred to as *armoring*.

RISK The likelihood of adverse consequences.

RISK ASSESSMENT As applied to dam safety, the process of identifying the likelihood and consequences of dam failure to provide the basis for informed decisions on a course of action.

RISK COST (EXPECTED COST OF FAILURE) The product of the risk and the monetary consequences of failure.

ROCK ANCHOR A steel rod or cable that is placed in a hole bored into rock and held in position by grout or a steel wedge. Usually the rock anchor is more than 6 meters long and is finally prestressed.

ROCK BOLT A steel rod usually less than 6 meters long that is placed in a hole drilled into rock and held in position by grout or a steel wedge.

ROCKFILL DAM *See* Embankment Dam.

ROLLCRETE A no-slump concrete that can be hauled in dump trucks, spread with a bulldozer or grader, and compacted with a vibratory roller.

ROLLED FILL DAM *See* Embankment Dam.

ROUND HEAD BUTTRESS DAM *See* Buttress Dam.

RUBBLE DAM A masonry dam in which the stones are unshaped or uncoursed.

SADDLE DAM A subsidiary dam of any type constructed across a saddle or low point on the perimeter of a reservoir.

SADDLE SPILLWAY *See* Spillway.

SEEPAGE The interstitial movement of water that may take place through a dam, its foundation, or its abutments.

SEEPAGE COLLAR A projecting collar usually of concrete built around the outside of a pipe, tunnel, or conduit, under an embankment dam, to lengthen the seepage path along the outer surface of the conduit.

SEISMIC INTENSITY *See* Intensity Scale.

SEMIPERVIOUS ZONE *See* Transition Zone.

SHAFT SPILLWAY *See* Spillway.

SHARP-CRESTED WEIR *See* Weir.

SHELL (SHOULDER) The upstream and downstream parts of the cross section of a zoned embankment dam on each side of the core or core wall; hence the expressions *upstream shoulder* and *downstream shoulder*.

SIDE CHANNEL SPILLWAY *See* Spillway.

SILL (a) A submerged structure across a river to control the water level upstream. (b) The crest of a spillway. (c) The horizontal gate seating, made of wood, stone, concrete or metal at the invert of any opening or gap in a structure. Hence the expressions *gate sill* and *stoplog sill*.

SLOPE (a) Side of a hill or a mountain. (b) The inclined face of a cutting or canal or embankment. (c) Inclination from the horizontal. In the United States, measured as the ratio of the number of units of the horizontal distance to the number of corresponding units of the vertical distance. Used in English for any inclination. Expressed in percent when the slope is gentle; in this case also termed *gradient*.

SLOPE PROTECTION The protection of a slope against wave action or erosion.

SLUICEWAY *See* Low-Level Outlet.

SLURRY TRENCH A narrow excavation whose sides are supported by a slurry made of mud, clay, or cement and mud filling the excavation. Sometimes used to describe the cutoff itself.

SOIL-CEMENT A well-compacted mixture of soil, Portland cement, and water that produces a hard pavement with more or less permanent cohesion. Used for road building and slope protection.

SOLID HEAD BUTTRESS DAM *See* Buttress Dam.

SPILLWAY A structure over or through which flood flows are discharged. If the flow is controlled by gates, it is considered a controlled spillway; if the elevation of the spillway crest is the only control, it is considered an uncontrolled spillway.

AUXILIARY SPILLWAY (EMERGENCY SPILLWAY) A secondary spillway designed to operate only during exceptionally large floods.

FUSE PLUG SPILLWAY A form of auxiliary or emergency spillway comprising a low embankment or a natural saddle designed to be overtopped and erroded away during a very rare and exceptionally large flood.

MORNING GLORY SPILLWAY *See* Shaft Spillway.

OGEE SPILLWAY (OGEE SECTION) An overflow weir in which in cross section the crest, downstream slope, and bucket have an S or ogee form of curve. The shape is intended to match the underside of the nappe at its upper extremities.

PRIMARY SPILLWAY (PRINCIPAL SPILLWAY) The principal or first-used spillway during flood flows.

SADDLE SPILLWAY A spillway constructed at a low saddle on the perimeter of a reservoir.

SERVICE SPILLWAY A principal spillway used to regulate reservoir releases additional to or in lieu of the outlet.

SHAFT SPILLWAY (MORNING GLORY SPILLWAY) A vertical or inclined shaft into which flood water spills and then is conducted through, under, or around a dam by means of a conduit or tunnel. If the upper part of the shaft is splayed out and terminates in a circular horizontal weir, it is termed a *bellmouth* or *morning glory spillway*.

SIDE CHANNEL SPILLWAY A spillway the crest of which is roughly parallel to the channel immediately downstream of the spillway.

SIPHON SPILLWAY A spillway with one or more siphons built at crest level. This type of spillway is sometimes used for providing automatic surface-level regulation within narrow limits or when considerable discharge capacity is necessary within a short period of time.

SPILLWAY CHANNEL (SPILLWAY TUNNEL) A channel or tunnel conveying water from the spillway to the river downstream.

SPILLWAY CHUTE A sloping spillway channel.

SPILLWAY DESIGN FLOOD (SDF) The largest flood that a given project is designed to pass safely. The reservoir inflow-discharge hydrograph used to estimate the spillway discharge capacity requirements and corresponding maximum surcharge elevation in the reservoir.

STILLING BASIN A basin constructed so as to dissipate the energy of fast-flowing water, e.g., from a spillway or bottom outlet, and to protect the river bed from erosion.

STOPLOGS Large logs or timbers or steel beams placed on top of each other with their ends held in guides on each side of a channel or conduit so as to provide a cheaper or more easily handled means of temporary closure than a bulkhead gate.

STORAGE The retention of water or delay of runoff either by planned operation, as in a reservoir, or by temporary filling of overflow areas, as in the progression of a flood crest through a natural stream channel.

STORAGE RESERVOIR A reservoir that is operated with changing water level for the purpose of storing and releasing water.

STRUCTURAL HEIGHT The distance between the lowest point in the excavated foundation (excluding narrow fault zones) and the top of the dam.

SUBMERGED WEIR *See* Weir.

SURCHARGE *See* Flood Surcharge.

TAILRACE The tunnel, channel, or conduit that conveys the discharge from the turbine to the river; hence the terms *tailrace tunnel* and *tailrace canal*.

TAILWATER LEVEL The level of water in the tailrace at the nearest free surface to the turbine or in the discharge channel immediately downstream of the dam.

THALWEG The line connecting the deepest or lowest points along the stream valley. Sometimes referred to as "the third of the stream."

THRUST BLOCK *See* Block.

TOE OF DAM The junction of the downstream face of a dam with the ground surface. Also referred to as *downstream toe*. For an embankment dam the junction of the upstream face with ground surface is called the *upstream toe*.

TOE WEIGHT Additional material placed at the toe of an embankment dam to increase its stability.

TOP OF DAM The elevation of the uppermost surface of a dam, usually a road or walkway, excluding any parapet wall, railings, etc.

TOP THICKNESS (TOP WIDTH) The thickness or width of a dam at the level of the top of the dam. In general, the term *thickness* is used for gravity and arch dams and *width* is used for other dams.

TRAINING WALL A wall built to confine or guide the flow of water.

TRANSITION ZONE (SEMIPERVIOUS ZONE) A part of the cross section of a zoned embankment dam comprising material whose grading is of intermediate size between that of an impervious zone and that of a permeable zone.

TRASH RACK A screen comprising metal or reinforced concrete bars located in the waterway at an intake so as to prevent the ingress of floating or submerged debris.

TRIBAR A precast concrete shape consisting essentially of three cylinders connected by beams. They are placed in interlocking patterns and are effective for high-energy dissipation.

TUNNEL A long underground excavation usually having a uniform cross section; hence the terms *headrace tunnel*, *pressure tunnel*, *collecting tunnel*, *diversion tunnel*, *power tunnel*, *tailrace tunnel*, *navigation tunnel*, *access tunnel*, *scour tunnel*, *drawoff tunnel*, and *spillway tunnel*.

UNDERSEEPAGE The interstitial movement of water through a foundation.

UPLIFT (a) The upward water pressure in the pores of a material (interstitial pressure) or on the base of a structure. (b) An upward force on a structure caused by frost heave or wind force.

UPSTREAM BLANKET An impervious layer placed on the reservoir floor upstream of a dam. In the case of an embankment dam the blanket may be connected to the impermeable element in the dam.

VALVE In general, a device fitted to a pipeline or orifice in which the closure member is either rotated or moved transversely or longitudinally in the waterway so as to control or stop the flow.

GUARD VALVE *See* Guard Gate.

REGULATING VALVE *See* Regulating Gate and Energy-Dissipating Valve.

VOLUME OF DAM The total space occupied by the materials forming the dam structure computed between abutments and from the top to the bottom of a dam. No deduction is made for small openings such as galleries, adits, tunnels, and operating chambers within the dam structure. Portions of power houses, locks, spillway, etc., may be included only if they are necessary for the structural stability of the dam.

WATERSHED DIVIDE The divide or boundary between catchment areas (or drainage areas).

WATERSTOP A strip of metal, rubber, or other material used to prevent leakage through joints between adjacent sections of concrete.

WAVE WALL A solid wall built along the upstream side at the top of a dam and designed to reflect waves.

WEIGHTING OF A SLOPE (WEIGHTING BERM) Additional material placed on the slope of an embankment.

WEIR A low dam or wall built across a stream to raise the upstream water level. Termed *fixed-crest weir* when uncontrolled. A structure built across a stream or channel for the purpose of measuring flow. Sometimes described as measuring weir or gauging weir. Types of weir include broad-crested weir, sharp-crested weir, drowned weir, and submerged weir.

ZONED EARTHFILL *See* Embankment Dam.

APPENDIX **A**

# Biographical Sketches of Committee Members

**ROBERT B. JANSEN** (*Chairman*) is a consulting civil engineer specializing in the engineering of dams. He has directed the design, construction, and dam safety programs for both the California Department of Water Resources and the U.S. Bureau of Reclamation. During 1963–1964 he was chairman of the California State Engineering Board of Inquiry, which investigated the failure of the Baldwin Hills Dam. In 1976 he was executive director for the Independent Panel to investigate the Teton Dam failure in Idaho. Mr. Jansen holds an M.S.C.E. degree from the University of Southern California. He was chairman of the U.S. Committee of the International Commission on Large Dams in 1979–1981. Mr. Jansen is the author of the book *Dams and Public Safety*, published by the U.S. Bureau of Reclamation in 1980 (reprinted in 1983).

**HARL P. ALDRICH** is president and cofounder of the consulting engineering firm of Haley and Aldrich, Inc. He has been in charge of geotechnical engineering investigations for more than 1,000 projects, concentrating on the design and construction aspects of buildings and earth dams. He received a Sc.D. degree in civil engineering from the Massachusetts Institute of Technology and taught undergraduate and graduate courses at MIT in soil mechanics, foundations, and seepage and groundwater flow. Dr. Aldrich chaired the earlier National Research Council study on dam safety that produced the 1977 report *A Review of the U.S. Bureau of Reclamation Program on the Safety of Existing Dams*.

**ROBERT A. BURKS** is chief civil engineer for the Southern California Edison Company, where he is responsible for all civil engineering performed for and by the company. He is also responsible for the surveillance of the company's 33 major dams. He received a B.E. and a M.S. degree in civil engineering from the University of Southern California and is a registered professional engineer in five states. Mr. Burks has worked for the U.S. Army Corps of Engineers and the California Department of Water Resources. He is a member of the U.S. Committee on Large Dams.

**CLIFFORD J. CORTRIGHT** holds a B.S.C.E. degree from North Dakota State University and is a consulting civil engineer specializing in engineering for dams and appurtenant structures. He was staff engineer for the Independent Panel to investigate the Teton Dam failure and was division engineer of the California Division of Safety of Dams, where he was involved in the design, construction, and operation of dams.

**JAMES J. DOODY** is chief of the Division of Safety of Dams of the California Department of Water Resources. The division supervises the safety of over 1,100 dams and reservoirs in California. His division deals with owners of dams and reservoirs and with representatives of local, state, and federal agencies concerning dam safety. He holds a B.S. degree in civil engineering from the University of California and previously served with the U.S. Army Corps of Engineers and the U.S. Naval Civil Engineer Corps. In his career Mr. Doody has held various positions involved in the design of multipurpose reservoir projects, including concrete dams and tunnels; earthfill dams; and associated structures, such as spillways and outlet works. He is a registered civil engineer in California, a fellow of the American Society of Civil Engineers, and a member of the U.S. Committee on Large Dams.

**JACOB H. DOUMA** has extensive experience in the hydraulic design divisions of both the U.S. Bureau of Reclamation and the U.S. Army Corps of Engineers. Before going into private consulting practice, he was chief hydraulic design engineer with the Office of the Chief of Engineers of the Corps in Washington, D.C., and was responsible for the final review of all hydraulic designs accepted in the 12 divisions and 41 districts of the Corps. Mr. Douma is a member of the National Academy of Engineering, the International Commission on Large Dams, and the U.S. Committee on Large Dams.

**JOSEPH J. ELLAM** is chief of the Division of Dam Safety of the Pennsylvania Department of Environmental Resources. He is responsible for administering the Pennsylvania dam safety program and supervises the review of the design and construction of new dams and the inspection program for existing dams in Pennsylvania. Mr. Ellam received a B.S. degree in civil engineering from the University of Notre Dame and holds an M.S. degree in government administration. He is a member of the U.S. Committee on Large Dams and the USCOLD Committee on Maintenance Operation and Public Safety. He was also a member of the earlier National Research Council committee on dam safety that produced the 1977 report *A Review of the U.S. Bureau of Reclamation Program on the Safety of Existing Dams*.

**CHARLES H. GARDNER** received an M.S. in geology from Emory University in 1961. He has studied civil and mining engineering and is a registered professional engineer and a certified professional geologist. He was chief geologist for International Minerals and Chemicals, Florida Phosphate Operations, and the chief geologist for Law Engineering Testing Company in Atlanta and Raleigh, which included responsibility for dam design and inspection of projects. Since 1976 he has been chief of the Land Quality Section of the North Carolina Department of Natural Resources and has overall responsibility for the state's programs in dam safety, mining, and sedimentation. Mr. Gardner is responsible for an inspection program covering 3,700 dams and has reviewed over 400 dam design and repair plans. Mr. Gardner is a member of the U.S. Committee of the International Commission on Large Dams.

**WILLIAM R. JUDD** has been a professor of rock mechanics at Purdue University since 1967 and was made head of the Geotechnical Engineering Area in 1976. Formerly, he held various positions within the U.S. Bureau of Reclamation as engineer/geologist and was head of the Geology Section I. He has been chairman of the National Research Council's U.S. National Committee for Rock Mechanics and has taken part in a National Research Council study on the safety of existing dams for the U.S. Bureau of Reclamation. Mr. Judd is also a member of the U.S. Committee of the International Commission on Large Dams Executive Council and the Committee on Earthquakes and assisted in the preparation of their publication, *Lessons from Dam Incidents, USA*.

**DAN R. LAWRENCE** received a B.S.C.E. degree from Utah State University and has had 14 years' experience in state-level dam safety engineering activities. He currently serves as chief of the Division of Safety of Dams of the Arizona Department of Water Resources. Mr. Lawrence is a registered professional engineer in Arizona, California, and Utah and is a member of the U.S. Committee on Large Dams.

ROBERT J. LEVETT has worked for 25 years in the hydraulics and structural engineering departments of Niagara Mohawk Power Corporation and has been involved in calculations of maximum and design flood potentials, spillway capacities, and the stability of structures at various hydro stations. He holds a B.S. in civil engineering from the Drexel Institute of Technology.

ARTHUR G. STRASSBURGER is with Pacific Gas and Electric Company, where he is presently project manager for the Helms Pumped Storage Project. He has 30 years' experience with Pacific Gas and Electric in the hydroelectric field, including engineering, construction, and project management. He has been involved with or responsible for most of Pacific Gas and Electric's major hydro projects in the last 25 years. This includes engineering and safety responsibility for about 200 dams. He has written a number of papers on dam rehabilitation and safety. Mr. Strassburger is a member of the U.S. Committee on Large Dams Executive Committee and has participated in several committees of the National Research Council, ASCE, EPRI, and EEI. He holds a B.S.C.E. degree from the University of Wisconsin, Madison.

BRUCE A. TSCHANTZ is a professor of civil engineering at the University of Tennessee. In 1980, on leave for one year from the university, he was chief of federal dam safety for FEMA. He holds a Sc.D. degree in civil engineering from New Mexico State University. From 1977 to 1979, Dr. Tschantz coordinated the executive office review of federal agency dam safety procedures, which in 1979 resulted in new federal guidelines for dam safety. His principal expertise is in dam safety, flood plain management, and the hydrologic impacts of strip mining.

ERIK H. VANMARCKE is professor of civil engineering at Massachusetts Institute of Technology. He holds a Ph.D. degree in civil engineering from MIT and organized MIT's Risk and Decisions in Geotechnical Engineering (1976) and New Perspectives on Dam Safety (1979) programs. His principal expertise is in risk analysis of dams and other structures. Dr. Vanmarcke was a committee member on the 1977 National Research Council study of dam safety. He is the author of *Random Fields: Analysis and Synthesis* and the editor of *The Journal of Structural Safety*.

HOMER B. WILLIS is a consulting engineer in private practice. He holds a B.S. degree in civil engineering from Ohio University. In over 38 years as an employee of the U.S. Army Corps of Engineers, he was involved in many aspects of engineering for dams. In his last assignment with the

Corps (1973–1979) he directed the technical engineering activities for the Civil Works program for the development of water resources, including the nationwide program for inspection of nonfederal dams.

### Technical Consultant

**CHARLES F. CORNS** is a consulting engineer specializing in dam safety and the structural engineering of all types of water resource projects. He holds a B.S. degree in civil engineering from Akron University. In January 1977 he retired from the U.S. Army Corps of Engineers, where he had been the chief structural engineer for the National Water Resources Development Program (civil works). Mr. Corns also served as the technical consultant for a previous National Research Council study of dam safety.

APPENDIX **B**

# Biographical Sketches of Workshop Participants

---

**GEORGE L. BUCHANAN** is chief of the Civil Engineering and Design Branch of the Tennessee Valley Authority. His principal work there has been in design and related areas of Tennessee Valley Authority's hydroelectric program. In addition to being a registered engineer in Tennessee, he is a member of the U.S. Committee on Large Dams and serves as Tennessee Valley Authority's representative to the Federal Emergency Management Agency/ Interagency Committee on Dam Safety concerned with federal dam safety. Previously, he was a member of the FCCSET/ICODS group during the development of federal guidelines for dam safety and was chairman of the Subcommittee on Site Investigation and Design.

**CATALINO B. CECILIO** is a senior civil engineer in charge of the hydrologic engineering group in the Civil Engineering Department of the Pacific Gas and Electric Company in San Francisco, California. He holds a B.S. degree in civil engineering and is registered as a professional engineer in the state of California. His job responsibilities and work experience since 1969 with Pacific Gas and Electric, which owns some 200 dams, has been in the hydraulics and hydrology of dam safety, with principal expertise on design floods up to and including the probable maximum flood and dam break analysis. His expertise includes flood plain evaluation and the impact on the hydrologic environment of plant construction. He is principal codeveloper of the ANS 2.8 *American National Standards for Determining Design Basis Flooding at Power Reactor Sites*, first issued in 1976 by the American National Standard Institute and revised in 1981.

LLEWELLYN L. CROSS, JR., is the chief hydrologist for Chas. T. Main, Inc., in charge of all hydrometeorological and related work. He has over 30 years of experience in hydrologic and hydrometeorological studies and designs. Mr. Cross holds a B.S. in civil engineering from Tufts University and is a member of the U.S. Committee on Large Dams and of the USCOLD Committee on Hydraulics of Spillways. At Chas. T. Main, Inc., he is responsible for hydrologic studies integral to the determination of spillway design floods, diversion floods, and reservoir yield studies for hydroelectric projects in the United States and abroad.

RAY F. DeBRUHL is director of the Division of State Construction, North Carolina Department of Administration. He is a civil engineering extension specialist at North Carolina State University and holds a B.S. degree in civil engineering from the University of South Carolina and an M.C.E. from North Carolina State University. As a consultant, he has provided structural engineering services for architects, engineers, steel fabricators, and contractors. As civil engineering extension specialist at North Carolina State University, in addition to teaching, he is responsible for developing and implementing short courses, seminars, conferences, etc., for the construction industry. He is a registered professional engineer in North and South Carolina.

JAMES M. DUNCAN has for 17 years been a professor of civil engineering at the University of California at Berkeley. He has taught courses and done research on shear strength, slope stability, and deformation of embankment dams. Mr. Duncan has also been a member of the Board of Consultants for the repair of the San Luis Dam in California. He has been involved in safety evaluations of six Corps of Engineers' dams and has participated in studies of the New Melones, Warm Springs, Wolf Creek, Henshaw, Birch, and Arcadia dams.

LLOYD C. FOWLER is general manager and chief engineer at the Goleta Water District, Goleta, California. He was with the Santa Clara Valley Water District for 17 years and supervised the operation and maintenance of 8 earth dams, including analyses for seismic stability. Mr. Fowler received his M.S. in civil engineering and the professional degree of civil engineer from the University of California. He has over 30 years' experience in irrigation, hydraulics, flood control, water supply, and river control works. Presently, he is also president of the Institute for Water Resources of the American Public Works Association.

VERNON K. HAGEN is chief of the Hydraulics and Hydrology Branch of the U.S. Army Corps of Engineers. His expertise involves hydraulic design, hydrologic engineering, and water control and quality. He received his B.S. in civil engineering at Montana State University and his M.S. in civil engineering at Catholic University. He is a registered professional engineer, State of Montana, and a member of USCOLD.

JOSEPH S. HAUGH is national planning engineer with the U.S. Department of Agriculture's Soil Conservation Service. He has worked at the USDA's national headquarters since 1973, with groups such as ICODS, and has served a one-year detail to the Water Resources Council on the development of principles and standards for water resource planning. His positions have encompassed engineering, planning, and design and construction for the Soil Conservation Service in West Virginia, Wisconsin, Kentucky, and South Carolina. Mr. Haugh holds a B.S.C.E. from West Virginia University and is a registered professional engineer in South Carolina.

DAVID S. LOUIE is senior associate and chief hydraulic engineer at Harza Engineering Company in Chicago, Illinois. He has been with Harza since 1950. Since 1967 he has been responsible for quality control of all aspects of work relating to complex problems in hydromechanics, hydraulic transients, hydraulic model experimentations, and prototype investigations. Mr. Louie is also the in-house consultant on all major hydraulic problems. He holds an Sc.D. in hydraulic engineering from Massachusetts Institute of Technology.

J. DAVID LYTLE is chief of the Instrumentation and Evaluation Section, U.S. Army Corps of Engineers, St. Louis district, where he is responsible for directing the programs for monitoring and evaluating the safety conditions of the U.S. Army Corps of Engineers' dams located within that district. He is a professional engineer in Missouri and a member of USCOLD and ICOLD. Mr. Lytle is coinventor of an instrument, the digital tri-axial inclinometer, that monitors the tilt of structures and internal movements within embankments.

MARTIN W. McCANN, JR., is an associate in the consulting firm of Jack R. Benjamin & Associates, Inc., and is an acting professor of civil engineering at Stanford University. He received his Ph.D. from Stanford in 1980, where his research work was in the area of seismic risk analysis. Currently, he is the director of a project sponsored by FEMA at Stanford University to develop procedures to conduct risk-based evaluations of existing dams. As a consulting engineer, Dr. McCann has participated in a number of projects involving the assessment of risk associated with nuclear power plants and dams. He has participated in the critical review of probabilistic safety studies for nuclear power plants in the area of seismic and flood risk analysis. His current research work involves investigating the attenuation and sources of variability of strong ground motion.

JEROME M. RAPHAEL is a consulting civil engineer specializing in dams and is a professor emeritus of civil engineering of the University of California at Berkeley. Previous to this, he served as a structural engineer on dams with the U.S. Army Corps of Engineers and with the U.S. Bureau of Reclamation. He holds an S.M. degree from Massachusetts Institute of Technology. He was an editor of the U.S. Bureau of Reclamation's *Design Manual* and is a member of its panel on criteria for the design of concrete dams. In addition to consulting on the design and analysis of concrete dams, Mr. Raphael has worked on problems of structural behavior, concrete technology, and temperature control. For many years he taught a graduate course on the design and analysis of concrete dams. He was chairman of the American Concrete Institute's Committee on Mass Concrete and of its Committee on Creep and Shrinkage and is a member of USCOLD's Committee on Concrete and of its Committee on Instrumentation. Mr. Raphael has participated in a number of investigations on the safety of existing dams to determine their safety under static and seismic loads and is the author of over 120 publications on the technology of dams.

HARESH SHAH is a professor of structural engineering and the director of the John A. Blume Earthquake Engineering Center at Stanford University. His fields of interest include structures, earthquake engineering, and statistical decision theory. He received his Ph.D. from Stanford University. Dr. Shah has worked with the Honduran, Guatemalan, and Algerian governments to help them develop an understanding of and expertise in the field of earthquake engineering.

THOMAS V. SWAFFORD holds a B.S. degree in civil engineering from the University of Tennessee. He is a registered professional engineer in Tennessee, Arkansas, and Arizona. As vice-president in charge of engineering, construction, and maintenance at Fairfield Glade, he is responsible for planning, designing, constructing, and maintaining several small- and medium-sized dams and lakes.

HARRY E. THOMAS is Chief, Inspections Branch, Division of Hydropower Licensing of the Federal Energy Regulatory Commission in Washington, D.C. Mr. Thomas's responsibilities include planning and coordination of technical direction and control of the Inspections Branch in establishing standards for inspection of licensed hydroelectric projects. He is also responsible for the overall monitoring of project construction and operation, with special emphasis on compliance with the Commission's Dam Safety Program. Mr. Thomas is FERC's representative to the Interagency Committee on Dam Safety. He holds an M.S. in geology from the University of Arkansas.

J. LAWRENCE VON THUN is a senior technical specialist in the Division of Design at the U.S. Bureau of Reclamation. He holds a B.S. in civil engineering from Colorado State University and has done graduate work in operations research at the Colorado School of Mines. He began federal service in 1967, working in Design and Analysis of Concrete Dams and Foundations. He developed specialized procedures using geologic data for the analysis of Auburn Dam's foundation. Mr. Von Thun is a member of the Stanford University advisory board on the risk analysis dam safety study being done for FEMA. He is also a liaison member of the National Academy of Science's panel on seismic hazards in the siting of critical facilities and of the U.S. Committee of the Interagency Committee on Large Dams.

JACK G. WULFF is senior vice-president and chief engineer of Wahler Associates, Consulting Geotechnical Engineers. For 11 years he has been responsible for the direction and management of the technical activities of the firm. These activities are heavily oriented toward earth dam design, consultation, and evaluation in seismic regions of the world. During his 31-year career in water resources and mine and mill waste disposal engineering, Mr. Wulff has specialized in the planning, design, evaluation, and construction aspects of earth dams and mine and mill waste impoundments. He previously held a position with the California Department of Water Resources as their first chief of Earth Dams Design and was directly in charge of the designs of most major dams of the California State Water Project, including Oroville Dam, the highest earthfill dam in the world at the time of its completion in 1967, and some 15 others. Mr. Wulff has been a member of U.S. Committee on Large Dams since 1966 and has coauthored numerous technical articles and papers. He received his B.S. in civil engineering from the University of Nevada and continued with postgraduate studies in soil mechanics and earthquake engineering at the University of California at Berkeley and Sacramento State College.

# Index

## A

## D

## E

## F

## G

## P



## Q



## R

## T



## U